Lecture Notes in Computer Science　13003

Sandy Engelhardt · Ilkay Oksuz · Dajiang Zhu ·
Yixuan Yuan · Anirban Mukhopadhyay ·
Nicholas Heller · Sharon Xiaolei Huang ·
Hien Nguyen · Raphael Sznitman ·
Yuan Xue (Eds.)

Deep Generative Models, and Data Augmentation, Labelling, and Imperfections

First Workshop, DGM4MICCAI 2021
and First Workshop, DALI 2021
Held in Conjunction with MICCAI 2021
Strasbourg, France, October 1, 2021
Proceedings

 Springer

Editors
Sandy Engelhardt (iD)
Universitätsklinikum Heidelberg
Heidelberg, Germany

Dajiang Zhu
The University of Texas at Arlington
Arlington, TX, USA

Anirban Mukhopadhyay (iD)
TU Darmstadt
Darmstadt, Germany

Sharon Xiaolei Huang (iD)
Pennsylvania State University
University Park, PA, USA

Raphael Sznitman (iD)
University of Bern
Bern, Switzerland

Ilkay Oksuz (iD)
Istanbul Technical University
Istanbul, Turkey

Yixuan Yuan (iD)
University of Hong Kong
Hong Kong, Hong Kong

Nicholas Heller (iD)
University of Minnesota
Minneapolis, MN, USA

Hien Nguyen (iD)
University of Houston
Houston, TX, USA

Yuan Xue (iD)
Johns Hopkins University
Baltimore, MD, USA

ISSN 0302-9743   ISSN 1611-3349 (electronic)
Lecture Notes in Computer Science
ISBN 978-3-030-88209-9   ISBN 978-3-030-88210-5 (eBook)
https://doi.org/10.1007/978-3-030-88210-5

LNCS Sublibrary: SL6 – Image Processing, Computer Vision, Pattern Recognition, and Graphics

This Springer imprint is published by the registered company Springer Nature Switzerland AG
The registered company address is: Gewerbestrasse 11, 6330 Cham, Switzerland

DGM4MICCAI 2021 Preface

It was our genuine honor and great pleasure to hold the inaugural Workshop on Deep Generative Models for Medical Image Computing and Computer Assisted Intervention (DGM4MICCAI 2021), a satellite event at the 24th International Conference on Medical Image Computing and Computer Assisted Intervention (MICCAI 2021). In addition to the workshop, we organized an associated challenge the AdaptOR: Deep Generative Model Challenge for Domain Adaptation in Surgery.

DGM4MICCAI was a single-track, half-day workshop consisting of high-quality, previously unpublished papers, presented orally (virtually), intended to act as a forum for computer scientists, engineers, clinicians and industrial practitioners to present their recent algorithmic developments, new results, and promising future directions in deep generative models. Deep generative models such as generative adversarial networks (GANs) and variational auto-encoders (VAEs), are currently receiving widespread attention from not only the computer vision and machine learning communities but also the MIC and CAI community. These models combine the advanced deep neural networks with classical density estimation (either explicit or implicit) for achieving state-of-the-art results. The AdaptOR challenge formulated a domain adaptation problem "from simulation to surgery", which was a clinically relevant technical problem due to data availability and data privacy concerns. As such, DGM4MICCAI provided an all-round experience for deep discussion, idea exchange, practical understanding, and community building around this popular research direction.

This year's DGM4MICCAI was held on October 1, 2021, virtually in Strasbourg, France. There was a very positive response to the call for papers for DGM4MICCAI 2021. We received 15 workshop papers and 2 challenge papers. Each paper was reviewed by at least two reviewers and we ended up with 10 accepted papers for the workshop and 2 for the AdaptOR challenge. The accepted papers present fresh ideas on broad topics ranging from methodology (image-to-image translation, synthesis) to applications (segmentation, classification).

The high quality of the scientific program of DGM4MICCAI 2021 was due first to the authors who submitted excellent contributions and second to the dedicated collaboration of the international Program Committee and the other researchers who reviewed the papers. We would like to thank all the authors for submitting their valuable contributions and for sharing their recent research activities.

We are particularly indebted to the Program Committee members and to all the external reviewers for their precious evaluations, which permitted us to set up this proceedings. We were also very pleased to benefit from the keynote lectures of the invited speakers: Andreas Maier, FAU Nürnberg, Germany, and Stefanie Speidel, NCT Dresden, Germany. We would like to express our sincere gratitude to these renowned experts for making the inaugural workshop a successful platform to rally deep generative models research within the MICCAI context.

August 2021

Sandy Engelhardt
Ilkay Oksuz
Dajiang Zhu
Yixuan Yuan
Anirban Mukhopadhyay

DGM4MICCAI 2021 Organization

Organizing Committee

Sandy Engelhardt	University Hospital Heidelberg, Germany
Ilkay Oksuz	Istanbul Technical University, Turkey
Dajiang Zhu	University of Texas at Arlington, USA
Yixuan Yuan	City University of Hong Kong, China
Anirban Mukhopadhyay	Technische Universität Darmstadt, Germany

Program Committee

Li Wang	University of Texas at Arlington, USA
Tong Zhang	Peng Cheng Laboratory, China
Ping Lu	University of Oxford, UK
Roxane Licandro	Medical University of Vienna, Austria
Chen Qin	University of Edinburgh, UK
Veronika Zimmer	TU Muenchen, Germany
Dwarikanath Mahapatra	Inception Institute of AI, UAE
Michael Sdika	CREATIS Lyon, France
Jelmer Wolterink	University of Twente, The Netherlands
Alejandro Granados	King's College London, UK
Liang Zhan	University of Pittsburgh, USA
Jinglei Lv	University of Sydney, Australia

Student Organizers

Lalith Sharan	University Hospital Heidelberg, Germany
Henry Krumb	Technische Universität Darmstadt, Germany
Moritz Fuchs	Technische Universität Darmstadt, Germany
Caner Özer	Istanbul Technical University, Turkey
Chen Zhen	City University of Hong Kong, China
Guo Xiaoqing	City University of Hong Kong, China

Additional Reviewers

Chen Chen
Shuo Wang
Matthias Perkonigg
Martin Menten
Alberto Gomez
Yanfu Zhang
Mariano Cabezas
Haoteng Tang
Jorge Cardoso

DALI 2021 Preface

This volume contains the proceedings of the 1st International Workshop on Data Augmentation, Labeling, and Imperfections (DALI 2021) which was held on October 1, 2021, in conjunction with the 24th International Conference on Medical Image Computing and Computer Assisted Intervention (MICCAI 2021). This event was originally planned for Strasbourg, France, but was ultimately held virtually due to the COVID-19 pandemic. While this is the first workshop under the "DALI" name, it is the result of a joining of forces between previous MICCAI workshops on Large Scale Annotation of Biomedical data and Expert Label Synthesis (LABELS 2016–2020) and on Medical Image Learning with Less Labels and Imperfect Data (MIL3ID 2019–2020).

Obtaining the huge amounts of labeled data that modern image analysis methods require is especially challenging in the medical imaging domain. Medical imaging data is heterogeneous and constantly evolving, and expert annotations can be prohibitively expensive and highly variable. Hard clinical outcomes such as survival are exciting targets for prediction but can be exceptionally difficult to collect. These challenges are especially acute in rare conditions, some of which stand to benefit the most from medical image analysis research. In light of this, DALI aims to bring together researchers in the MICCAI community who are interested in the rigorous study of medical data as it relates to machine learning systems.

This year's DALI workshop received 32 paper submissions from authors all over the world. Each paper was reviewed by at least three peer-experts, and in the end, 15 high-quality papers were selected for publication. The workshop day included presentations for each of these 15 papers as well as longer-form invited talks from Margrit Betke of Boston University, Ekin Dogus Cubuk of Google Brain, Jerry Prince of Johns Hopkins University, Adrian Dalca of Massachusetts Institute of Technology, and Stephen Wong of Weill Cornell Medical College.

No scientific program would be successful without a monumental effort on the part of its peer reviewers. We are deeply grateful to the more than 30 scientists who volunteered a substantial amount of their time to provide valuable feedback to the authors and to help our editorial team make final decisions. We would also like to thank Histosonics Inc. for its generous financial support of the DALI workshop.

August 2021

Nicholas Heller
Sharon Xiaolei Huang
Hien V. Nguyen

DALI 2021 Organization

Organizing Committee

Nicholas Heller   University of Minnesota, USA
Sharon Xiaolei Huang   Pennsylvania State University, USA
Hien V. Nguyen   University of Houston, USA

Editorial Chairs

Raphael Sznitman   University of Bern, Switzerland
Yuan Xue   Johns Hopkins University, USA

Award Committee

Dimitris N. Metaxas   Rutgers University, USA
Diana Mateus   Centrale Nantes, France

Advisory Board

Stephen Wong   Houston Methodist Hospital, USA
Jens Rittscher   University of Oxford, UK
Margrit Betke   Boston University, USA
Emanuele Trucco   University of Dundee, Scotland

Program Committee

Alison C. Leslie   University of Minnesota, USA
Amelia Jiménez-Sánchez   Pompeu Fabra University, Spain
Anjali Balagopal   UT Southwestern, USA
Brett Norling   University of Minnesota, USA
Chandra Kambhamettu   University of Delaware, USA
Chao Chen   Stony Brook University, USA
Christoph M. Friedrich   Dortmund University of Applied Sciences and Arts, Germany
Devante F. Delbrune   University of Minnesota, USA
Edward Kim   Drexel University, USA
Emanuele Trucco   University of Dundee, UK
Filipe Condessa   Instituto Superior Tecnico, Portugal/Carnegie Mellon University, USA

Haomiao Ni	Pennsylvania State University, USA
Hui Qu	Adobe Inc., USA
Jiarong Ye	Pennsylvania State University, USA
Jue Jiang	Memorial Sloan Kettering Cancer Center, USA
Kelvin Wong	Houston Methodist Hospital Research Institute, USA
Li Xiao	Chinese Academy of Science, China
Michael Goetz	German Cancer Research Center (DFKZ), Germany
Nicha Dvornek	Yale University, USA
Niklas E. P. Damberg	University of Minnesota, USA
Pengyu Yuan	University of Houston, USA
Pietro Antonio Cicalese	University of Houston, USA
Rafat Solaiman	University of Minnesota Medical School, USA
Samira Zare	University of Houston, USA
Ti Bai	UT Southwestern, USA
Weidong Cai	University of Sydney, Australia
Wen Hui Lei	University of Electronic Science and Technology of China, China
Xiao Liang	UT Southwestern, USA
Xiaoxiao Li	Princeton University, USA
Xiaoyang Li	Adobe Research, USA

Contents

Image-to-Image Translation, Synthesis

Frequency-Supervised MR-to-CT Image Synthesis

Zenglin Shi[1(✉)], Pascal Mettes[1], Guoyan Zheng[2], and Cees Snoek[1]

[1] University of Amsterdam, Amsterdam, The Netherlands
z.shi@uva.nl
[2] Shanghai Jiao Tong University, Shanghai, China

Abstract. This paper strives to generate a synthetic computed tomography (CT) image from a magnetic resonance (MR) image. The synthetic CT image is valuable for radiotherapy planning when only an MR image is available. Recent approaches have made large strides in solving this challenging synthesis problem with convolutional neural networks that learn a mapping from MR inputs to CT outputs. In this paper, we find that all existing approaches share a common limitation: reconstruction breaks down in and around the high-frequency parts of CT images. To address this common limitation, we introduce frequency-supervised deep networks to explicitly enhance high-frequency MR-to-CT image reconstruction. We propose a frequency decomposition layer that learns to decompose predicted CT outputs into low- and high-frequency components, and we introduce a refinement module to improve high-frequency reconstruction through high-frequency adversarial learning. Experimental results on a new dataset with 45 pairs of 3D MR-CT brain images show the effectiveness and potential of the proposed approach. Code is available at https://github.com/shizenglin/Frequency-Supervised-MR-to-CT-Image-Synthesis.

Keywords: Deep learning · CT synthesis · Frequency supervision

1 Introduction

Magnetic resonance (MR) image is widely used in clinical diagnosis and cancer monitoring, as it is obtained through a non-invasive imaging protocol, and it delivers excellent soft-tissue contrast. However, MR image does not provide electron density information that computed tomography (CT) image can provide, which is essential for applications like dose calculation in radiotherapy treatment planning [2,5,10,18] and attenuation correction in positron emission tomography reconstruction [16,22,24]. To overcome this limitation, a variety of approaches have been proposed to recreate a CT image from the available MR images [9,11,25,26,31]. Recently, deep learning-based synthesis methods [9,11,19,25,26,31] have shown superior performance over alternatives such as segmentation-based [1,5,17] and atlas-based methods [3,4,7,30].

© Springer Nature Switzerland AG 2021
S. Engelhardt et al. (Eds.): DGM4MICCAI 2021/DALI 2021, LNCS 13003, pp. 3–13, 2021.
https://doi.org/10.1007/978-3-030-88210-5_1

A typical approach for deep learning-based synthesis is through 2D convolutional networks on 2D MR images [8,9,11,14,18,28,31]. A downside of this setup is that 2D approaches are applied to 3D MR images slice-by-slice, which can cause discontinuous prediction results across slices [26]. To take full use of 3D spatial information of volumetric data, 3D-based synthesis models have been explored using 3D convolutional networks [25,29] and 3D GANs [26]. In this work we adopt a similar setup, which uses paired MR and CT images during training, but we tackle a common limitation amongst existing 3D-based synthesis approaches: imperfect CT image synthesis in high-frequency parts of the volume.

The main motivation behind our work is visualized in Fig. 1. For MR (a) to CT (b) image synthesis using 3D networks [6], the reconstruction error (c) is most dominant in regions that directly overlap with the high-frequency parts of the CT image (d). This is a direct result of the used loss function, *e.g.*, an ℓ_1 or ℓ_2 loss, which results in blurring since they are minimized by averaging all possible outputs [13,23]. As a result, the low-frequency parts are reconstructed well, at the cost of the high-frequency parts. Interestingly, Lin *et al.* [21] also found CNN-based synthesis models tend to lose high-frequency image details for CT-to-MR image synthesis. To address this limitation, they propose the frequency-selective learning, where multiheads are used in the deep network for learning the reconstruction of different frequency components. Differently, in this work, we propose frequency-supervised networks that explicitly aim to enhance high-frequency reconstruction in MR-to-CT image synthesis.

We make three contributions in this work: *i)* we propose a frequency decomposition layer to decompose the predicted CT image into high-frequency and low-frequency parts. This decomposition is supervised by decomposing ground truth CT images using low-pass filters. In this way, we can focus on improving the quality of the high-frequency part, assisted by *ii)* a high-frequency refinement module. This module is implemented as a 3D symmetric factorization convolutional block to maximize reconstruction performance with minimal parameters; and *iii)* we outline a high-frequency adversarial learning to further improve the quality of the high-frequency CT image. Experimental results on a dataset with 45 pairs of 3D MR-CT brain images shows the effectiveness and potential of the proposed approach.

2 Method

We formulate the MR-to-CT image synthesis task as a 3D image-to-image translation problem. Let $\mathcal{X} = \{x_i\}_{i=1}^{H \times W \times L}$ be an input MR image of size $H \times W \times L$, and $\mathcal{Y} = \{y_i\}_{i=1}^{H \times W \times L}$ be the target CT image for this MR image, where y_i is the target voxel for the voxel of x_i. The transformation from input images to target images can be achieved by learning a mapping function, *i.e.*, $f : \mathcal{X} \mapsto \mathcal{Y}$. In this paper, we learn the mapping function by way of voxel-wise nonlinear regression, implemented through a 3D convolutional neural network. Let $\Psi(\mathcal{X}) : \mathbb{R}^{H \times W \times L} \mapsto \mathbb{R}^{H \times W \times L}$ denote such a mapping given an arbitrary 3D convolutional neural network Ψ for input MR image $\mathcal{X}$.

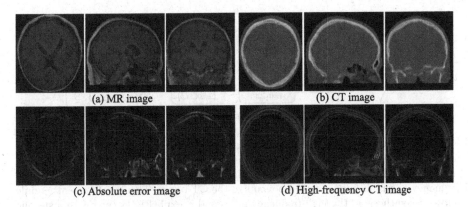

(a) MR image   (b) CT image

(c) Absolute error image   (d) High-frequency CT image

Fig. 1. High-frequency supervision motivation for MR-to-CT image synthesis. For MR (a) to CT (b) image synthesis using 3D networks [6], the reconstruction error (c) is most dominant in regions that directly overlap with the high-frequency parts of the CT image (d). For computing the high-frequency CT image, we first obtain a low-frequency CT image through a Gaussian low-pass filter. Then, we subtract the low-frequency CT image from the raw CT image to generate a high-frequency CT image. In the error image (c), the brighter the voxel, the bigger the error.

2.1 Frequency-Supervised Synthesis Network

We propose a 3D network that specifically emphasizes the high-frequency parts of CT images. Standard losses for 3D networks, such as the ℓ_1 loss, perform well in the low-frequency parts of CT images, at the cost of loss in precision for the high-frequency parts. Our approach is agnostic to the base 3D network and introduces two additional components to address our desire for improved synthesis in the high-frequency parts: a decomposition layer and a refinement module, which is explicitly learned with a specific high-frequency supervision. The overall network is visualized in Fig. 2.

Decomposition Layer. We account for a specific focus on high-frequency CT image parts through a decomposition layer. This layer learns to split the output features of a 3D base network into low-frequency and high-frequency components in a differentiable manner. Let $V = \Psi(\mathcal{X}) \in \mathbb{R}^{C \times H \times W \times L}$ denote the output of the penultimate layer of the base network for input $\mathcal{X}$. We add a $3 \times 3 \times 3$ convolution layer with parameters $\theta_d \in \mathbb{R}^{C \times 2 \times 3 \times 3 \times 3}$, followed by a softmax function to generate probability maps $P = [p_l, p_h] = \text{softmax}(\theta_d V) \in \mathbb{R}^{2 \times H \times W \times L}$. The probability scores for each voxel denote the likelihood of the voxel belonging to the low- or high-frequency parts of the CT images. Using the probabilities, we obtain low-frequency features $V_l = p_l * V$ and high-frequency features $V_h = p_h * V$, where p_l and p_h are first tiled to be the same size as V. For the low-frequency part, we use a $3 \times 3 \times 3$ convolution layer with the parameters $\theta_l \in \mathbb{R}^{C \times 1 \times 3 \times 3 \times 3}$ to generate the low-frequency CT image $\hat{\mathcal{Y}}_l = \theta_l V_l$.

Refinement Module. To generate the high-frequency CT image $\hat{\mathcal{Y}}_h$, we introduce a refinement module to improve the quality of the high-frequency features

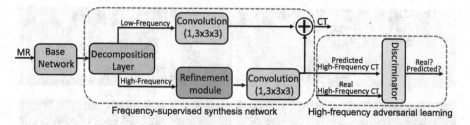

Fig. 2. Frequency-supervised network architecture. Our approach is agnostic to the base 3D MR-to-CT image synthesis network. The decomposition layer splits the output features of the 3D base network into two parts that generate the low-frequency and high-frequency components of the CT image. Then a refinement module improves synthesis in the high-frequency parts and is explicitly learned with a specific high-frequency supervision. Finally, the predicted high-frequency CT image is further enhanced by means of adversarial learning. $\oplus$ denotes the element-wise sum.

V_h. Since the high-frequency features are close to zero in most regions, its learning usually requires a large receptive field for capturing enough context information. The enhancement is performed on top of a base network, *e.g.*, a 3D U-Net [6], thus we should limit the amount of extra parameters and layers to avoid making the network hard to optimize. To this end, we introduce a 3D symmetric factorization module. The module explicitly factorizes a 3D convolution into three 1D convolutions along three dimensions. To process each dimension equally, the module employs a symmetric structure with the combination of $(k \times 1 \times 1)+(1 \times k \times 1)+(1 \times 1 \times k)$, $(1 \times k \times 1)+(1 \times 1 \times k)+(k \times 1 \times 1)$, and $(1 \times 1 \times k)+(k \times 1 \times 1)+(1 \times k \times 1)$ convolutions. Specifically, the input of this module is convolved with three 1D convolutions for each dimension, the output of each 1D convolution is convolved two more times over the remaining dimensions. Then the outputs of last three 1D convolutions is summed as the output of this module. Compared to a standard 3D $k \times k \times k$ convolution layer, parameters is reduced from k^3 to $9k$. In this paper, we use $k = 13$ for relatively large receptive field. The module is denoted as ϕ with the parameters θ_e. Then we use a $3 \times 3 \times 3$ convolution layer with the parameters $\theta_h \in \mathbb{R}^{C \times 1 \times 3 \times 3 \times 3}$ to generate the high-frequency CT image $\hat{\mathcal{Y}}_h = \theta_h \phi(V_h)$.

Optimization. We use a specific loss for high-frequency CT image synthesis to explicitly learn the high-frequency refinement module. Another loss is used for overall CT image synthesis. Empirically, low-frequency CT image can be synthesised correctly with only an overall loss. Thus, we minimize the difference between predicted high-frequency CT image $\hat{\mathcal{Y}}_h$ and its ground-truth $\mathcal{Y}_h$, and the difference between predicted CT image $(\hat{\mathcal{Y}}_h + \hat{\mathcal{Y}}_l)$ and its ground-truth $\mathcal{Y}$ using the L_1-norm, which is defined as:

$$\mathcal{L} = \| \hat{\mathcal{Y}}_h - \mathcal{Y}_h \|_1 + \| (\hat{\mathcal{Y}}_l + \hat{\mathcal{Y}}_h) - \mathcal{Y} \|_1 . \tag{1}$$

During training, the ground truth high-frequency CT image $\mathcal{Y}_h$ is obtained fully automatically without *any* manual labeling. Specifically, we first obtain a low-frequency CT image through a Gaussian low-pass filter with filtering size $\sigma = 15$. Then we subtract the low-frequency CT image from the CT image to obtain the high-frequency CT image. During inference, we input a MR $\mathcal{X}$, and output $(\hat{\mathcal{Y}}_l + \hat{\mathcal{Y}}_h)$ as the synthesised CT image.

2.2 High-Frequency Adversarial Learning

Lastly, we enhance the predicted high-frequency CT image $\hat{\mathcal{Y}}_h$ by means of adversarial learning. Adversarial learning has shown its benefits in MR-to-CT image synthesis by Nie *et al.* [26]. We have observed that the low-frequency CT image can be reconstructed well. Thus, we propose to apply the discriminator only on the high frequencies. This reduces the complexity of the problem, making it easier for the discriminator to focus on the relevant image features. We use the relativistic discriminator introduced by [15]. The discriminator makes adversarial learning considerably more stable and generates higher quality images.

3 Experiments and Results

3.1 Experimental Setup

Dataset and Pre-processing. We evaluate our approach on a dataset with 45 pairs of 3D MR-CT brain images. When comparing the size of our data to previous supervised works, our data set size is reasonable. Such as, Nie *et al.* [26] report on 38 data pairs and Han *et al.* [11] use 33 data pairs. Our images are acquired in the head region for the clinical indications of dementia, epilepsy and grading of brain tumours. The MR images have a spacing of $0.8 \times 0.8 \times 0.8$ mm^3 while the CT images have a spacing of $0.9 \times 0.9 \times 2.5$ mm^3. Registration is performed to align the two modalities and to sample the aligned images with a spacing of $1.0 \times 1.0 \times 1.5$ mm^3. The gray values of the CT were uniformly distributed in the range of $[-1024, 2252.7]$ Hounsfield unit. We resample all the training data to isotropic resolution and normalized each MR image as zero mean and unit variance. We also normalize each CT image into the range of $[0, 1]$ and we expect that the output synthetic values are also in the same range. The final synthetic CT will be obtained by multiplying the normalized output with the range of Hounsfield unit in our training data.

Implementation Details. Implementation is done with Python using TensorFlow. Network parameters are initialized with He initialization and trained using Adam with a mini-batch of 1 for 5,000 epochs. We set β_1 to 0.9, β_2 to 0.999 and the initial learning rate to 0.001. Data augmentation is used to enlarge the training samples by rotating each image with a random angle in the range of $[-10°, 10°]$ around the z axis. Randomly cropped $64 \times 64 \times 64$ sub-volumes serve as input to train our network. During testing, we adopt sliding window and overlap-tiling stitching strategies to generate predictions for the whole volume. We use MAE (Mean Absolute Error), PSNR (Peak Signal-to-Noise Ratio) and SSIM (Structural Similarity Index Measure) as evaluation measurements. These measurements are reported by 5-fold cross-validation.

Table 1. Effect of frequency-supervised learning. For all base networks, our high-frequency supervised learning results in improved synthesis across all metrics.

	3D FC-Net [25]			3D Res-Net [20]			3D U-Net [6]		
	MAE↓	PSNR↑	SSIM↑	MAE↓	PSNR↑	SSIM↑	MAE↓	PSNR↑	SSIM↑
Base network	94.55	24.43	0.681	81.26	25.89	0.724	79.09	26.10	0.726
w/Boundary refinement [27]	90.15	25.03	0.697	82.54	25.76	0.723	82.83	25.63	0.723
w/ *This paper*	**84.31**	**25.72**	**0.736**	**73.61**	**26.72**	**0.741**	**72.71**	**26.86**	**0.747**

Table 2. Effect of refinement module. L denotes layer number, C denotes channel number and K denotes kernel size. Our proposed refinement module introduces relatively few parameters with large receptive fields, leading to improved performance.

Stacking			Large kernel size			This paper		
3D Convolutions	MAE↓	#params	3D Convolutions	MAE↓	#params	1D Convolutions	MAE↓	#params
($L=3, C=32, K=3$)	76.38	82.9k	($L=3, C=32, K=3$)	76.38	82.9k	($L=9, C=32, K=3$)	74.08	64.5k
($L=6, C=32, K=3$)	79.60	165.9k	($L=3, C=32, K=5$)	75.78	384.0k	($L=9, C=32, K=13$)	72.71	119.8k
($L=9, C=32, K=3$)	81.89	248.8k	($L=3, C=32, K=7$)	79.09	1053.7k	($L=9, C=32, K=19$)	72.69	175.1k

3.2   Results

Effect of Frequency-Supervised Learning. We first analyze the effect of the proposed frequency-supervised learning. We compare it to two baselines. The first performs synthesis using the base network only. Here, we compare three widely used network architectures, *i.e.*, 3D fully convolutional network (3D FC-Net) [25], 3D residual network (3D Res-Net) [20] and 3D U-Net [6]. The second baseline adds a boundary refinement module introduced by [27] on top of the base network, which improves the structures of the predicted image by means of residual learning. The losses of these two baseline models are optimized with respect to the overall CT image estimation. As a result, low-frequency parts of predicted CT image are reconstructed well, at the cost of high-frequency parts. By contrast, we first decompose the predicted CT image into low-frequency and high-frequency parts, and then improve the quality of high-frequency parts with a high-frequency refinement module, which is learned by a specific high-frequency loss. The results are shown in Table 1. With the base network only, 3D U-Net obtains the best performance. With the addition of the boundary refinement module, only the performance of the 3D FC-Net is improved. By contrast, our method improves over all three base networks. Figure 3 highlights our ability to better reduce synthesised errors, especially in high-frequency parts. We conclude our frequency-supervised learning helps MR-to-CT image synthesis, regardless of the 3D base network, and we will use 3D U-Net as basis for further experiments.

Effect of High-Frequency Refinement Module. In the second experiment, we demonstrate the effect of the proposed refinement module. To capture more context information by enlarging the receptive field, standard approaches include stacking multiple $3 \times 3 \times 3$ convolutional layers or use convolutions with larger kernel, *e.g.*, $7 \times 7 \times 7$. The experimental results in Table 2 show that neither

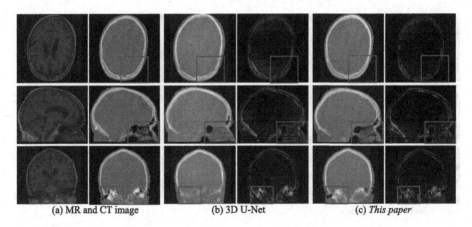

(a) MR and CT image   (b) 3D U-Net   (c) *This paper*

Fig. 3. Effect of frequency-supervised learning. The right images of (b) and (c) are the error images, where the brighter the voxel, the bigger the error. Our method better reduces the synthesised errors than 3D U-Net, as highlighted by the red squares. (Color figure online)

Table 3. Effect of high-frequency adversarial learning. Our approach, with both frequency-supervision and high-frequency adversarial learning, outperforms 3D U-Net with standard adversarial learning.

	MAE↓	PSNR↑	SSIM↑
3D U-Net	79.09	26.10	0.726
3D GAN	76.83	26.55	0.742
This paper: Frequency-supervised synthesis	72.71	26.86	0.747
This paper*: *Frequency-supervised synthesis and adversarial learning	**69.57**	**27.39**	**0.758**

approach works well. Such as, the MAE of stacking three $3 \times 3 \times 3$ convolutional layers is 76.38, while the MAE of stacking six is 79.60. The MAE of stacking three $3 \times 3 \times 3$ convolutional layers is 76.38, while the MAE of stacking three $7 \times 7 \times 7$ is 79.09. By contrast, our proposed module with $k = 13$ introduces relatively few parameters with large convolution kernels, leading to a better MAE of 72.71.

Effect of High-Frequency Adversarial Learning. In this experiment, we show the effectiveness of the proposed high-frequency adversarial learning. For the generator network, we use the 3D U-Net for standard adversarial learning, and frequency-supervised synthesis network for our high-frequency adversarial learning, as shown in Fig. 2, where 3D U-Net is the base network. For the discriminator, we both use the relativistic average discriminator introduced in [15] (See Sect. 2.2). The network architecture of the discriminator is the same as the encoder of 3D U-Net. 3D U-Net combined with standard adversarial learning leads to a 3D GAN based synthesis model, as the work of Nie *et al.* [26]. As shown in Table 3, the 3D GAN achieves better synthesis performance than 3D

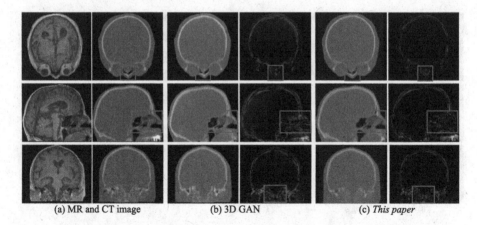

(a) MR and CT image   (b) 3D GAN   (c) *This paper*

Fig. 4. Effect of high-frequency adversarial learning. From the figure, one can observe that the proposed method reduces the synthesized error and yields synthesized CT images with better perceptive quality, as highlighted by the red squares. In the error images, the brighter the voxel, the bigger the error. (Color figure online)

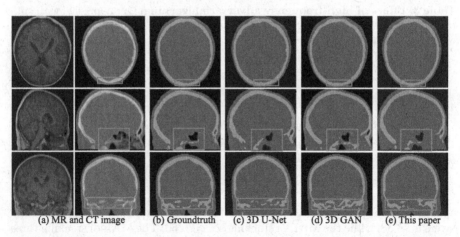

(a) MR and CT image   (b) Groundtruth   (c) 3D U-Net   (d) 3D GAN   (e) This paper

Fig. 5. Evaluations by segmentation. The regions with black color represent background and air, the regions with red color represent soft tissue, the regions with green color represent bone tissue. Our method achieves better synthetic bone tissues, as highlighted by the blue squares. (Color figure online)

U-Net only. Our high-frequency adversarial learning further improves the performance of our frequency-supervised synthesis network. From Fig. 4, we observe the proposed method yields synthesized CT images with better perceptive quality, in particular higher structural similarity and more anatomical details.

Evaluation by Segmentation. To further evaluate the quality of synthesised CT images generated by various methods. Following Hangartner [12], we use a simple thresholding to segment ground-truth and synthesised CT images into three classes: 1) background and air; 2) soft tissue; and 3) bone tissue. As shown in Fig. 5, our proposed methods mainly improve the model's capability on bone tissue synthesis compared to other methods.

4 Conclusion

In this paper, we have shown that existing deep learning based MR-to-CT image synthesis methods suffer from high-frequency information loss in the synthesized CT image. To enhance the reconstruction of high-frequency CT images, we present a method. Our method contributes a frequency decomposition layer, a high-frequency refinement module and a high-frequency adversarial learning, which are combined to explicitly improve the quality of the high-frequency CT image. Our experimental results demonstrate the effectiveness of the proposed methods.

References

1. Berker, Y., et al.: MRI-based attenuation correction for hybrid PET/MRI systems: a 4-class tissue segmentation technique using a combined ultrashort-echo-time/Dixon MRI sequence. J. Nucl. Med. **53**(5), 796–804 (2012)
2. Burgos, N., Guerreiro, F., et al.: Iterative framework for the joint segmentation and CT synthesis of MR images: application to MRI-only radiotherapy treatment planning. Phys. Med. Biol. **62**, 4237–4253 (2017)
3. Burgos, N., et al.: Robust CT synthesis for radiotherapy planning: application to the head and neck region. In: Navab, N., Hornegger, J., Wells, W.M., Frangi, A.F. (eds.) MICCAI 2015. LNCS, vol. 9350, pp. 476–484. Springer, Cham (2015). https://doi.org/10.1007/978-3-319-24571-3_57
4. Burgos, N., et al.: Joint segmentation and CT synthesis for MRI-only radiotherapy treatment planning. In: Ourselin, S., Joskowicz, L., Sabuncu, M.R., Unal, G., Wells, W. (eds.) MICCAI 2016. LNCS, vol. 9901, pp. 547–555. Springer, Cham (2016). https://doi.org/10.1007/978-3-319-46723-8_63
5. Burgos, N., et al.: Iterative framework for the joint segmentation and CT synthesis of MR images: application to MRI-only radiotherapy treatment planning. Phys. Med. Biol. **62**(11), 4237 (2017)
6. Çiçek, Ö., Abdulkadir, A., Lienkamp, S.S., Brox, T., Ronneberger, O.: 3D U-net: learning dense volumetric segmentation from sparse annotation. In: Ourselin, S., Joskowicz, L., Sabuncu, M.R., Unal, G., Wells, W. (eds.) MICCAI 2016. LNCS, vol. 9901, pp. 424–432. Springer, Cham (2016). https://doi.org/10.1007/978-3-319-46723-8_49
7. Degen, J., Heinrich, M.P.: Multi-atlas based pseudo-CT synthesis using multimodal image registration and local atlas fusion strategies. In: CVPR Workshops, pp. 160–168 (2016)
8. Emami, H., Dong, M., Nejad-Davarani, S.P., Glide-Hurst, C.K.: Generating synthetic CTs from magnetic resonance images using generative adversarial networks. Med. Phys. **45**(8), 3627–3636 (2018)
9. Ge, Y., Wei, D., Xue, Z., Wang, Q., Zhou, X., Zhan, Y., Liao, S.: Unpaired MR to CT synthesis with explicit structural constrained adversarial learning. In: ISBI (2019)

10. Guerreiro, F., et al.: Evaluation of a multi-atlas CT synthesis approach for MRI-only radiotherapy treatment planning. Physica Med. **35**, 7–17 (2017)
11. Han, X.: MR-based synthetic CT generation using a deep convolutional neural network. Med. Phys. **44**(4), 1408–1419 (2017)
12. Hangartner, T.N.: Thresholding technique for accurate analysis of density and geometry in QCT, PQCT and muCT images. J. Musculoskelet. Neuronal Interact. **7**(1), 9 (2007)
13. Isola, P., Zhu, J.Y., Zhou, T., Efros, A.A.: Image-to-image translation with conditional adversarial networks. In: CVPR (2017)
14. Jin, C.B., et al.: Deep CT to MR synthesis using paired and unpaired data. Sensors **19**(10), 2361 (2019)
15. Jolicoeur-Martineau, A.: The relativistic discriminator: a key element missing from standard GAN. In: ICLR (2019)
16. Kläser, K., et al.: Improved MR to CT synthesis for PET/MR attenuation correction using imitation learning. In: SSMI Workshop, pp. 13–21 (2019)
17. Korhonen, J., Kapanen, M., et al.: A dual model HU conversion from MRI intensity values within and outside of bone segment for MRI-based radiotherapy treatment planning of prostate cancer. Med Phys **41**, 011704 (2014)
18. Largent, A., et al.: Pseudo-CT generation for MRI-only radiation therapy treatment planning: comparison among patch-based, atlas-based, and bulk density methods. Int. J. Radiat. Oncol.* Biol.* Phys. **103**(2), 479–490 (2019)
19. Largent, A., et al.: Comparison of deep learning-based and patch-based methods for pseudo-CT generation in MRI-based prostate dose planning. Int. J. Radiat. Oncol.* Biol.* Phys. **105**(5), 1137–1150 (2019)
20. Li, W., Wang, G., Fidon, L., Ourselin, S., Cardoso, M.J., Vercauteren, T.: On the compactness, efficiency, and representation of 3D convolutional networks: brain parcellation as a pretext task. In: Niethammer, M., et al. (eds.) IPMI 2017. LNCS, vol. 10265, pp. 348–360. Springer, Cham (2017). https://doi.org/10.1007/978-3-319-59050-9_28
21. Lin, Z., Zhong, M., Zeng, X., Ye, C.: Frequency-selective learning for CT to MR synthesis. In: Burgos, N., Svoboda, D., Wolterink, J.M., Zhao, C. (eds.) SASHIMI 2020. LNCS, vol. 12417, pp. 101–109. Springer, Cham (2020). https://doi.org/10.1007/978-3-030-59520-3_11
22. Liu, F., Jang, H., et al.: Deep learning MR imaging-based attenuation correction for PET/MR imaging. Radiology **286**(2), 676–684 (2017)
23. Mathieu, M., Couprie, C., LeCun, Y.: Deep multi-scale video prediction beyond mean square error. In: ICLR (2016)
24. Navalpakkam, B., Braun, H., et al.: Magnetic resonance-based attenuation correction for PET/MR hybrid imaging using continuous valued attenuation maps. Invest. Radiol. **48**, 323–332 (2013)
25. Nie, D., Cao, X., Gao, Y., Wang, L., Shen, D.: Estimating CT image from MRI data using 3D fully convolutional networks. In: Carneiro, G., et al. (eds.) LABELS/DLMIA -2016. LNCS, vol. 10008, pp. 170–178. Springer, Cham (2016). https://doi.org/10.1007/978-3-319-46976-8_18
26. Nie, D., et al.: Medical image synthesis with context-aware generative adversarial networks. In: Descoteaux, M., Maier-Hein, L., Franz, A., Jannin, P., Collins, D.L., Duchesne, S. (eds.) MICCAI 2017. LNCS, vol. 10435, pp. 417–425. Springer, Cham (2017). https://doi.org/10.1007/978-3-319-66179-7_48
27. Peng, C., Zhang, X., Yu, G., Luo, G., Sun, J.: Large kernel matters-improve semantic segmentation by global convolutional network. In: CVPR (2017)

28. Ronneberger, O., Fischer, P., Brox, T.: U-net: convolutional networks for biomedical image segmentation. In: Navab, N., Hornegger, J., Wells, W.M., Frangi, A.F. (eds.) MICCAI 2015. LNCS, vol. 9351, pp. 234–241. Springer, Cham (2015). https://doi.org/10.1007/978-3-319-24574-4_28

29. Roy, S., Butman, J.A., Pham, D.L.: Synthesizing CT from ultrashort echo-time MR images via convolutional neural networks. In: SSMI Workshop, pp. 24–32 (2017)

30. Sjölund, J., Forsberg, D., Andersson, M., Knutsson, H.: Generating patient specific pseudo-CT of the head from MR using atlas-based regression. Phys. Med. Biol. **60**(2), 825 (2015)

31. Wolterink, J.M., Dinkla, A.M., Savenije, M.H., Seevinck, P.R., van den Berg, C.A., Išgum, I.: MR-to-CT synthesis using cycle-consistent generative adversarial networks. In: NIPS (2017)

Ultrasound Variational Style Transfer to Generate Images Beyond the Observed Domain

Alex Ling Yu Hung[✉] and John Galeotti

Carnegie Mellon University, Pittsburgh, PA 15213, USA
{lingyuh,jgaleotti}@andrew.cmu.edu

Abstract. The use of deep learning in medical image analysis is hindered by insufficient annotated data and the inability of models to generalize between different imaging settings. We address these problems using a novel variational style-transfer neural network that can sample various styles from a computed latent space to generate images from a broader domain than what was observed. We show that using our generative approach for ultrasound data augmentation and domain adaptation during training improves the performance of the resulting deep learning models, even when tested within the observed domain.

Keywords: Style transfer · Variational autoencoder · Generative models · Data augmentation · Domain generalization

1 Introduction

Deep learning in medical imaging has immense potential but the challenge is that images taken by different machines or in different settings follow different distributions [4]. As a result, models are usually not generalized enough to be used across datasets. Especially in ultrasound, images captured by different imaging settings on different subjects can be so different that the model trained on one dataset could completely fail on another dataset. Besides, since annotation of medical images is laborious work which requires experts, the number of annotated images is limited. It is extremely difficult to get a large dataset of annotated medical images that are from different distributions. Therefore, we propose a continuous neural style transfer algorithm, which is capable of generating ultrasound images with known content from unknown style latent space.

In the first attempt at style transfer with convolutional neural networks (CNNs), the content and style features were extracted using pretrained VGGs [21] and used to iteratively optimize the output image on-the-fly [6]. Feed-forward frameworks were then proposed to get rid of the numerous iterations during inference [10,23]. The gram matrix, which describes the style information in images, were further explained in [14]. To control the style of the output, ways to manipulate the spatial location, color information and spatial scale were introduced in

S. Engelhardt et al. (Eds.): DGM4MICCAI 2021/DALI 2021, LNCS 13003, pp. 14–23, 2021.
https://doi.org/10.1007/978-3-030-88210-5_2

[7]. Adaptive Instance Normalization (AdaIN) was proposed as a way to perform arbitrary style transfer in real-time [8]. Further improvements were made where the image transformation network was directly generated from a style image via meta network [20]. Some works are dedicated to using style transfer as a data augmentation tool [17,25], which showed that style transfer on natural images can improve the performance of classification and segmentation.

In medical image analysis, [22] directly applied style transfer to fundus images for augmentation, while [3] analyzed the style of ultrasound images encoded by VGG encoders. [16] improved the segmentation results of cardiovascular MR images by style transfer. [5] built their network upon StyleGAN [11] to generate high-resolution medical images. [24] showed that generated medical images based on style transfer can improve the results of semantic segmentation on CT scans. [15] proposed a method to do arbitrary style transfer on ultrasound images.

However, current works can only generate one result given a content and a style image, and few can sample the style from a latent space. Works in medical domain are mostly tested on images with similar styles, limiting the ability of generalization. As the style should follow a certain distribution instead of some specific values, we intent to generate multiple plausible output images given a content and a style image. Furthermore, we also want to sample the style in a continuous latent space so that we would be able to generate images from unseen styles. We propose a variational style transfer approach on medical images, which has the following contribution: (1) To the best of our knowledge, our method is the first variational style transfer approach that to explicitly sample the style from a latent space without giving the network a style reference image. (2) Our approach can be used to augment the data in ultrasound images, which results in better segmentation. (3) The method that we propose is able to transform the ultrasound images taken in one style to an unobserved style.

2   Methods

Our style transfer network consists of three parts: style encoder E_s, content encoder E_c, and decoder D. The network structure is shown in Fig. 1, where I_s is the style image, I_c is the content image and $\hat{I}$ is the output image. During training, the decoder D learns to generate $\hat{I}$ with the Gaussian latent variables z conditioned on the content image I_c, while the style encoder E_s learns the distribution of z given style image I_s. Before putting the three parts together, we pre-train the style encoder and the content encoder separately to provide better training stability. When generating images, our method can either use certain given style, or sample the style from the latent space.

2.1   Style Encoder

The style encoder E_s is the encoder part of a U-Net [19] based Variational Autoencoder (VAE) [13]. The difference between our VAE and traditional ones

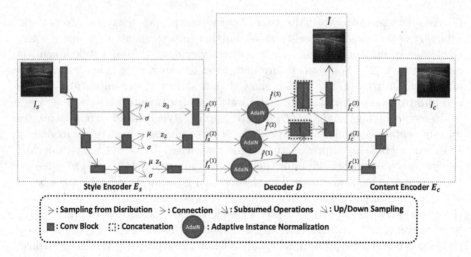

Fig. 1. The network structure of the proposed method, which consists of two encoders that process the style image and content image separately, and a decoder. The style encoder generates distributions for some latent variables given the style image $q_\phi(\mathbf{z}|I_s)$, to create style that could change continuously.

is that our latent variables are at different scales to generate images with better resolutions. The style encoder E_s approximates the distribution of latent variables at different scale. In other words, it learns distributions $q_\phi(z_i|I)$ that approximates the intractable true distribution $p_\theta(z_i|I)$, where ϕ is the variational parameters, while θ is the generative model parameters. Therefore the variational lower bound could be written as:

$$\mathcal{L}(\theta, \phi; I_s) = -KL(q_\phi(\mathbf{z}|I_s)||p_\theta(\mathbf{z})) + \mathbb{E}_{q_\phi(\mathbf{z}|I_s)} \log p_\theta(I_s|\mathbf{z}) \qquad (1)$$

where $KL(\cdot)$ is the Kullback–Leibler (KL) divergence between two distributions, and we further assume $\mathbf{z}|\theta \sim \mathcal{N}(0, 1)$.

The encoder is further incorporated into the style transfer network while the decoder here is only used during initial training. The structure of the style encoder is shown in Fig. 1, while the decoder part is a traditional U-Net decoder.

2.2 Content Encoder

The structure of the content encoder E_c is shown in Fig. 1, and like the style encoder E_s, it is the encoder of a U-Net autoencoder, where the decoder is only used in the initial training with the exact opposite structure of the encoder.

2.3 Decoder

The decoder takes in the encoded style and content before generating a new image, and is also designed based on a traditional U-Net. The structure is shown in Fig. 1. Denote the content features as $f_c = E_c(I_c)$ and the style feature as $f_s = E_s(I_s)$. Also denote the content and style features at scale i as $f_c^{(i)}$ and $f_s^{(i)}$ respectively. The features of the output images at scale i, $\hat{f}^{(i)}$, can be treated as the input into the decoder of a normal U-Net. Inspired by [8], we utilize AdaIN to perform style transfer in feature space at each scale to calculate $\hat{f}^{(i)}$ based on $f_c^{(i)}$ and $f_s^{(i)}$. The calculation can be expressed as:

$$\hat{f}^{(i)} = AdaIN(f_c^{(i)}, f_s^{(i)}) = \sigma(f_s^{(i)}(\frac{f_c^{(i)} - \mu(f_c^{(i)})}{\sigma(f_c^{(i)})})) + \mu(f_s^{(i)}) \tag{2}$$

2.4   Loss Functions

The training objective of the network is to generate an output image $\hat{I}$ containing the contents in the content image I_c and having the style of the style image I_s, all while maximizing the variational lower bound. Therefore, the loss function is made up of three parts: perceptual loss, style loss, and KL divergence loss.

Perceptual and style losses, are based on high level features extracted by pre-trained VGGs [21] and were first utilized in [6]. Perceptual loss is the difference between two feature maps encoded by a CNN. Since spatial correlation is considered in perceptual loss L_p, it is deemed as an expression of the content similarity between two images, which can be expressed as follows:

$$L_p = \sum_{i=1}^{N_{VGG}} w_i^p ||\psi_i(I_c) - \psi_i(\hat{I})|| \tag{3}$$

where $\psi_i(x)$ extracts the layer i of VGG from x, w_i^p is the weight at layer i for perceptual loss, and N_{VGG} as the total number of layers in VGG.

Denote the number of channels, height, and width of the i th layer of the feature map as C_i, H_i, and W_i respectively. We also denote the $C_i \times C_i$ gram matrix of i th layer of the feature map of image x as $G_i(x)$. The gram matrix can be described as:

$$G_i(x)(u,v) = \frac{\sum_{h=1}^{H_i} \sum_{w=1}^{W_i} \psi_i(x)(h,w,u)\psi_i(x)(h,w,v)}{C_i H_i W_i} \tag{4}$$

Since the gram matrix only records the relationship between different channels rather than the spatial correlation, it is considered to be the representation of the general textures and patterns of an image. Let w_i^s be the weight of the loss at layer i for style loss, the style loss L_s is the distance of two gram matrices:

$$L_s = \sum_{i=1}^{N_{VGG}} w_i^s ||G_i(I_s) - G_i(\hat{I})|| \tag{5}$$

To maximize the variational lower bound, we need to minimize the KL divergence between $q_\phi(\mathbf{z}|I_s)$ and $p_\theta(\mathbf{z})$. Under the assumption that all the latent variables are i.i.d., we calculate the KL divergence on each scale and sum over all the

KL divergence to get the KL loss. Since we already assume that $\mathbf{z}|\theta \sim \mathcal{N}(0,1)$, we can derive the KL divergence at each scale as:

$$KL(q_\phi(\mathbf{z}|I_s)\|p_\theta(\mathbf{z})) = \frac{1}{2}(\frac{\mu^2}{\sigma^2} + \sigma^2 - \log \sigma^2 - 1) \tag{6}$$

where we assume $\mathbf{z}|I_s; \phi \sim N(\mu,\sigma)$.

2.5 Implementation Details

Encoders and decoders in the network follow the architecture below. There are 2 convolutional layers in each ConvBlock, followed by batch normalization [9] and swish activation [18]. We use 5 ConvBlocks in the encoder side which are followed by a max pooling layer, except for the last one. The number of convolutional filters are 64, 128, 256, 512, 1024 respectively. The decoders follow the inverse structure of the encoder. Note that in the style encoder, shown in Fig. 1, there is an additional ConvBlock before generating the distribution $q_\phi(z_i|I_s)$ after the normal U-Net [19] encoder at each scale. There is also another ConvBlock after sampling z_i from $q_\phi(z_i|I_s)$ to calculate $f_s^{(i)}$ at each scale. The weights for perceptual and style loss are set to 0, 0, 0, 0, 0.01 and 0.1, 0.002, 0.001, 0.01, 10 for *block1_conv1, block2_conv1, block3_conv1, block4_conv1, block5_conv1* of VGG respectively. The network is optimized via Adam optimizer [12] with a learning rate of 5×10^{-5} and trained for 20 epochs on a single Nvidia Titan RTX GPU.

3 Experiments

The images in the experiments are a combination of numerous datasets: (1) lung images on clinical patients by Sonosite ultrasound machine with HFL38xp linear transducer, (2) chicken breast images by UF-760AG Fukuda Denshi using a linear transducer with 51 mm scanning width (FDL), (3) live-pig images by FDL, (4) blue-gel phantom images by FDL, (5) leg phantom images by FDL, (6) Breast Ultrasound Images Dataset [2], (7) Ultrasound Nerve Segmentation dataset from Kaggle [1], (8) arteries and veins in human subjects by aVisualsonics Vevo 2100 UHFUS machine with ultrahigh frequency scanners. In total, there are 18308, 2285, 2283 images for training, validation, and testing respectively in the combined dataset. During training, we randomly select a pair of content and style images from the combined dataset without any additional restrictions.

3.1 Qualitative Results

In the first experiment, we directly generate the outputs given the content and style images. We transfer the style of images across (1)–(8). Shown in Fig. 2, the visual results are good in each combination of content and style images. Moreover, all the results still have the anatomy in the content images, including but not limited to vessels and ligaments, while looking like the style images. On

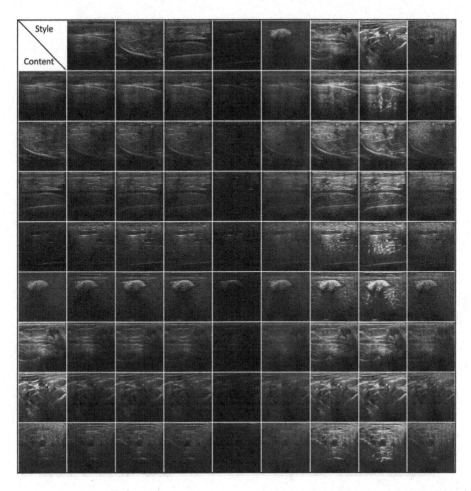

Fig. 2. Results of the algorithm given style and content images. We transfer content images from the first column to the style in the corresponding image in the first row. The content images from top to bottom and the style images from left to right are images from dataset (1)–(8).

Fig. 3. Results of the method by Huang et al. [8]. The left is the content image (same as the last one in Fig. 2), whose style is transferred to styles of (1)–(8) (right 8 images).

the contrary, shown in Fig. 3, method proposed by Huang et al. [8], is not able to capture the fine details in the content and generate realistic textures.

Another qualitative experiment is that we directly sample the style from the latent space without giving the model a style image. Figure 4 shows the

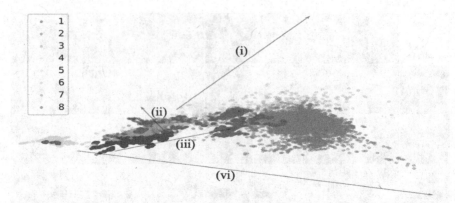

Fig. 4. The distribution (flattened to fit the paper) of styles in the latent space.

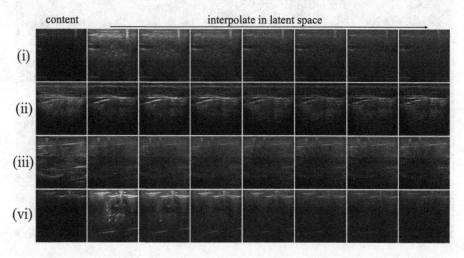

Fig. 5. Results of directly sampling from the latent space and interpolation. The first column are the content image. We sampled the second and the last column (starting point and ending point of the arrows in Fig. 4), then interpolate in between in the latent space in the direction of the corresponding arrows.

distribution of our training style images in the latent space. We then randomly sampled two styles (at the end points of the arrow in Fig. 4) from the latent space and interpolate between the two, and the results can be found in Fig. 5. We observe that even without a style image, the model can still generate visually reasonable ultrasound images given the content image without losing significant details while sampling from latent space that is not covered by the training data.

3.2   Quantitative Results

To show that our approach is effective in augmentation, assume that we only have limited live-pig data while having more leg-phantom data for veins and

Table 1. Comparison between style-transfer-based augmentation and the traditional augmentation

	Veins				Arteries			
	Dice	IoU	Pre.	Rec.	Dice	IoU	Pre.	Rec.
vt-ph/pig-99/aug.	0.96	0.92	0.93	0.99	0.93	0.86	0.87	**0.99**
vt-ph/pig-99	**0.97**	**0.95**	**0.96**	**0.99**	0.95	0.91	0.92	0.99
dt-ph/pig-99/aug.	0.94	0.90	0.92	0.97	0.92	0.84	0.88	0.96
dt-ph/pig-99	0.89	0.81	0.82	0.98	0.83	0.71	0.71	**0.99**
pig-99/aug.	0.87	0.77	0.84	0.90	0.86	0.75	0.81	0.91
pig-99	0.83	0.71	0.73	0.96	0.81	0.68	0.70	0.95
ph/pig-99/aug.	0.07	0.04	0.04	0.59	0.02	0.01	0.01	0.97
ph/pig-99	0.05	0.03	0.36	0.03	0.14	0.07	0.13	0.14
vt-ph/pig-77/aug.	**0.95**	**0.91**	**0.95**	0.96	**0.93**	**0.87**	0.91	0.95
vt-ph/pig-77	0.93	0.88	0.91	0.96	0.90	0.81	**0.95**	0.90
dt-ph/pig-77/aug.	0.93	0.88	0.89	**0.98**	0.87	0.77	0.78	0.98
dt-ph/pig-77	0.94	0.90	0.92	0.97	0.91	0.84	0.85	**0.99**
pig-77/aug.	0.92	0.86	0.87	**0.98**	0.91	0.84	0.85	0.98
pig-77	0.88	0.78	0.90	0.86	0.68	0.51	0.71	0.65
ph/pig-77/aug.	0.07	0.04	0.04	0.55	0.03	0.02	0.02	0.06
ph/pig-77	0.38	0.23	0.68	0.25	0.21	0.12	0.22	0.19
vt-ph/pig-55/aug.	0.74	0.59	0.80	0.69	**0.79**	**0.65**	**0.73**	**0.87**
vt-ph/pig-55	**0.80**	**0.66**	0.87	**0.73**	0.59	0.41	0.65	0.54
dt-ph/pig-55/aug.	0.70	0.53	0.71	0.69	0.53	0.36	0.59	0.48
dt-ph/pig-55	0.76	0.61	**0.88**	0.67	0.66	0.60	0.56	0.81
pig-55/aug.	0.67	0.50	0.87	0.54	0.43	0.27	0.37	0.27
pig-55	0.15	0.01	0.09	0.43	0.01	0	0.01	0
ph/pig-55/aug.	0.04	0.02	0.02	0.07	0.02	0.02	0.01	0.81
ph/pig-55	0.22	0.12	0.45	0.14	0.1	0.05	0.05	0.53
vt-ph/pig-33/aug.	**0.54**	**0.37**	0.63	0.47	0.37	0.23	0.42	0.33
vt-ph/pig-33	0.43	0.28	0.38	**0.49**	**0.56**	**0.39**	**0.43**	**0.82**
dt-ph/pig-33/aug.	0.11	0.06	0.13	0.10	0.06	0.03	0.03	0.37
dt-ph/pig-33	0.53	**0.37**	**0.68**	0.44	0.34	0.20	0.36	0.32
pig-33/aug.	0.15	0.08	0.09	0.66	0.01	0	0	0.01
pig-33	0.06	0.03	0.06	0.06	0	0	0.01	0
ph/pig-33/aug.	0.07	0.04	0.04	0.58	0.02	0.01	0.01	0.37
ph/pig-33	0.01	0	0.02	0	0	0	0	0

arteries (405 images). Denote *vt-ph*, *dt-ph* and *ph* as variational-style-transferred (ours), deterministic-style-transferred [8] and original phantom images respectively, *aug.* as traditional augmentation methods including gamma transform, Gaussian blurring, and image flipping. We further evaluate the effects of the number of the real images have on the final results. We transfer the leg-phantom images into the style of pig images. For live-pig data, we set 99 images as training set while having 11 images in each of the validation set and test set. In the experiments, we train U-Nets with the same network implementation and training settings as [19], on 99, 77, 55, 33 live-pig images denoted as *pig-99*, *pig-77* etc. to show that our method works with very limited number of images. Note that, we balance the phantom data and pig data as roughly 1:1 ratio in each epoch. Shown in Table 1, the segmentation performs better by the networks trained on variational-style-transferred phantom images and pig images than the ones trained on other images in general. Additional augmentation on top of our variational-style-transfer augmentation sometimes improve the performance but our variational-style-transfer augmentation is an upgrade over the traditional augmentation and deterministic style transfer by Huang et al. [8]. Besides, when the number of real live-pig data is really limited, only style-transfer augmentation can produce a decent result. In any case, it can be seen that style-transfer-based approach has a significant improvement over traditional methods and our variantional approach is superior than Huang et al.'s method [8].

4 Conclusion

We demonstrated that our method is capable of transferring the style of one ultrasound image to another style, e.g. from the style of phantom data to that of pig data, from the style of normal ultrasound machines to the style of high frequency ultrasound machines, etc.. Besides, it is also able to sample arbitrary and continuous style from a latent space. Our method can generate ultrasound images from both observed and unobserved domains, which helps address the insufficiency of data and labels insufficiency in medical imaging.

References

1. Ultrasound Nerve Segmentation: Identify nerve structures in ultrasound images of the neck (2016). https://www.kaggle.com/c/ultrasound-nerve-segmentation
2. Al-Dhabyani, W., Gomaa, M., Khaled, H., Fahmy, A.: Dataset of breast ultrasound images. Data Brief **28**, 104863 (2020)
3. Byra, M.: Discriminant analysis of neural style representations for breast lesion classification in ultrasound. Biocybern. Biomed. Eng. **38**(3), 684–690 (2018)
4. Ching, T., et al.: Opportunities and obstacles for deep learning in biology and medicine. J. R. Soc. Interface **15**(141), 20170387 (2018)
5. Fetty, L., et al.: Latent space manipulation for high-resolution medical image synthesis via the styleGAN. Z. Med. Phys. **30**(4), 305–314 (2020)
6. Gatys, L.A., Ecker, A.S., Bethge, M.: A neural algorithm of artistic style. arXiv preprint arXiv:1508.06576 (2015)
7. Gatys, L.A., Ecker, A.S., Bethge, M., Hertzmann, A., Shechtman, E.: Controlling perceptual factors in neural style transfer. In: Proceedings of the IEEE Conference on Computer Vision and Pattern Recognition, pp. 3985–3993 (2017)
8. Huang, X., Belongie, S.: Arbitrary style transfer in real-time with adaptive instance normalization. In: Proceedings of the IEEE International Conference on Computer Vision, pp. 1501–1510 (2017)
9. Ioffe, S., Szegedy, C.: Batch normalization: accelerating deep network training by reducing internal covariate shift. arXiv preprint arXiv:1502.03167 (2015)

10. Johnson, J., Alahi, A., Fei-Fei, L.: Perceptual losses for real-time style transfer and super-resolution. In: Leibe, B., Matas, J., Sebe, N., Welling, M. (eds.) ECCV 2016. LNCS, vol. 9906, pp. 694–711. Springer, Cham (2016). https://doi.org/10.1007/978-3-319-46475-6_43

11. Karras, T., Laine, S., Aila, T.: A style-based generator architecture for generative adversarial networks. In: Proceedings of the IEEE Conference on Computer Vision and Pattern Recognition, pp. 4401–4410 (2019)

12. Kingma, D.P., Ba, J.: Adam: a method for stochastic optimization. arXiv preprint arXiv:1412.6980 (2014)

13. Kingma, D.P., Welling, M.: Auto-encoding variational bayes. arXiv preprint arXiv:1312.6114 (2013)

14. Li, Y., Wang, N., Liu, J., Hou, X.: Demystifying neural style transfer. arXiv preprint arXiv:1701.01036 (2017)

15. Liu, Z., et al.: Remove appearance shift for ultrasound image segmentation via fast and universal style transfer. In: 2020 IEEE 17th International Symposium on Biomedical Imaging (ISBI), pp. 1824–1828. IEEE (2020)

16. Ma, C., Ji, Z., Gao, M.: Neural style transfer improves 3D cardiovascular MR image segmentation on inconsistent data. In: Shen, D., et al. (eds.) MICCAI 2019. LNCS, vol. 11765, pp. 128–136. Springer, Cham (2019). https://doi.org/10.1007/978-3-030-32245-8_15

17. Mikołajczyk, A., Grochowski, M.: Style transfer-based image synthesis as an efficient regularization technique in deep learning. In: 2019 24th International Conference on Methods and Models in Automation and Robotics (MMAR), pp. 42–47. IEEE (2019)

18. Ramachandran, P., Zoph, B., Le, Q.V.: Swish: a self-gated activation function. arXiv preprint arXiv:1710.05941 **7** (2017)

19. Ronneberger, O., Fischer, P., Brox, T.: U-net: convolutional networks for biomedical image segmentation. In: Navab, N., Hornegger, J., Wells, W.M., Frangi, A.F. (eds.) MICCAI 2015. LNCS, vol. 9351, pp. 234–241. Springer, Cham (2015). https://doi.org/10.1007/978-3-319-24574-4_28

20. Shen, F., Yan, S., Zeng, G.: Neural style transfer via meta networks. In: Proceedings of the IEEE Conference on Computer Vision and Pattern Recognition, pp. 8061–8069 (2018)

21. Simonyan, K., Zisserman, A.: Very deep convolutional networks for large-scale image recognition. arXiv preprint arXiv:1409.1556 (2014)

22. Singh, K., Drzewicki, D.: Neural style transfer for medical image augmentation. https://scholar.google.com/scholar?hl=en&as_sdt=0%2C5&q=Singh%2C+K.%2C+Drzewicki%2C+D.%3A+Neural+style+transfer+for+medical+image+augmentation&btnG=

23. Ulyanov, D., Lebedev, V., Vedaldi, A., Lempitsky, V.S.: Texture networks: feed-forward synthesis of textures and stylized images. In: ICML, vol. 1, p. 4 (2016)

24. Xu, Y., Li, Y., Shin, B.-S.: Medical image processing with contextual style transfer. Hum.-Cent. Inf. Sci. **10**(1), 1–16 (2020). https://doi.org/10.1186/s13673-020-00251-9

25. Zheng, X., Chalasani, T., Ghosal, K., Lutz, S., Smolic, A.: STaDA: style transfer as data augmentation. arXiv preprint arXiv:1909.01056 (2019)

3D-StyleGAN: A Style-Based Generative Adversarial Network for Generative Modeling of Three-Dimensional Medical Images

Sungmin Hong[1]([envelope]), Razvan Marinescu[2], Adrian V. Dalca[2,3],
Anna K. Bonkhoff[1], Martin Bretzner[1,4], Natalia S. Rost[1], and Polina Golland[2]

[1] JPK Stroke Research Center, Department of Neurology,
Massachusetts General Hospital, Harvard Medical School, Boston, MA, USA
`shong20@mgh.harvard.edu`
[2] Computer Science and Artificial Intelligence Laboratory,
Massachusetts Institute of Technology, Cambridge, MA, USA
[3] A.A. Martinos Center for Biomedical Imaging, Department of Radiology,
Massachusetts General Hospital, Harvard Medical School, Cambridge, MA, USA
[4] Univ. Lille, Inserm, CHU Lille, U1172 - LilNCog (JPARC) - Lille Neurosciences
and Cognition, 59000 Lille, France

Abstract. Image synthesis via Generative Adversarial Networks (GANs) of three-dimensional (3D) medical images has great potential that can be extended to many medical applications, such as, image enhancement and disease progression modeling. Current GAN technologies for 3D medical image synthesis must be significantly improved to be suitable for real-world medical problems. In this paper, we extend the state-of-the-art StyleGAN2 model, which natively works with two-dimensional images, to enable 3D image synthesis. In addition to the image synthesis, we investigate the behavior and interpretability of the 3D-StyleGAN via style vectors inherited form the original StyleGAN2 that are highly suitable for medical applications: (i) the latent space projection and reconstruction of unseen real images, and (ii) style mixing. The model can be applied to any 3D volumetric images. We demonstrate the 3D-StyleGAN's performance and feasibility with ∼12,000 three-dimensional full brain MR T1 images. Furthermore, we explore different configurations of hyperparameters to investigate potential improvement of the image synthesis with larger networks. The codes and pre-trained networks are available online: https://github.com/sh4174/3DStyleGAN.

1 Introduction

Generative modeling via Generative Adversarial Networks (GAN) has achieved remarkable improvements with respect to the quality of generated images [3,4, 11,21,32]. StyleGAN2, a style-based generative adversarial network, has been recently proposed for synthesizing highly realistic and diverse natural images. It

© Springer Nature Switzerland AG 2021
S. Engelhardt et al. (Eds.): DGM4MICCAI 2021/DALI 2021, LNCS 13003, pp. 24–34, 2021.
https://doi.org/10.1007/978-3-030-88210-5_3

progressively accounts for multi-resolution information of images during training and controls image synthesis using style vectors that are used at each block of a style-based generator network [18–21]. StyleGAN2 achieves the outstanding quality of generated images, and enhanced control and interpretability for image synthesis compared to previous works [3,4,20,21]. Despite of its great potential for medical applications due to its enhanced performance and interpretability, it is yet to be extended to the generative modeling of 3D medical images.

One of the challenges of using generative models such as GANs for medical applications is that medical images are often three-dimensional (3D) and have a significantly higher number of voxels compared to two-dimensional natural images. Due to the high-memory requirements of GANs, it is often not feasible to directly apply large networks for 3D image synthesis [4,22,30]. To address this challenge, some models for 3D image synthesis generate the image in successive 2D slices, which are then combined to render a 3D image while accounting for relationships among slices [30]. However, these methods often result in discontinuity between slices, and the interpretation and manipulation of the slice-specific latent vectors is complicated [30].

In this paper, we present 3D-StyleGAN to enable synthesis of high-quality 3D medical images by extending the StyleGAN2. We made several changes to the original StyleGAN2 architecture: (1) we replaced 2D operations, layers, and noise inputs with their 3D counterparts, and (2) significantly decreased the depths of filter maps and latent vector sizes. We trained different configurations of 3D-StyleGAN on a collection of ∼12,000 T1 MRI brain scans [6]. We additionally demonstrate (i) the possibility of synthesizing realistic 3D brain MRIs at 2mm resolution, corresponding to $80 \times 96 \times 12$ voxels, (ii) projection of unseen test images to the latent space that results in reconstructions of high fidelity to the input images by a projection function suitable for medical images and (iii) StyleGAN2's style-mixing used to "transfer" anatomical variation across images. We discuss the performance and feasibility of the 3D-StyleGAN with limited filter depths and latent vector sizes for 1mm isotropic resolution full brain T1 MR images. The source code and pre-trained networks are publicly available in https://github.com/sh4174/3DStyleGAN.

2   Methods

Figure 1 illustrates the architecture of the 3D-StyleGAN. A style-based generator was first suggested in [20] and updated in [21]. We will briefly introduce the style-based generator suggested in [21] and then summarize the changes we made for enabling 3D image synthesis.

StyleGAN2: A latent vector $z \in \mathcal{Z}$ is first normalized and then mapped by a mapping network m to $\mathbf{w} = m(\mathbf{z})$, $\mathbf{w} \in \mathcal{W}$. A synthesis network g starts with a constant layer with a base size $\mathbf{B} \in \mathbb{R}^d$, where d is the input dimension (two for natural images). The transformed $A(\mathbf{w})$ is modulated (Mod) with trainable convolution weights w and then demodulated ($Demod$), which act as instance normalization to reduce artifacts caused by arbitrary amplification of certain

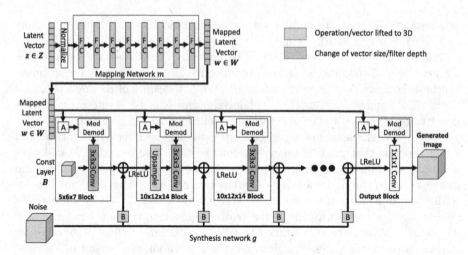

Fig. 1. The style-based generator architecture of 3D-StyleGAN. The mapping network m, made of 8 fully-convolutional layers, maps a normalized latent vector $\mathbf{z}$ to an intermediate latent vector $\mathbf{w}$. This is then transformed by a learned feature-wise affine transform A at each layer, then further used by modulation (Mod) and demodulation (Demod) operations that are applied on the weights of each convolution layer [8,17]. The synthesis network starts with a constant $5 \times 6 \times 7$ block and applies successive convolutions with modulated weights, up-sampling, and activations with Leaky ReLU. 3D noise maps, downscaled to the corresponding layer resolutions, are added before each nonlinear activation by LReLU. At the final layer, the output is convolved by a 1 $\times$ 1 $\times$ 1 convolution to generate the final image. $\mathbf{B}$ is the size of base layer ($5 \times 6 \times 7$). The discriminator network (3D-ResNet) was omitted due to space constraint.

feature maps. Noise inputs scaled by a trainable factor B and a bias b are added to the convolution output at each style block. After the noise addition, the Leaky Rectifier Linear Unit (LReLU) is applied as nonlinear activation [23]. At the final layer in the highest resolution block, the output is fed to the 1×1 convolution filter to generate an image.

Loss Functions and Optimization: For the generator loss function, Style-GAN2 uses the logistic loss function with the path length regularization [21]:

$$\mathbb{E}_{\mathbf{w},\mathbf{y} \sim \mathcal{N}(0,\mathbf{I})}(\|\nabla_{\mathbf{w}}(g(\mathbf{w}) \cdot \mathbf{y})\|_2 - a)^2, \tag{1}$$

where $\mathbf{w}$ is a mapped latent vector, g is a generator, $\mathbf{y}$ is a random image following a normal intensity distribution with the identity covariance matrix, and a is the dynamic constant learned as the running exponential average of the first term over iterations [21]. It regularizes the gradient magnitude of a generated image $g(\mathbf{w})$ projected on $\mathbf{y}$ to be similar to the running exponential average to smooth the (mapped) latent space $\mathcal{W}$. The generator regularization was applied once every 16 minibatches following the lazy regularization strategy. For the discriminator loss function, StyleGAN2 uses the standard logistic loss function with $\mathbf{R}_1$ or $\mathbf{R}_2$ regularizations.

2.1   3D-StyleGAN

We modified StyleGAN2 for 3D image synthesis by replacing the 2D convolutions, upsampling and downsampling with 3D operations. We started from Style-GAN2 configuration F (see [21]), but switched back to the original StyleGAN generator with $(De)Mod$ operators, as it achieved the best performance in the preliminary experiments for 3D images [13,14,21]. We used the 3D residual network for the discriminator and the standard logistic loss without regularization for the discriminator loss function that showed the best empirical results.

Image Projection: An image (either generated or real) can be projected to the latent space by finding $\mathbf{w}$ and stochastic noise $\mathbf{n}$ at each layer that minimize a distance between an input image I and a generated image $g(\mathbf{w}, \mathbf{n})$. In the original StyleGAN2, LPIPS distance was used [31]. However, the LPIPS distance uses the VGG16 network trained with 2D natural images, which is not straightforwardly applicable to 3D images [29]. Instead, we used two mean squared error (MSE) losses, one computed at full resolution, and the other at an (x8) downsampled resolution. The second loss was added to enhance the optimization stability.

Configurations: We tested different resolutions, filter depths, latent vector sizes, minibatch sizes summarized in Table 1. Because of the high dimensionality of 3D images, the filter map depths of the 3D-StyleGAN needed to be significantly lower than in the 2D version. We tested five different feature depths: 16, 32, 64, and 96, with 2mm isotropic resolution brain MR images to investigate how different filter depths affect the quality of generated images. For 1mm isotropic images, we used the filter depth of 16 for the generator and discriminator, 32 filter depth for the mapping network, and 32 latent vector size that our computational resource allowed.

Slice-specific Fréchet Inception Distance Metric: Conventionally, the Fréchet Inception Distance (FID) score is measured by comparing the distributions of randomly sampled real images from a training set and generated images [15]. Because FID relies on the Inception V3 network that was pre-trained with 2D natural images, it is not directly applicable to 3D images [29]. Instead, we measured the Fréchet Inception Distance (FID) scores on the middle slices on axial, coronal, and sagittal planes of the generated 3D images [15].

Additional Evaluation Metrics: In addition to the slice-specific FID metric, we evaluated the quality of generated images with the batch-wise squared Maximum Mean Discrepancy (bMMD^2) as suggested in [22] and [30]. Briefly, MMD^2 measures the distance between two distributions with finite sample estimates with kernel functions in the reproducing kernel Hilbert space [12]. The bMMD^2 measures the discrepancy between images in a batch with a dot product as a kernel [22,30]. To measure the diversity of generated images, we employed the pair-wise Multi-Scale Structural Similarity (MS-SSIM) [22,30]. It captures the perceptual diversity of generated images as the mean of the MS-SSIM scores of pairs of generated image [28].

Table 1. The list of configurations of 3D-StyleGAN with different filter map depths, latent vector sizes and minibatch sizes.

Config.	Filter depth	Latent vector size	Minibatch size {B, 2B, 2^2B, 2^3B, 2^4B}	# trainable params
2mm-fd96	96	96	{32, 32, 16, 16}	6.6M
2mm-fd64	64	64	{32, 32, 16, 16}	3.0M
2mm-fd32	32	128	{32, 32, 16, 16}	0.9M
2mm-fd16	16	64	{32, 32, 16, 16}	0.2M
1mm-fd16	16	64	{64, 64, 32, 16, 16}	0.2M

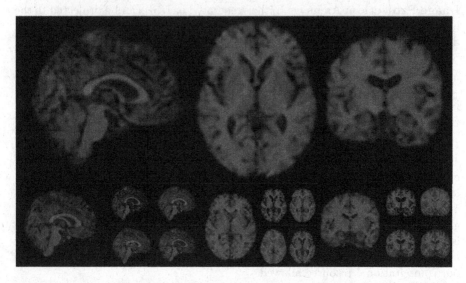

Fig. 2. Uncurated randomly generated 3D images by the 3D-StyleGAN. The images are generated by the network trained with configuration 2mm-fd96. The middle slices of sagittal, axial, and coronal axes of five 3D images were presented.

3   Results

For all experiments, we used 11,820 brain MR T1 images from multiple publicly available datasets: ADNI, OASIS, ABIDE, ADHD2000, MCIC, PPMI, HABS, and Harvard GSP [5–7,10,16,24–27]. All images were skull-stripped, affine-aligned, resampled to 1mm isotropic resolution, and trimmed to 160 × 192 × 224 size using FreeSurfer [6,9]. For the experiments with 2mm isotropic resolution, the images were resampled to an image size of 80 × 96 × 112. Among those images, 7,392 images were used for training. The remaining 4,329 images were used for evaluation. The base layer was set to $\mathbf{B} = 5 \times 6 \times 7$ to account for the input image size. For each experiment, we used four NVIDIA Titan Xp GPUs for the training with 2mm-resolution images, and eight for the training with 1mm-resolution images. The methods were implemented in Tensorflow 1.15 and Python 3.6 [1].

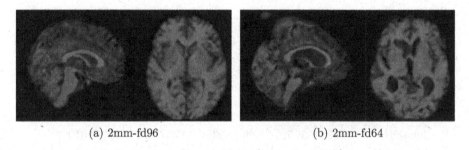

(a) 2mm-fd96   (b) 2mm-fd64

Fig. 3. The qualitative comparison between the networks trained with the filter map depths of (a) 96 (2mm-fd96) and (b) 64 (2mm-fd64). While the generated images of the 2mm-fd64 are sharper, e.g., cerebellum and sulci, the overall brain structures are not as coherent as those in the images generated with 2mm-fd96.

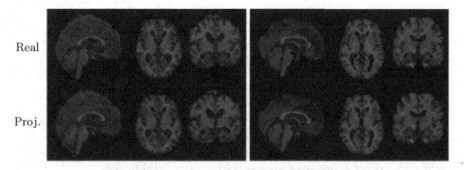

Real

Proj.

Fig. 4. Results of projecting real, unseen 3D images to the latent space. Top row shows the real image, and bottom row shows the reconstruction from the projected embedding. Configuration used was 2mm-fd96. The middle sagittal, axial and coronal slices of the 3D images are displayed. The differences between the real and reconstructed images are indistinguishable other than lattice-like artifacts caused by the generator.

3D Image Synthesis at 2 mm Isotropic Resolution: Figure 2 shows randomly generated 2mm-resolution 3D images using 3D-StyleGAN that was trained with the configuration fd-96 with the filter map depth of 96 (see Table 1 for all configurations). Each model was trained for the average of $\sim$2 days. We observe that anatomical structures are generated correctly and are coherently located.

In Table 2, we show the $bMMD^2$, MS-SSIM, and FIDs for the middle sagittal (FID-Sag), axial (FID-Ax) and coronal (FID-Cor) slices calculated using 4,000 generated images with respect to the unseen test images. One interesting observation was that the slice-specific FIDs of middle sagittal and axial slices of the network trained with the filter map depth of 64 were lower than the one with the filter map depth of 96. Figure 3 provides a qualitative comparison between images generated by networks with a filter map depth of 64 *vs.* 96. Networks with a filter depth of 64 produce sharper boundaries in the cerebellum and sulci, but the overall brain structures are not as coherent as in the images with filter

Table 2. The quantitative evaluations of generated images with different configurations. The Maximum Mean Discrepancy (bMMD^2) and Multi-Scale Structural Similarity Index (MS-SSIM) were calculated on each set of 3D generated and training real images. The slice-specific Fréchet Inception Distance (FID) of the middle slices (Sag: Sagittal, Ax: Axial, and Cor: Coronal) of generated and training real images with respect to test unseen real images were summarized.

Configurations	bMMD^2	MS-SSIM	FID-Sag	FID-Ax	FID-Cor
2mm-fd16	7026 (555.79)	0.96 (0.03)	145.6	153.6	164.1
2mm-fd32	**4470** (532.65)	0.94 (0.07)	129.3	144.3	128.8
2mm-fd64	4497 (898.53)	**0.93** (0.12)	**106.9**	**71.3**	90.2
2mm-fd96	4475 (539.38)	0.96 (0.04)	138.3	83.2	**88.5**
2mm-Real	449 (121.43)	0.85 (0.002)	3.0	2.1	2.9

depth of 96. This is possibly caused by the characteristics of the FID score that focuses on the texture of the generated images, rather than taking into account the overall structures of the brain [31]. The network with filter map depth of 32 showed the best bMMD^2 result while the distributions of the metrics largely overlapped between fd-32, fd-64, and fd-96. The MS-SSIM showed that the network with filter map depth of 64 showed the most diversity in generated images. This may indicate that the better FID metrics of the fd-64 configuration was due to the diversity of generated images and the fd-96 configuration possibly overfit and did not represent the large variability of real images.

Image Projection: Figure 4 shows the results of projecting unseen images onto the space of latent vector **w** and noise map **n**. Top row shows the real, unseen image, while the bottom row shows the reconstruction from the projected embedding. Reconstructed images are almost identical to the input images, a result that has also been observed by [2] on other datasets, likely due to the over-parameterization of the latent space of StyleGAN2.

Style Mixing: Figure 5 shows the result of using 3D-StyleGAN's style mixing capability. The images at the top row and the first column are randomly generated from the trained 3D-StyleGAN with the configuration 2mm-fd96. The style vectors of the images at the top row were used as inputs into the high resolution style blocks (i.e., the 5th to 9th). The style vector of the image in the first column is provided to the low resolution style blocks (i.e., the 1st to 4th). We observed that high-level anatomical variations, such as the size of a ventricle and a brain outline, were controlled by the style vectors of the image at the first column mixed to the lower-resolution style blocks. These results indicate that the style vectors at different layers could potentially control different levels of anatomical variability although our results are preliminary and need further investigation.

Failure Cases: We trained the 3D-StyleGAN for image synthesis of 1mm isotropic resolution images. The depths of filter map and latent vectors sizes were set to 16 and 32, respectively, because of the limited GPU memory (12 GB).

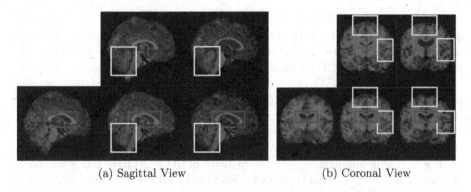

(a) Sagittal View   (b) Coronal View

Fig. 5. Style-mixing example. The style vectors of two 3D images at the top row at the high resolution (5th–9th) style blocks were mixed with the style vectors at at the low resolution (1st–4th) style blocks of an image at the first column were mixed and generated new images. Higher-level anatomical variations, such as the size of a ventricle and corpus callosum (red boxes), were controlled by the lower-resolution style blocks and the detailed variations, such as sulci and cerebellum structures (yellow boxes) by the higher-resolution style blocks. Three input images were randomly generated full 3D images and displayed on the (a) sagittal and (b) coronal views of the respective middle slices. (Color figure online)

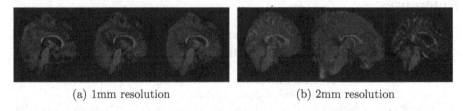

(a) 1mm resolution   (b) 2mm resolution

Fig. 6. Failure cases of 1 mm and 2 mm isotropic resolution with the filter map depth of 16. The low filter map depth resulted in the low quality of images.

The model was trained for ∼5 days. Figure 6 (a) shows the examples of generated 1mm-resolution images on the sagittal view after 320 iterations. The results showed the substantially lower quality of generated images compared to those from the networks trained with deeper filter depths for 2 mm-resolution images. Figure 6 (b) shows the image generation result for 2 mm-resolution images with the same filter depth, 16. Compared to the generated images shown in Fig. 2, the quality of generated images is also substantially lower. This experiment showed that the filter map depth of generator and discriminator networks need to be above certain thresholds to assure the high quality of generated images.

4   Discussion

We presented 3D-StyleGAN, an extension of StyleGAN2 for the generative modeling of 3D medical images. In addition to the high quality of the generated images, the latent space projection of unseen real images and the control of anatomical variability through style vectors can be utilized to tackle clinically important problems. One limitation of this work is in the use of limited evaluation metrics. We used FIDs that evaluated 2D slices due to their reliance on the 2D VGG network. We plan to address 3D quantitative evaluation metrics on perceptual similarity with a large 3D network in future work. Another limitation is the limited size of the training data and the trained networks. We plan to address this in future work through extensive augmentation of the training images and investigating memory-efficient network structures. We believe our work will enable important downstream tasks of 3D generative modeling, such as super-resolution, motion correction, conditional atlas estimation, and image progression modeling with respect to clinical attributes.

Acknowledgements. This research was supported by NIH-NINDS MRI-GENIE: R01NS086905; K23NS064052, R01NS082285, NIH NIBIB NAC P41EB015902, NIH NINDS U19NS115388, Wistron, and MIT-IBM Watson AI Lab. A.K.B. was supported by MGH ECOR Fund for Medical Discovery (FMD) Clinical Research Fellowship Award. M.B. was supported by the ISITE and Planiol foundations and the Sociétés Françaises de Neuroradiologie et Radiologie.

References

1. Abadi, M., et al.: TensorFlow: large-scale machine learning on heterogeneous systems (2015). https://www.tensorflow.org/
2. Abdal, R., Qin, Y., Wonka, P.: Image2styleGAN: how to embed images into the StyleGAN latent space? In: Proceedings of the IEEE/CVF International Conference on Computer Vision, pp. 4432–4441 (2019)
3. Arjovsky, M., Chintala, S., Bottou, L.: Wasserstein generative adversarial networks. In: International Conference on Machine Learning, pp. 214–223. PMLR (2017)
4. Brock, A., Donahue, J., Simonyan, K.: Large scale GAN training for high fidelity natural image synthesis. arXiv preprint arXiv:1809.11096 (2018)
5. Dagley, A., et al.: Harvard aging brain study: dataset and accessibility. Neuroimage **144**, 255–258 (2017)
6. Dalca, A.V., Guttag, J., Sabuncu, M.R.: Anatomical priors in convolutional networks for unsupervised biomedical segmentation. In: Proceedings of the IEEE Conference on Computer Vision and Pattern Recognition, pp. 9290–9299 (2018)
7. Di Martino, A., et al.: The autism brain imaging data exchange: towards a large-scale evaluation of the intrinsic brain architecture in autism. Mol. Psychiatry **19**(6), 659–667 (2014)
8. Dumoulin, V., et al.: Feature-wise transformations. Distill **3**(7), e11 (2018)
9. Fischl, B.: Freesurfer. Neuroimage **62**(2), 774–781 (2012)
10. Gollub, R.L., et al.: The MCIC collection: a shared repository of multi-modal, multi-site brain image data from a clinical investigation of schizophrenia. Neuroinformatics **11**(3), 367–388 (2013)
11. Goodfellow, I.J., et al.: Generative adversarial networks. arXiv preprint arXiv:1406.2661 (2014)
12. Gretton, A., Borgwardt, K.M., Rasch, M.J., Schölkopf, B., Smola, A.: A kernel two-sample test. J. Mach. Learn. Res. **13**(1), 723–773 (2012)

13. Hara, K., Kataoka, H., Satoh, Y.: Can spatiotemporal 3d CNNs retrace the history of 2D CNNs and ImageNet? In: Proceedings of the IEEE Conference on Computer Vision and Pattern Recognition (CVPR), pp. 6546–6555 (2018)
14. He, K., Zhang, X., Ren, S., Sun, J.: Deep residual learning for image recognition. In: Proceedings of the IEEE Conference on Computer Vision and Pattern Recognition, pp. 770–778 (2016)
15. Heusel, M., Ramsauer, H., Unterthiner, T., Nessler, B., Hochreiter, S.: GANs trained by a two time-scale update rule converge to a local nash equilibrium. arXiv preprint arXiv:1706.08500 (2017)
16. Holmes, A.J., et al.: Brain genomics superstruct project initial data release with structural, functional, and behavioral measures. Sci. Data 2(1), 1–16 (2015)
17. Huang, X., Belongie, S.: Arbitrary style transfer in real-time with adaptive instance normalization. In: Proceedings of the IEEE International Conference on Computer Vision, pp. 1501–1510 (2017)
18. Karras, T., Aila, T., Laine, S., Lehtinen, J.: Progressive growing of GANs for improved quality, stability, and variation. arXiv preprint arXiv:1710.10196 (2017)
19. Karras, T., Aittala, M., Hellsten, J., Laine, S., Lehtinen, J., Aila, T.: Training generative adversarial networks with limited data. arXiv preprint arXiv:2006.06676 (2020)
20. Karras, T., Laine, S., Aila, T.: A style-based generator architecture for generative adversarial networks. In: Proceedings of the IEEE/CVF Conference on Computer Vision and Pattern Recognition, pp. 4401–4410 (2019)
21. Karras, T., Laine, S., Aittala, M., Hellsten, J., Lehtinen, J., Aila, T.: Analyzing and improving the image quality of StyleGAN. In: Proceedings of the IEEE/CVF Conference on Computer Vision and Pattern Recognition, pp. 8110–8119 (2020)
22. Kwon, G., Han, C., Kim, D.: Generation of 3D brain MRI using auto-encoding generative adversarial networks. In: Shen, D., et al. (eds.) MICCAI 2019. LNCS, vol. 11766, pp. 118–126. Springer, Cham (2019). https://doi.org/10.1007/978-3-030-32248-9_14
23. Maas, A.L., Hannun, A.Y., Ng, A.Y.: Rectifier nonlinearities improve neural network acoustic models. In: Proceedings of ICML, vol. 30, p. 3. Citeseer (2013)
24. Marcus, D.S., Wang, T.H., Parker, J., Csernansky, J.G., Morris, J.C., Buckner, R.L.: Open access series of imaging studies (OASIS): cross-sectional MRI data in young, middle aged, nondemented, and demented older adults. J. Cogn. Neurosci. 19(9), 1498–1507 (2007)
25. Marek, K., et al.: The Parkinson progression marker initiative (PPMI). Prog. Neurobiol. 95(4), 629–635 (2011)
26. Milham, M.P., Fair, D., Mennes, M., Mostofsky, S.H., et al.: The ADHD-200 consortium: a model to advance the translational potential of neuroimaging in clinical neuroscience. Front. Syst. Neurosci. 6, 62 (2012)
27. Mueller, S.G., et al.: Ways toward an early diagnosis in Alzheimer's disease: the Alzheimer's disease neuroimaging initiative (ADNI). Alzheimer's Dement. 1(1), 55–66 (2005)
28. Odena, A., Olah, C., Shlens, J.: Conditional image synthesis with auxiliary classifier GANs. In: International Conference on Machine Learning, pp. 2642–2651. PMLR (2017)
29. Simonyan, K., Zisserman, A.: Very deep convolutional networks for large-scale image recognition. arXiv preprint arXiv:1409.1556 (2014)
30. Volokitin, A., et al.: Modelling the distribution of 3D brain MRI using a 2D slice VAE. In: Martel, A.L., et al. (eds.) MICCAI 2020. LNCS, vol. 12267, pp. 657–666. Springer, Cham (2020). https://doi.org/10.1007/978-3-030-59728-3_64

31. Zhang, R., Isola, P., Efros, A.A., Shechtman, E., Wang, O.: The unreasonable effectiveness of deep features as a perceptual metric. In: Proceedings of the IEEE Conference on Computer Vision and Pattern Recognition, pp. 586–595 (2018)
32. Zhu, J.Y., Park, T., Isola, P., Efros, A.A.: Unpaired image-to-image translation using cycle-consistent adversarial networks. In: Proceedings of the IEEE International Conference on Computer Vision, pp. 2223–2232 (2017)

Bridging the Gap Between Paired and Unpaired Medical Image Translation

Pauliina Paavilainen[1,2]($\boxtimes$), Saad Ullah Akram[1,2]iD, and Juho Kannala[1]iD

[1] Aalto University, Espoo, Finland
[2] MVision AI, Helsinki, Finland
{pauliina.paavilainen,saad.akram}@mvision.ai

Abstract. Medical image translation has the potential to reduce the imaging workload, by removing the need to capture some sequences, and to reduce the annotation burden for developing machine learning methods. GANs have been used successfully to translate images from one domain to another, such as MR to CT. At present, paired data (registered MR and CT images) or extra supervision (e.g. segmentation masks) is needed to learn good translation models. Registering multiple modalities or annotating structures within each of them is a tedious and laborious task. Thus, there is a need to develop improved translation methods for unpaired data. Here, we introduce modified pix2pix models for tasks CT→MR and MR→CT, trained with unpaired CT and MR data, and MRCAT pairs generated from the MR scans. The proposed modifications utilize the paired MR and MRCAT images to ensure good alignment between input and translated images, and unpaired CT images ensure the MR→CT model produces realistic-looking CT and CT→MR model works well with real CT as input. The proposed pix2pix variants outperform baseline pix2pix, pix2pixHD and CycleGAN in terms of FID and KID, and generate more realistic looking CT and MR translations.

Keywords: Medical image translation · Generative adversarial network

1 Introduction

Each medical imaging modality captures specific characteristics of the patient. In many medical applications, complimentary information from multiple modalities can be combined for better diagnosis or treatment. However, due to limited time, cost and patient safety, not all desired modalities are captured for every patient. Medical image translation can play a vital role in many of these scenarios as it can be used to generate non-critical (i.e. the ones which are not needed for fine pattern matching) missing modalities. One such clinical application is in MR-based radiotherapy, where MR images are used for delineating targets (e.g. tumors) and organs-at-risk (OARs). However, the clinicians still need CT scans for calculating the dose delivered to OARs and targets. Since this CT scan is

© Springer Nature Switzerland AG 2021
S. Engelhardt et al. (Eds.): DGM4MICCAI 2021/DALI 2021, LNCS 13003, pp. 35–44, 2021.
https://doi.org/10.1007/978-3-030-88210-5_4

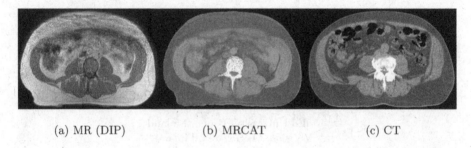

(a) MR (DIP)   (b) MRCAT   (c) CT

Fig. 1. MR and MRCAT are voxel-aligned, while CT is from a different patient.

used to compute dose delivered to various structures, the alignment of the patient anatomy is more important than the precise voxel intensities. There exist few commercial solutions which can be used to generate synthetic CT scans, such as Philips' MRCAT. However, these translated images do not look very realistic due to quantization artifacts, failure to reduce blurring caused by patient breathing and lack of air cavities, etc. (see Fig. 1b).

Medical image translation can also play a critical role in development of machine learning based image analysis methods. In many applications, same tasks are performed in different modalities, e.g. in radiotherapy, OAR contouring is done in either CT or MR. Medical image translation can significantly speed-up the development of these automated methods by reducing the annotation requirements for new modalities/sequences. A large dataset can be annotated in one modality and then image translation can be used to generate scans of new modalities and annotations copied to synthetic scans.

During recent years, there has been increased interest in using Generative Adversarial Networks (GANs) [8] in medical image generation. The research in the field includes low-dose CT denoising [28], MR to CT translation [7,19,20,27], CT to MR translation [13], applications in deformable image registration [24] and segmentation [12], data augmentation [23], PET to CT translation, MR motion correction [1], and PET denoising [2] (for review, see [29]).

pix2pix [11] and CycleGAN [32] are two popular general-purpose image-to-image translation methods. pix2pix requires paired and registered data, while CycleGAN can be trained with unpaired images. pix2pix and its variants can produce high quality realistic-looking translations, however, capturing voxel-aligned scans or registering scans is a time-consuming task. In the medical imaging field, there is often lack of paired data, making CycleGAN more suitable method. However, without any additional constraints, it is difficult to optimize and the translated images may not always have good alignment with input scans. Many variants have been proposed which impose additional constraints, e.g. mask alignment between both input and output scans [31].

Besides CycleGAN, another relatively popular unpaired image-to-image translation method is UNsupervised Image-to-image Translation (UNIT) [18], which is a Variational Autoencoder (VAE)-GAN-based method [8,15,16]. UNIT

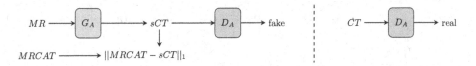

Fig. 2. pix2pix$_{M \rightarrow C}$ has a generator G_A, which generates sCT from MR, and a discriminator D_A, which distinguishes between real CT and sCT. L_1 loss between MRCAT (pair of MR) and sCT is used as an auxiliary supervision.

has been utilized for T1 to T2 MR and T2 to T1 MR translation [26], PET to CT translation, and MR motion correction [1]. Other potential generative models for unpaired image translation include Multimodal UNIT (MUNIT) [10], Disentangled Representation for Image-to-Image Translation++ (DRIT++) [17], Multi-Season GAN (MSGAN) [30], and StarGAN v2 [4].

We propose modifications to original pix2pix model for tasks CT→MR and MR→CT for situations where there is only unpaired CT and MR data. MR scans can be used to generate voxel-aligned (paired) Magnetic Resonance for Calculating ATtenuation (MRCAT) scans, which look somewhat like CT but are not very realistic and thus not suitable for many tasks (see Fig. 1). Our proposed models utilize the alignment information between MR and MRCAT as an auxiliary supervision and produce more realistic CT and MR translations.

2 Methods

We propose two pix2pix variants, pix2pix$_{M \rightarrow C}$ and pix2pix$_{C \rightarrow M}$, for tasks MR→CT and CT→MR, respectively. These models are trained with unpaired CT and MR, and paired MR and MRCAT, which are used as auxiliary supervision to preserve anatomic alignment between input and translated images. We use U-Net-based [22] generators and PatchGAN discriminators as in [11].

pix2pix$_{M \rightarrow C}$ (MR→CT): pix2pix$_{M \rightarrow C}$ consists of one generator G_A and an unconditional discriminator D_A (Fig. 2). The generator is trained to generate synthetic CT (sCT=$G_A(MR)$) from real MR input, while the discriminator D_A is trained to classify between the sCT and real CT. pix2pix$_{M \rightarrow C}$ has an unconditional GAN objective, $\mathcal{L}_{GAN}(G_A, D_A)$, as the conditional GAN (cGAN) objective cannot be used due to lack of paired CT and MR. Following [8,11], D_A is trained to maximize this objective, while G_A is trained to maximize $\log(D_A(G_A(MR)))$.

$$\mathcal{L}_{GAN}(G_A, D_A) = \mathbb{E}_{CT}[\log D_A(CT)] + \mathbb{E}_{MR}[\log(1 - D_A(G_A(MR)))] \quad (1)$$

In addition, G_A is trained to minimize an L_1 loss between MRCAT and the generated sCT, $\mathcal{L}_{L_1}(G_A)$. Note that we do not have ground truth CTs for the MRs, which is why we use MRCATs in the L_1 loss. The L_1 loss plays similar role as cGAN objective in original pix2pix [11] as it encourages the sCTs to be aligned with the input MRs. Our target distribution is the distribution of

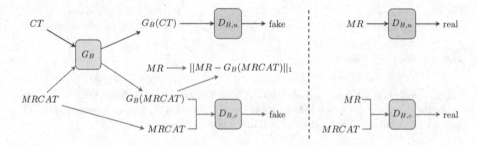

Fig. 3. pix2pix$_{C \to M}$ has a generator G_B, for producing sMR, and two discriminators, $D_{B,u}$ and $D_{B,c}$. $D_{B,c}$, is conditioned on the input MRCAT, and learns to classify between $G_B(MRCAT)$ and real MR. $D_{B,u}$ distinguishes between $G_B(CT)$ and real MR. $\to$: path for CT input. $\to$: path for MRCAT input.

real CTs instead of MRCATs, which is why the MRCATs are only used as an auxiliary supervision.

$$\mathcal{L}_{L_1}(G_A) = \mathbb{E}_{MRCAT,MR}[||MRCAT - G_A(MR)||_1] \tag{2}$$

The full training objective of pix2pix$_{M \to C}$ is

$$\mathcal{L}(G_A, D_A) = \mathcal{L}_{GAN}(G_A, D_A) + \lambda \mathcal{L}_{L_1}(G_A) \tag{3}$$

pix2pix$_{C \to M}$ (CT→MR): pix2pix$_{C \to M}$ consists of one generator G_B and two discriminators $D_{B,u}$ and $D_{B,c}$. The generator takes either real CT, with probability of 0.5, or MRCAT as input during training, and it is trained to generate synthetic MR (sMR) images from both input modalities. The unconditional discriminator $D_{B,u}$ is trained to classify between the sMR generated from real CT, $G_B(CT)$, and real MR. The conditional discriminator $D_{B,c}$ is conditioned on MRCAT, and is trained to classify between the synthetic MR generated from MRCAT, $G_B(MRCAT)$, and real MR.

pix2pix$_{C \to M}$ has two GAN objectives: $\mathcal{L}_{GAN}(G_B, D_{B,u})$ and $\mathcal{L}_{cGAN}(G_B, D_{B,c})$. Following [8,11], $D_{B,u}$ and $D_{B,c}$ are trained to maximize $\mathcal{L}_{GAN}(G_B, D_{B,u})$ and $\mathcal{L}_{cGAN}(G_B, D_{B,c})$, respectively, while G_B is trained to maximize $\log(D_{B,u}(G_B(CT)))$ and $\log(D_{B,c}(MRCAT, G_B(MRCAT)))$.

$$\mathcal{L}_{GAN}(G_B, D_{B,u}) = \mathbb{E}_{MR}[\log D_{B,u}(MR)] + \mathbb{E}_{CT}[\log(1 - D_{B,u}(G_B(CT)))] \tag{4}$$

$$\mathcal{L}_{cGAN}(G_B, D_{B,c}) = \mathbb{E}_{MRCAT,MR}[\log D_{B,c}(MRCAT, MR)] + \mathbb{E}_{MRCAT}[\log(1 - D_{B,c}(MRCAT, G_B(MRCAT)))] \tag{5}$$

In addition, G is trained to minimize the L_1 loss between MR and the sMR generated from MRCAT, $\mathcal{L}_{L_1}(G_B)$. Since we do not have paired CT and MR, we cannot compute L_1 loss between $G_B(CT)$ and MR. While our primary goal is CT→MR translation, the introduction of MRCAT inputs allows us to use the alignment information between the MRCAT and MR pairs to encourage the generator to produce sMR that is aligned with the input.

$$\mathcal{L}_{L_1}(G_B) = \mathbb{E}_{MR,MRCAT}[|||MR - G_B(MRCAT)||_1] \tag{6}$$

The full training objective of pix2pix$_{C \to M}$ is as follows

$$\mathcal{L}(G_B, D_{B,u}, D_{B,c}) = \mathcal{L}_{GAN}(G_B, D_{B,u}) + \mathcal{L}_{cGAN}(G_B, D_{B,c}) + \lambda \mathcal{L}_{L_1}(G_B) \tag{7}$$

Training Details: We use random horizontal flip, random zoom (scale: 0.6 − 1.4) and random crop as data augmentations. In ablation experiments, we use down-sampled (by a factor of 4) images. We utilize the code [6, 21] provided by the authors [11, 25, 32]. We use instance normalization, and a batch size of 8 for pix2pixHD, and batch size of 16 for the other models. All models are trained for 50 epochs, and 13K iterations per epoch. We use Adam optimizer with constant learning rate (LR) (0.0002) for the first 30 epochs and with linearly decaying LR from 0.0002 to zero over the last 20 epochs.

For pix2pix$_{M \to C}$ and pix2pix$_{C \to M}$, we use the vanilla GAN loss (the cross-entropy objective) like in original pix2pix [11] (Table 1). For pix2pix$_{C \to M}$ we use $\lambda = 100$, and discriminator receptive field (RF) size 70×70 in ablation (low resolution) experiments and 286×286 in our final model. For pix2pix$_{M \to C}$, we use $\lambda = 50$, and discriminator RF size 70×70. Since CycleGAN with the default 70×70 discriminators fails, we use a stronger baseline CycleGAN with discriminator RF size 142×142.

3   Experiments

Dataset: The dataset contains 51 pairs of Dixon-In-Phase (DIP) MR and MRCAT scans, and 220 unpaired CT scans of prostate cancer patients, treated with radiotherapy, from Turku University Hospital. Scans were randomly split into training and evaluation set. The number of training/evaluation scans is 41/10 for MR and MRCAT, and 179/41 for CT.

Evaluation Metrics: We use Kernel Inception Distance (KID) [3] and Fréchet Inception Distance (FID) [9] between real and translated images as the evaluation metrics. They measure the distance between the distribution of the generated images and the target distribution of real images, and lower values indicate better performance. In addition, we use DICE coefficient between automatically segmented [14] structures (e.g. Body, Femurs, Prostate, etc.) of input and translated images to evaluate the anatomic alignment and quality of translations. For task MR→CT, we also compute the mean absolute HU intensity difference (HU-Dif) between mean HU value for each segmented ROI in real and translated CT.

Table 1. Overview of the models. Paired data refers to MR and MRCAT pairs. CTs are unpaired with MRs/MRCATs.

Model	Generator	GAN mode	Training data	#CT	#MR	#MRCAT
CycleGAN	unet_512	lsgan	unpaired	179	41	0
pix2pixHD	global	lsgan	paired	0	41	41
pix2pix	unet_512	vanilla	paired	0	41	41
pix2pix$_{M \to C}$/pix2pix$_{C \to M}$	unet_512	vanilla	unpaired & paired	179	41	41

Table 2. MR→CT translation: Performance comparison.

Model	Training data			$FID_{CT,sCT}$	$KID_{CT,sCT}$	DICE	HU-Dif
	CT	MR	MRCAT				
CycleGAN	✓	✓		42.8	0.019	0.80±0.20	**15.5**
pix2pixHD		✓	✓	121.9	0.118	**0.91±0.10**	23.2
pix2pix		✓	✓	122.6	0.119	**0.91±0.12**	23.0
pix2pix$_{M \to C}$	✓	✓	✓	**34.3**	**0.009**	0.88±0.14	16.1

3.1 Comparison with Baselines

We compare our models, pix2pix$_{M \to C}$ and pix2pix$_{C \to M}$, with pix2pix [11] and pix2pixHD [25], and CycleGAN [32]. Table 1 provides the details of the models and training data.

MR→CT: Table 2 shows that pix2pix$_{M \to C}$ outperformed all other methods in terms of FID and KID, indicating that its translated images better resembled real CT. It had slightly worse DICE (standard deviation is much higher than the difference) compared to pix2pix and pix2pixHD, which can partly be explained by the fewer artifacts produced by these models. However, since pix2pix and pix2pixHD were trained using MRCAT as their target, their predictions had similar limitations as the MRCAT, e.g. clear quantization artifacts were present (see bones in Fig. 4). pix2pix and pix2pixHD had high FID and KID, primarily due to absence of the patient couch in the translated CT. CycleGAN translations had small misalignment with inputs and some moderate artifacts, lowering its DICE score. pix2pix$_{M \to C}$ had some prominent artifacts, primarily in the bottom few slices, these might have been caused by the slight difference in the field-of-view of MR and CT datasets. pix2pix$_{M \to C}$ and CycleGAN generated air cavities and hallucinated patient tables.

CT→MR: pix2pix$_{C \to M}$ had the best performance in terms of FID, KID and DICE, as shown in Table 3. pix2pix and pix2pixHD produced relatively good translations of the patient anatomy but due to their failure to ignore the couch (see Fig. 5), visible in CT, their FID and KID values were high. CycleGAN had the worst translations, with large artifacts in some parts of the body (see Fig. 5e), frequently causing large segmentation failures and leading to very low DICE. pix2pix$_{C \to M}$ produced less artifacts and more realistic sMRs, however, some sMR slices had a small misalignment with inputs near the couch.

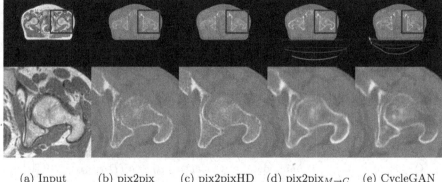

(a) Input   (b) pix2pix   (c) pix2pixHD   (d) pix2pix$_{M\to C}$ (e) CycleGAN

Fig. 4. sCTs produced by pix2pix$_{M\to C}$ and the baselines. First row: Slices from complete scans. Second row: cropped slices.

Table 3. CT→MR translation: Performance comparison.

Model	Training data			$FID_{MR,sMR}$	$KID_{MR,sMR}$	DICE
	CT	MR	MRCAT			
CycleGAN	✓	✓		55.9	0.029	0.58±0.29
pix2pixHD		✓	✓	127.8	0.122	0.81±0.17
pix2pix		✓	✓	90.8	0.086	0.81±0.16
pix2pix$_{C\to M}$	✓	✓	✓	**45.3**	**0.021**	**0.83±0.15**

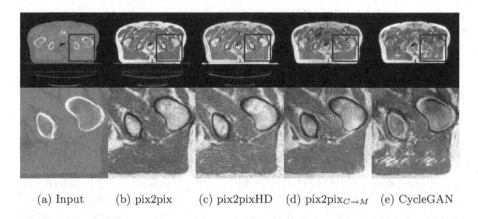

(a) Input   (b) pix2pix   (c) pix2pixHD   (d) pix2pix$_{C\to M}$ (e) CycleGAN

Fig. 5. sMRs generated by pix2pix$_{C\to M}$ and the baselines. First row: Slices from complete scans. Second row: cropped slices.

Table 4. Training objectives for pix2pix$_{M \to C}$ in low resolution. $\lambda = 50$.

Objective	Training data			$FID_{CT,sCT}$	$KID_{CT,sCT}$	DICE
	CT	MR	MRCAT			
L_1		✓	✓	147.4	0.146	0.72
GAN	✓	✓		67.3	0.032	0.23
GAN + λL_1	✓	✓	✓	**39.4**	**0.014**	**0.73**

Table 5. Training objectives for pix2pix$_{C \to M}$ in low resolution. $\lambda = 100$.

Objective	Training data			$FID_{MR,sMR}$	$KID_{MR,sMR}$	DICE
	CT	MR	MRCAT			
GAN	✓	✓		40.0	0.015	0.24
GAN+cGAN	✓	✓	✓	32.8	0.013	0.52
GAN+cGAN+λL_1	✓	✓	✓	**30.0**	**0.012**	**0.57**

3.2 Ablation Studies

pix2pix$_{M \to C}$ Objective: When the training objective included both L_1 and GAN loss, the translations looked realistic and the performance of pix2pix$_{M \to C}$ was better in terms of FID, KID and DICE scores compared to the experiment with only L_1 objective or only GAN objective (Table 4). When only GAN objective was used, the generated sCTs had poor alignment with the input MRs.

pix2pix$_{C \to M}$ Objective: When only CT inputs were used with a GAN objective (see $\to$ path in Fig. 3), the generated sMR images were not well aligned with the inputs. The translations were better aligned when the cGAN objective was included, i.e. with a probability of 0.5 either CT or MRCAT input was used. Inclusion of the L_1 objective (between sMR generated from MRCAT and real MR) with GAN+cGAN produced the best results in terms of FID, KID and DICE (see Table 5).

4 Conclusion

Our results show that CT$\to$MR and MR$\to$CT translation with unpaired CT and MR, using the MR and MRCAT pairs as an auxiliary supervision, produces more realistic translated CT and MR images. This additional supervision reduces artifacts and improves alignment between the input and the translated images. The proposed pix2pix$_{M \to C}$ and pix2pix$_{C \to M}$, outperformed the baseline pix2pix, pix2pixHD and CycleGAN, in terms of FID and KID scores.

pix2pix$_{M \to C}$ and pix2pix$_{C \to M}$, like other GAN-based methods, may be useful in producing realistic-looking translated images for research purposes. Since these methods can hallucinate features in images [5], they require extensive validation of their image quality and fidelity before clinical use. It remains as task for future research to develop improved translation methods and design quantitative metrics which better capture the quality of translations.

References

1. Armanious, K., Jiang, C., Abdulatif, S., Küstner, T., Gatidis, S., Yang, B.: Unsupervised medical image translation using cycle-MedGAN. In: 2019 27th European Signal Processing Conference (EUSIPCO), pp. 1–5. IEEE (2019)
2. Armanious, K., et al.: MedGAN: medical image translation using GANs. Comput. Med. Imaging Graph. **79**, 101684 (2020)
3. Bińkowski, M., Sutherland, D.J., Arbel, M., Gretton, A.: Demystifying MMD GANs. arXiv preprint arXiv:1801.01401 (2018)
4. Choi, Y., Uh, Y., Yoo, J., Ha, J.W.: StarGAN v2: diverse image synthesis for multiple domains. In: Proceedings of the IEEE/CVF Conference on Computer Vision and Pattern Recognition, pp. 8188–8197 (2020)
5. Cohen, J.P., Luck, M., Honari, S.: Distribution matching losses can hallucinate features in medical image translation. In: Frangi, A.F., Schnabel, J.A., Davatzikos, C., Alberola-López, C., Fichtinger, G. (eds.) MICCAI 2018. LNCS, vol. 11070, pp. 529–536. Springer, Cham (2018). https://doi.org/10.1007/978-3-030-00928-1_60
6. CycleGAN and pix2pix (2021). https://github.com/junyanz/pytorch-CycleGAN-and-pix2pix
7. Emami, H., Dong, M., Nejad-Davarani, S.P., Glide-Hurst, C.K.: Generating synthetic CTS from magnetic resonance images using generative adversarial networks. Med. Phys. **45**(8), 3627–3636 (2018)
8. Goodfellow, I.J., et al.: Generative adversarial networks. arXiv preprint arXiv:1406.2661 (2014)
9. Heusel, M., Ramsauer, H., Unterthiner, T., Nessler, B., Hochreiter, S.: GANs trained by a two time-scale update rule converge to a local Nash equilibrium. arXiv preprint arXiv:1706.08500 (2017)
10. Huang, X., Liu, M.-Y., Belongie, S., Kautz, J.: Multimodal unsupervised image-to-image translation. In: Ferrari, V., Hebert, M., Sminchisescu, C., Weiss, Y. (eds.) ECCV 2018. LNCS, vol. 11207, pp. 179–196. Springer, Cham (2018). https://doi.org/10.1007/978-3-030-01219-9_11
11. Isola, P., Zhu, J.Y., Zhou, T., Efros, A.A.: Image-to-image translation with conditional adversarial networks. In: Proceedings of the IEEE Conference on Computer Vision and Pattern Recognition, pp. 1125–1134 (2017)
12. Jiang, J., et al.: PSIGAN: joint probabilistic segmentation and image distribution matching for unpaired cross-modality adaptation-based mri segmentation. IEEE Trans. Med. Imaging **39**(12), 4071–4084 (2020)
13. Jin, C.B., et al.: Deep CT to MR synthesis using paired and unpaired data. Sensors **19**(10), 2361 (2019)
14. Kiljunen, T., et al.: A deep learning-based automated CT segmentation of prostate cancer anatomy for radiation therapy planning-a retrospective multicenter study. Diagnostics **10**(11), 959 (2020). https://doi.org/10.3390/diagnostics10110959
15. Kingma, D.P., Welling, M.: Auto-encoding variational Bayes. arXiv preprint arXiv:1312.6114 (2013)
16. Larsen, A.B.L., Sønderby, S.K., Larochelle, H., Winther, O.: Autoencoding beyond pixels using a learned similarity metric. In: International Conference on Machine Learning, pp. 1558–1566. PMLR (2016)

17. Lee, H.Y., et al.: DRIT++: diverse image-to-image translation via disentangled representations. Int. J. Comput. Vision **128**(10), 2402–2417 (2020). https://doi.org/10.1007/s11263-019-01284-z

18. Liu, M.Y., Breuel, T., Kautz, J.: Unsupervised image-to-image translation networks. arXiv preprint arXiv:1703.00848 (2017)

19. Nie, D., et al.: Medical image synthesis with context-aware generative adversarial networks. In: Descoteaux, M., Maier-Hein, L., Franz, A., Jannin, P., Collins, D.L., Duchesne, S. (eds.) MICCAI 2017. LNCS, vol. 10435, pp. 417–425. Springer, Cham (2017). https://doi.org/10.1007/978-3-319-66179-7_48

20. Peng, Y., et al.: Magnetic resonance-based synthetic computed tomography images generated using generative adversarial networks for nasopharyngeal carcinoma radiotherapy treatment planning. Radiother. Oncol. **150**, 217–224 (2020)

21. pix2pixHD (2021). https://github.com/NVIDIA/pix2pixHD

22. Ronneberger, O., Fischer, P., Brox, T.: U-Net: convolutional networks for biomedical image segmentation. In: Navab, N., Hornegger, J., Wells, W.M., Frangi, A.F. (eds.) MICCAI 2015. LNCS, vol. 9351, pp. 234–241. Springer, Cham (2015). https://doi.org/10.1007/978-3-319-24574-4_28

23. Sandfort, V., Yan, K., Pickhardt, P.J., Summers, R.M.: Data augmentation using generative adversarial networks (CycleGAN) to improve generalizability in CT segmentation tasks. Sci. Rep. **9**(1), 1–9 (2019)

24. Tanner, C., Ozdemir, F., Profanter, R., Vishnevsky, V., Konukoglu, E., Goksel, O.: Generative adversarial networks for MR-CT deformable image registration. arXiv preprint arXiv:1807.07349 (2018)

25. Wang, T.C., Liu, M.Y., Zhu, J.Y., Tao, A., Kautz, J., Catanzaro, B.: High-resolution image synthesis and semantic manipulation with conditional GANs. In: Proceedings of the IEEE Conference on Computer Vision and Pattern Recognition, pp. 8798–8807 (2018)

26. Welander, P., Karlsson, S., Eklund, A.: Generative adversarial networks for image-to-image translation on multi-contrast MR images-a comparison of CycleGAN and unit. arXiv preprint arXiv:1806.07777 (2018)

27. Wolterink, J.M., Dinkla, A.M., Savenije, M.H.F., Seevinck, P.R., van den Berg, C.A.T., Išgum, I.: Deep MR to CT synthesis using unpaired data. In: Tsaftaris, S.A., Gooya, A., Frangi, A.F., Prince, J.L. (eds.) SASHIMI 2017. LNCS, vol. 10557, pp. 14–23. Springer, Cham (2017). https://doi.org/10.1007/978-3-319-68127-6_2

28. Wolterink, J.M., Leiner, T., Viergever, M.A., Išgum, I.: Generative adversarial networks for noise reduction in low-dose CT. IEEE Trans. Med. Imaging **36**(12), 2536–2545 (2017)

29. Yi, X., Walia, E., Babyn, P.: Generative adversarial network in medical imaging: a review. Med. Image Anal. **58**, 101552 (2019)

30. Zhang, F., Wang, C.: MSGAN: generative adversarial networks for image seasonal style transfer. IEEE Access **8**, 104830–104840 (2020)

31. Zhang, Z., Yang, L., Zheng, Y.: Translating and segmenting multimodal medical volumes with cycle- and shape-consistency generative adversarial network. In: CVPR (2018). http://arxiv.org/abs/1802.09655

32. Zhu, J.Y., Park, T., Isola, P., Efros, A.A.: Unpaired image-to-image translation using cycle-consistent adversarial networks. In: Proceedings of the IEEE International Conference on Computer Vision, pp. 2223–2232 (2017)

Conditional Generation of Medical Images via Disentangled Adversarial Inference

Mohammad Havaei[1(✉)], Ximeng Mao[2], Yipping Wang[1], and Qicheng Lao[1,2]

[1] Imagia, Montreal, Canada
mohammad@imagia.com
[2] Montréal Institute for Learning Algorithms (MILA), Université de Montréal, Montreal, Canada

Abstract. We propose DRAI—a dual adversarial inference framework with augmented disentanglement constraints—to learn from the image itself, disentangled representations of style and content, and use this information to impose control over conditional generation process. We undergo two novel regularization steps to ensure content-style disentanglement. First, we minimize the shared information between content and style by introducing a novel application of the gradient reverse layer (GRL); second, we introduce a self-supervised regularization method to further separate information in the content and style variables. We conduct extensive qualitative and quantitative assessments on two publicly available medical imaging datasets (LIDC and HAM10000) and test for conditional image generation and style-content disentanglement. We also show that our proposed model (DRAI) achieves the best disentanglement score and has the best overall performance.

1 Introduction

In recent years, conditional generation of medical images has become a popular area for research using conditional Generative Adversarial Networks (cGAN) [20, 36]. One common pitfall of cGAN is that the conditioning codes are extremely high-level and do not cover nuances of the data. This challenge is exacerbated in the medical imaging domain where insufficient label granularity is a common occurrence. We refer to the factors of variation that depend on the conditioning vector as *content*. Another challenge in conditional image generation is that the image distribution also contains factors of variation that are agnostic to the conditioning code. These types of information are shared among different classes or different conditioning codes. In this work we refer to such information as *style*, which depending on the task, could correspond to position, orientation,

---

M. Havaei and X. Mao—Equal contribution.

Electronic supplementary material The online version of this chapter (https://doi.org/10.1007/978-3-030-88210-5_5) contains supplementary material, which is available to authorized users.

S. Engelhardt et al. (Eds.): DGM4MICCAI 2021/DALI 2021, LNCS 13003, pp. 45–66, 2021.
https://doi.org/10.1007/978-3-030-88210-5_5

location, background information, etc. Learning disentangled representation of content and style allows us to control the detailed nuances of the generation process.

In this work, we consider two types of information to preside over the image domain: *content* and *style*, which by definition, are independent and this independence criteria should be taken into account when training a model. By explicitly constraining the model to disentangle content and style, we ensure their independence and prevent information leakage between them. To achieve this goal, we introduce Dual Regularized Adversarial Inference (DRAI), a conditional generative model that leverages unsupervised learning and novel disentanglement constraints to learn disentangled representations of content and style, which in turn enables more control over the generation process.

We impose two novel disentanglement constraints to facilitate this process: Firstly, we introduce a novel application of the *Gradient Reverse Layer* (GRL) [16] to minimize the shared information between the two variables. Secondly, we present a new type of self-supervised regularization to further enforce disentanglement; using content-preserving transformations, we *attract* matching content information, while *repelling* different style information.

We compare the proposed method with multiple baselines on two datasets. We show the advantage of using two latent variables to represent style and content for conditional image generation. To quantify style-content disentanglement, we introduce a disentanglement measure and show the proposed regularizations can improve the separation of style and content information. The contributions of this work can be summarized as follows:

- To the best of our knowledge, this is the first time disentanglement of content and style has been explored in the context of medical image generation.
- We introduce a novel application of GRL that penalizes shared information between content and style in order to achieve better disentanglement.
- We introduce a self-supervised regularization that encourages the model to learn independent information as content and style.
- we introduce a quantitative content-style disentanglement measure that does not require any content or style labels. This is especially useful in real world scenarios where attributes contributing to content and style are not available.

2 Method

2.1 Overview

Let t be the conditioning vector associated with image x. Using the pairs $\{(t_i, x_i)\}, i = 1, \ldots, N$, where N denotes the size of the dataset, we train an inference model $G_{c,z}$ and a generative model G_x such that (i) the inference model $G_{c,z}$ infers content c and style z in a way that they are disentangled from each other and (ii) the generator G_x can generate realistic images that not only visually respect the conditioning vector t but also the style/content disentanglement. An Illustration of DRAI is made in Fig. 1

It is worth noting that our generative module is *not* constrained to require a style image. Having a probabilistic generative model allows us to sample the style code from the style prior distribution and generate images with random style attributes. The framework also allows us to generate hybrid images by mixing style and content from various sources (details can be found in Sect. B.2).

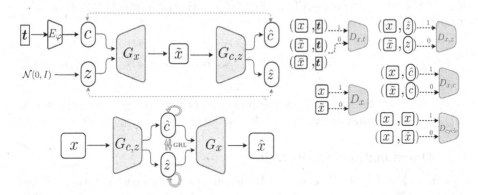

Fig. 1. Overview of DRAI. The dashed purple arrows mark the cycle consistency between features implemented via ℓ_1 norm, while the solid purple arrows show the imposed disentanglement constrains. On the right hand side of the figure we show all the discriminators used for training. $\hat{c}$ represents the inferred content, $\hat{z}$ the inferred style, $\hat{x}$ the reconstructed input image and $\bar{x}$ the image with mismatched conditioning.

2.2 Dual Adversarial Inference (DAI)

We follow the formulation of [30] for Dual Adversarial Inference (DAI) which is a conditional generative model that uses bidirectional adversarial inference [14,15] to learn content and style variables from the image data. To impose alignment between conditioning vector t and the generated image $\tilde{x}$, we seek to match $p(\tilde{x}, t)$ with $p(x, t)$. To do so, we adopt the matching-aware discriminator proposed by [40]. For this discriminator—denoted as $D_{x,t}$—the positive sample is the pair of real image and its corresponding conditioning vector (x, t), whereas the negative sample pairs consist of two groups; the pair of real image with mismatched conditioning $(\bar{x}, t)$, and the pair of synthetic image with corresponding conditioning $(G_x(z, c), t)$. In order to retain the fidelity of the generated images, we also train a discriminator D_x that distinguishes between real and generated images. The loss function imposed by $D_{x,t}$ and D_x is as follows:

$$\min_G \max_D V_{t2i}(D_x, D_{x,t}, G_x) = \mathbb{E}_{p_{\text{data}}}[\log D_x(x)] + \mathbb{E}_{p(z), q(c)}[\log(1 - D_x(G_x(z, c)))] +$$

$$\mathbb{E}_{p_{\text{data}}}[\log D_{x,t}(x, t)] + \tfrac{1}{2}\{\mathbb{E}_{p_{\text{data}}}[\log(1 - D_{x,t}(\bar{x}, t))] + \mathbb{E}_{p(z), q(c), p_{\text{data}}}[\log(1 - D_{x,t}(G_x(z, c), t))]\},$$

where $\tilde{x} = G_x(z, c)$ is the generated image and $(\bar{x}, t)$ designates a mis-matched pair.

We use adversarial inference to infer style and content codes from the image. Using the adversarial inference framework, we are interested in matching the conditional $q(z, c|x)$ to the posterior $p(z, c|x)$. Given the Independence assumption of c and z, can use the bidirectional adversarial inference formulation individually for style and content. This dual adversarial inference objective is thus formulated as:

$$\min_G \max_D V_{\text{dALI}}(D_{x,z}, D_{x,c}, G_x, G_{c,z}) = \mathbb{E}_{q(x),q(z,c|x)}[\log D_{x,z}(x, \hat{z}) + \log D_{x,c}(x, \hat{c})] +$$

$$\mathbb{E}_{p(x|z,c),p(z),p(c)}[\log(1 - D_{x,z}(\tilde{x}, z)) + \log(1 - D_{x,c}(\tilde{x}, c))]. \quad (1)$$

To improve the stability of training, we include image-cycle consistency ($V_{\text{image-cycle}}$) [51] and latent code cycle consistency ($V_{\text{code-cycle}}$) objectives [12].

2.3   Disentanglement Constrains

The dual adversarial inference (DAI) encourages disentanglement through the independence assumption of style and content. However, it does not explicitly penalize entanglement. We introduce two constraints to impose style-content disentanglement. Refer to the Appendix for details.

Content-Style Information Minimization: We propose a novel application of the Gradient Reversal Layer (GRL) strategy [16] to *explicitly* minimize the shared information between style and content. We train an encoder F_c to predict the content from style and use GRL to minimize the information between the two. The same process is done for predicting style from content through F_z. This constrains the content feature generation to disregard style features and the style feature generation to disregard content features.

Self-supervised Regularization: We incorporate a self-supervised regularization such that the content is invariant to content-preserving transformations (such as a rotation, horizontal or vertical flip) while the style is sensitive to such transformations. More formally, we maximize the similarity between the inferred contents of x and the transformed x' while minimizing the similarity between their inferred styles. This constrains the content feature generation to focus on the content of the image reflected in the conditioning vector and the style feature generation to focus on the transformation attributes.

DRAI is a probabilistic model that requires reparameterization trick to sample from the approximate posteriors $q(z|x)$, $q(c|x)$ and $q(c|t)$. We use KL divergence in order to regularize these posteriors to follow the normal distribution $\mathcal{N}(0, I)$. Taking that into account, the complete objective criterion for DRAI is:

$$\min_G \max_{D,F} V_{\text{t2i}} + V_{\text{dALI}} + V_{\text{image-cycle}} + V_{\text{code-cycle}} + V_{\text{GRL}} + V_{\text{self}} +$$

$$\lambda D_{KL}(q(z|x) \,\|\, \mathcal{N}(0, I)) + \lambda D_{KL}(q(c|x) \,\|\, \mathcal{N}(0, I)) + \lambda D_{KL}(q(c|t) \,\|\, \mathcal{N}(0, I)). \quad (2)$$

3   Experiments

We conduct experiments on two publicly available medical imaging datasets: LIDC [4] and HAM10000 [46] (see Appendix for details on these datasets). To evaluate the quality of generation, inference, and disentanglement, we consider two types of baselines. To show the effectiveness of dual variable inference, we compare our framework with single latent variable models. For this, we introduce a conditional adaptation of InfoGan [12] referred to as cInfoGAN and a conditional adversarial variational Autoencoder (cAVAE). We also compare DRAI to Dual Adversarial Inference (DAI) [30] and show how using our proposed disentanglement constraints together with latent code cycle-consistency can significantly boost performance. See Appendix for more details on various baselines. Finally, we conduct rigorous ablation studies to evaluate the impact of each component in DRAI.

3.1   Generation Evaluation

To evaluate the quality and diversity of the generated images, we measure FID and IS (see Appendix Sect. D.3) for the proposed DRAI model and various double and single latent variable baselines described in Appendix Sect. D. The results are reported in Table 1 for both LIDC and HAM10000 datasets. For the LIDC dataset, we observe all methods have comparable IS score while DRAI and DAI have significantly lower FID compared to other baselines, with DRAI having better performance. For the HAM10000 dataset, DRAI once again achieves the best FID score while D-cInfoGAN achieves the best IS.

Table 1. Comparison of image generation metrics (FID, IS) and disentanglement metric(CIFC) on HAM10000 and LIDC datasets for single and double variable baselines. CIFC is only evaluated for double variable baselines.

Method	HAM10000			LIDC		
	FID ($\downarrow$)	IS ($\uparrow$)	CIFC ($\downarrow$)	FID ($\downarrow$)	IS ($\uparrow$)	CIFC ($\downarrow$)
cInfoGAN	1.351 ± 0.33	1.326 ± 0.03	–	0.283 ± 0.06	1.366 ± 0.02	–
cAVAE	3.566 ± 0.56	1.371 ± 0.01	–	0.181 ± 0.03	$\mathbf{1.424 \pm 0.01}$	–
D-cInfoGAN	1.684 ± 0.42	$\mathbf{1.449 \pm 0.03}$	1.201 ± 0.17	0.333 ± 0.06	1.342 ± 0.09	1.625 ± 0.11
D-cAVAE	4.893 ± 0.99	1.321 ± 0.01	1.354 ± 0.03	0.378 ± 0.03	1.371 ± 0.04	1.944 ± 0.02
DAI [30]	1.327 ± 0.06	1.304 ± 0.01	0.256 ± 0.01	$\mathbf{0.106 \pm 0.02}$	1.423 ± 0.05	1.096 ± 0.28
DRAI	$\mathbf{1.224 \pm 0.05}$	1.300 ± 0.01	$\mathbf{0.210 \pm 0.01}$	$\mathbf{0.089 \pm 0.02}$	1.422 ± 0.03	$\mathbf{0.456 \pm 0.06}$

We highlight that while FID and IS are the most common metrics for the evaluation of GAN based models, they do not provide the optimum assessment [5] and thus qualitative assessment is needed. We use the provided conditioning vector for the generation process and only sample the style variable z. The generated samples are visualized in Fig. 2. In every sub-figure, the first column represents the reference image corresponding to the conditioning vector used for the image generation, and the remaining columns represent synthesized images.

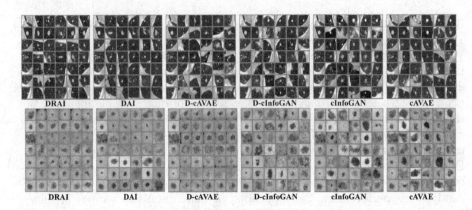

Fig. 2. Conditional generations on LIDC and HAM10000. The images are generated by keeping the content code (c) fixed and only sampling the style codes (z).

By fixing the content and sampling the style variable, we can discover the types of information that are encoded as style and content for each dataset. We observe that the learned content information are color and lesion size for HAM10000, and nodule size for LIDC; while the learned style information are location, orientation and lesion shape for HAM10000 and background for LIDC. We also observe that DRAI is very successful in preserving the content information when there is no stochasticity in the content variable (*i.e.*, c is fixed). As for other baselines, sampling style results in changing the content information of the generated images, which indicates information leak from the content variable to the style variable. The results show that compared to DAI and other baselines, DRAI achieves better separation of style and content.

3.2 Style-Content Disentanglement

Achieving good style-content disentanglement in both inference and generation phases is the main focus of this work. We conduct multiple quantitative and qualitative experiments to asses the quality of disentanglement in DRAI (our proposed method) as well as the competing baselines.

As a quantitative metric, we introduce the disentanglement error CIFC (refer to Appendix for details). Table 1 shows results on this metric. As seen from this table, in both HAM10000 and LIDC datasets, DRAI improves over DAI by a notable margin, which demonstrates the advantage of the proposed disentanglement regularizations; on one hand, the information regularization objective through GRL minimizes the shared information between style and content variables, and on the other hand, the self-supervised regularization objective not only allows for better control of the learned features but also facilitates disentanglement. In the ablation studies (Sect. 3.3), we investigate the effect of the individual components of DRAI on disentanglement.

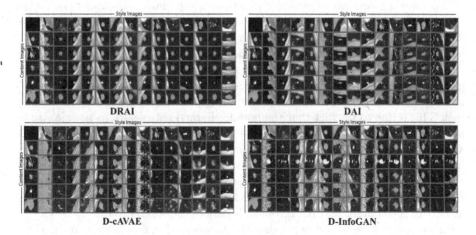

Fig. 3. Qualitative evaluation of style-content disentanglement through hybrid image generation on LIDC dataset. In every sub-figure, images in the first row present style image references and those in the first column present content image references. Hybrid images are generated by using the style and content codes inferred from style and content reference images respectively.

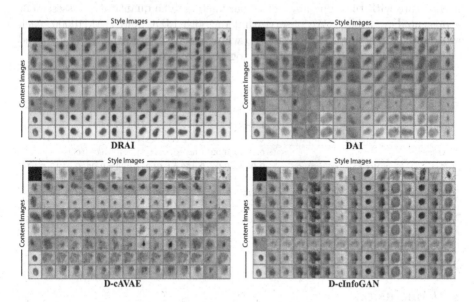

Fig. 4. Qualitative evaluation of style-content disentanglement through hybrid image generation on HAM10000 dataset. In every sub-figure, images in the first row represent style image references and those in the first column represent content image references. Hybrid images are generated by using the style and content codes inferred from style and content reference images respectively.

To have a more interpretable evaluation, we qualitatively assess the style-content disentanglement through generating hybrid images by combining style and content information from different sources (See Appendix for details on hybrid images). We can then evaluate the extent to which the style and content of the generated images respect the corresponding style and content of the source images. Figure 3 and Fig. 4 show these results on the two datasets. For the LIDC dataset, DAI and DRAI learn CT image background as style and nodule as content. This is due to the fact that the nodule characteristics such as nodule size is included in the conditioning factor and thus the content tends to focus on those attributes. Thanks to the added disentanglement regularizations, DRAI has the best content-style separation compared to all other baselines and demonstrates clear decoupling of the two variables. Because of the self-supervised regularization objective, DRAI assigns more emphases on capturing nodule characteristics as part of the content and background as part of the style. Overall, it is evident from the qualitative experiments that the proposed disentanglement regularizations help to decouple the style and content variables.

3.3   Ablation Studies

In this section, we perform ablation studies to evaluate the effect of each component on disentanglement using the CIFC metric. Ablated models use the same architecture with the same amount of parameters. The quantitative assessment is presented in Table 2. We observe that on both LIDC and HAM10000, each added component improves over DAI, while the best performance is achieved when these components are combined together to form DRAI.

Table 2. Quantitative ablation study on LIDC and HAM10000 datasets

Method	LIDC		HAM10000	
	FID ($\downarrow$)	CIFC ($\downarrow$)	FID ($\downarrow$)	CIFC ($\downarrow$)
DAI [30]	**0.106 ± 0.02**	1.096 ± 0.284	1.327 ± 0.06	0.256 ± 0.01
DRAI = DAI+selfReg+MIReg+featureCycle	**0.089 ± 0.02**	**0.456 ± 0.069**	**1.224 ± 0.05**	**0.210 ± 0.01**
DAI+selfReg+MIReg	0.176 ± 0.06	0.554 ± 0.185	1.350 ± 0.12	0.233 ± 0.01
DAI+featureCycle	0.221 ± 0.07	0.913 ± 0.074	1.367 ± 0.12	0.311 ± 0.01
DAI+MIReg	0.154 ± 0.04	0.747 ± 0.226	**1.298 ± 0.12**	**0.228 ± 0.01**
DAI+selfReg	0.208 ± 0.05	0.781 ± 0.203	1.347 ± 0.14	**0.219 ± 0.04**

4   Conclusion

We introduce DRAI, a frame work for generating synthetic medical images which allows control over the style and content of the generated images. DRAI uses adversarial inference together with conditional generation and disentanglement constraints to learn content and style variables from the dataset. We compare

DRAI quantitatively and qualitatively with multiple baselines and show its superiority in image generation in terms of quality, diversity and style-content disentanglement. Through ablation studies and comparisons with DAI [30], we show the impact of imposing the proposed disentanglement constraints over the content and style variables.

A   Disentanglement Constrains

Lao et al. [30] use double variable ALI as a criterion for disentanglement. However, ALI does approximate inference and does not necessarily guarantee disentanglement between variables. To further impose disentanglement between style and content, we propose additional constrains and regularization measures.

A.1   Content-Style Information Minimization

The content should not include any information of the style and vice versa. We seek to *explicitly* minimize the shared information between style and content. For this, we propose a novel application of the Gradient Reversal Layer (GRL) strategy. First introduced in [16], the GRL strategy is used in domain adaptation methods to learn domain-agnostic features, where it acts as the identity function in the forward pass but reverses the direction of the gradients in the backward pass. In domain adaptation literature, GRL is used with a domain classifier. Reversing the direction of the gradients coming from the domain classification loss has the effect of minimizing the information between the representations and domain identity, thus, learning domain invariant features. Inspired by the literature on domain adaptation, we use GRL to minimize the information between style and content. More concretely, for a given example x, we train an encoder F_c to predict the content from style and use GRL to minimize the information between the two. The same process is done for predicting style from content through F_z, resulting in the following objective function:

$$\min_G \max_F V_{\text{GRL}}(F_z, F_c, G_{c,z}) \tag{3}$$
$$= -\mathbb{E}_{x \sim q(x), (\hat{z}, \hat{c}) \sim q(z, c|x)}[\|\hat{z} - F_z(\hat{c})\| + \|\hat{c} - F_c(\hat{z})\|].$$

This constrains the content feature generation to disregard style features and the style feature generation to disregard content features. Figure 5b shows a visualization of this module.

We can show that Eq. (3) minimizes the mutual information between the style variable and the content variable. Here, we only provide the proof for using GRL with F_z to predict style from content. Similar reasoning can be made for using GRL with F_c. Let $I(z; c)$ denote the mutual information between the inferred content and the style variables, where

$$I(z; c) = H(z) - H(z|c). \tag{4}$$

Once again, following [2], we define a variational lower bound on $I(z;c)$ by rewriting the conditional entropy in (4) as:

$$-H(z|c) = \mathbb{E}_{\hat{c}\sim q(c|x)}[\log q(z|\hat{c}) + D_{KL}(p(z|\hat{c})||q(z|\hat{c}))]],$$

and by extension:

$$I(z;c) = H(z) + \max_{F_z} \mathbb{E}_{\hat{c}\sim q(c|x)}[\log q(z|\hat{c})], \qquad (5)$$

where the maximum is achieved when $D_{KL}(p(z|\hat{c})||q(z|\hat{c}))] = 0$. Since $H(z)$ is constant for F_z and $||\hat{z} - F_z(\hat{c})||$ corresponds to $-\log q(z|\hat{c})$, minimization of mutual information can be written as:

$$\min_G I(z;c) = \min_G \max_{F_z} -\mathbb{E}_{\hat{c}\sim q(c|x),\hat{z}\sim q(z|x)}[||\hat{z} - F_z(\hat{c})||], \qquad (6)$$

which corresponds to Eq. (3).

A.2 Self-supervised Regularization

Self-supervised learning has shown great potential in unsupervised representation learning [11,21,39]. To provide more control over the latent variables c and z, we incorporate a self-supervised regularization such that the content is invariant to content-preserving transformations while the style is sensitive to such transformations. The proposed self-supervised regularization constraints the feature generator $G_{c,z}$ to encode different information for content and style. More formally, let $\mathcal{T}$ be a random content-preserving transformation such as a rotation, horizontal or vertical flip. For every example $x \sim q(x)$, let x' be its transformed version; $x' = T_i(x)$ for $T_i \sim p(\mathcal{T})$. We would like to maximize the similarity between the inferred contents of x and x' and minimize the similarity between their inferred styles. This constrains the content feature generation to focus on the content of the image reflected in the conditioning vector and the style feature generation to focus on other attributes. This regularization procedure is visualized in Fig. 5a. The objective function for the self-supervised regularization is defined as:

$$\min_G V_{\text{self}}(G_{c,z}) = \mathbb{E}_{x\sim q(x)}[||\hat{c} - \hat{c}'|| - ||\hat{z} - \hat{z}'||], \qquad (7)$$

where $(\hat{z}, \hat{c}) \sim q(z,c|x)$ and $(\hat{z}', \hat{c}') \sim q(z,c|x')$.

B   Implementation Details

B.1   Implementation Details

In this section, we provide the important implementation details of DRAI. Firstly, to reduce the risk of information leak between style and content, we use completely separate encoders to infer the two variables. For the same reason, the

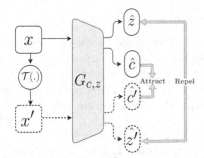 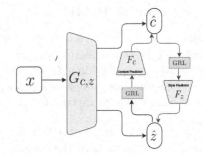

(a) Self-Supervised regularization. Given x and its transformed version x', their corresponding content codes c and $\hat{c}$ form a positive pair and the disparity between them is minimized (*i.e.*, attract each other) while their corresponding style codes z and $\hat{z}$ form a negative pair and the disparity between them is maximized (*i.e.*, repel each other).

(b) Content-Style information minimization. For a given image x, F_c is trained to predict the content $\hat{c}$ from the style $\hat{z}$. By reversing the direction of the gradients, the GRL penalizes $G_{c,z}$ to minimize the content information in the style variable z. The same procedure is carried out to minimize style information in the content variable c.

Fig. 5. Proposed disentanglement constraints.

dual adversarial discriminators are also implemented separately for style and content. The data augmentation includes random flipping and cropping. To enable self-supervised regularization, each batch is trained twice, first with the original images and then with the transformed batch. The transformations include rotations of 90, 180, and 270 degrees, as well as horizontal and vertical flipping. LSGAN (Least Square GAN) [34] loss is used for all GAN generators and discriminators, while ℓ_1 loss is used for the components related to disentanglement constraints, *i.e.*, GRL strategy and self-supervised regularization. In general, we found that "Image cycle-consistency" and "Latent code cycle-consistency" objectives improve the stability of training. This is evident by DRAI achieving lower prediction intervals (*i.e.*, standard deviation across multiple runs with different seeds) in our experiments.

We did not introduce any coefficients for the loss components in Equation (2) since other than the KL terms, they were all relatively on the same scale. As for the KL co-efficients λ, we tried multiple values and qualitatively evaluated the results. Since the model was not overly sensitive to KL, we used a coefficient of 1 for all KL components.

All models including the baselines are implemented in TensorFlow [1] version 2.1, and the models are optimized via Adam [27] with initial learning rate $1e^{-5}$.

For IS and FID computation, we fine-tune the inception model on a 5 way classification on nodule size for LIDC and a 7 way classification on lesion type for HAM10000. FID and IS are computed over a set of 5000 generated images.

B.2   Generating Hybrid Images

Thanks to our encoder that is able to infer disentangled codes for style and content and also our generator that does not have a hard constraint on requiring the conditioning embedding t, we can generate hybrid images where we mix style and content from different image sources. Let i and j be the indices of two different images. There are two ways in which DRAI can generate hybrid images:

1. Using a conditioning vector t_i and a style image x_j: In this setup, we use the conditioning factor t_i as the content and the inferred $\hat{z}_j$ from the style image x_j as the style:

$$c_i = E_\varphi(t_i)$$
$$\hat{z}_j, \hat{c}_j = G_{c,z}(x_j)$$
$$\tilde{x}_{ij} = G_x(\hat{z}_j, c_i).$$

2. Using a content image x_i and a style image x_j: In this setup we do not rely on the conditioning factor t. Instead, we infer codes for both style and content (i.e., $\hat{z}_j$ and $\hat{c}_i$) from style and content source images respectively.

$$\hat{z}_i, \hat{c}_i = G_{c,z}(x_i)$$
$$\hat{z}_j, \hat{c}_j = G_{c,z}(x_j)$$
$$\tilde{x}_{ij} = G_x(\hat{z}_j, \hat{c}_i)$$

The generation of hybrid images is graphically explained in Fig. 6 for the aforementioned two scenarios.

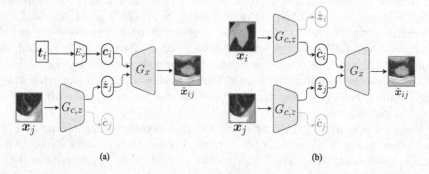

(a)   (b)

Fig. 6. Hybrid image generation: (a) via the conditioning factor t_i (representing content) and the style code $\hat{z}_j$ inferred from the style reference image. (b) via the content code $\hat{c}_i$ inferred from the content reference image and the style code $\hat{z}_j$ inferred from the style reference image.

C   Datasets

C.1   HAM10000

Human Against Machine (HAM10000) [46], contains approximately 10000 training images, includes 10015 dermatoscopic images of seven types of skin lesions
and is widely used as a classification benchmark. One of the lesion types,
"Melanocytic nevi" (nv), occupies around 67% of the whole dataset, while the
two lesion types that have the smallest data size, namely, "Dermatofibroma"
(df) and "Vascular skin lesions" (vasc), have only 115 and 143 images respectively. Such data imbalance is undesirable for our purpose since limitations on
the data size lead to severe lack of image diversity of the minority classes. For
our experiments, we select the three largest skin lesion types, which in order of
decreasing size are: "nv" with 6705 images; "Melanoma" (mel) with 1113 images;
and "Benign keratosis-like lesions" (bkl) with 1099 images. Patches of size 48×48
centered around the lesion are extracted and then resized to 64×64. To balance
the dataset, we augment mel and bkl three times with random flipping. We follow the train-test split provided by the dataset, and the data augmentation is
done only on the training data.

C.2   LIDC

The Lung Image Database Consortium image collection (LIDC-IDRI) consists
of lung CT scans from 1018 clinical cases [4]. In total, 7371 lesions are annotated
by one to four radiologists, of which 2669 are given ratings on nine nodule characteristics: "malignancy", "calcification", "lobulation", "margin", "spiculation",
"sphericity", "subtlety", "texture" and "internal structure". We take the following pre-processing steps for LIDC: a) We normalize the data such that it respects
the Hounsfield units (HU), b) the volume size is converted to $256 \times 256 \times 256$,
c) areas around the lungs are cropped out. For our experiments, we extract a
subset of 2D patches composing nodules with consensus from at least three radiologists. Patches of size 48×48 centered around the nodule are extracted and
then resized to 64×64. Furthermore, we compute the inter-observer median of
the malignancy ratings and exclude those with malignancy median of 3 (out of
5). This is to ensure a clear separation between benign and malignant classes presented in the dataset. The conditioning factor for each nodule is a 17-dimensional
vector, coming from six of its characteristic ratings, as well as the nodule size.
Note that "lobulation" and "spiculation" are removed due to known annotation
inconsistency in their ratings [3], and "internal structure" is removed since it
has a very imbalanced distribution. We quantize the remaining characteristics
to binary values following the same procedure of Shen et al. [43] and use the one-
hot encoding to generate a 12-dimensional vector for each nodule. The remaining
five dimensions are reserved for the quantization of the nodule size, ranging from
2 to 12 with an interval of 2. Following the above described procedure, the nodules with case index less than 899 are included in the training dataset while the
nodules of the remaining cases are considered as the test set. By augmenting the

label in such way, we exploit the richness of each nodule in LIDC, which proves to be beneficial for training.

D   Baselines

To evaluate the quality of generation, inference, and disentanglement, we consider two types of baselines. To show the effectiveness of dual variable inference, we compare our framework with single latent variable models. For this, we introduce a conditional adaptation of InfoGan [12] referred to as cInfoGAN and a conditional adversarial variational Autoencoder (cAVAE), both of which are explained in this section.

To compare our approach to dual latent variable inference methods, we extend InfoGAN and cAVAE to dual variables which we denote as D-cInfoGAN and D-cAVAE respectively.

We also compare DRAI to Dual Adversarial Inference (DAI) [30] and show how using our proposed disentanglement constraints together with latent code cycle-consistency can significantly boost performance. Finally, we conduct rigorous ablation studies to evaluate the impact of each component in DRAI.

D.1   Conditional InfoGAN

InfoGAN is a variant of generative adversarial network that aims to learn unsupervised disentangled representations. In order to do so, InfoGAN modifies the original GAN in two ways. First, it adds an additional input c to the generator. Second, using an encoder network Q, it predicts c from the generated image and effectively maximizes a lower bound on the mutual information between the input code c and the generated image $\tilde{x}$. The final objective is the combination of the original GAN objective plus that of the inferred code $\hat{c} \sim Q(c|x)$:

$$\min_{G,Q} \max_D V_{\text{InfoGAN}}(D,G,Q) = V_{\text{GAN}}(D,G) - \lambda(\mathbb{E}_{G(z,c),p(c)}[\log Q(c|x)] + H(c)). \quad (8)$$

The variable c can follow a discrete categorical distribution or a continuous distribution such as the normal distribution. InfoGAN is an unsupervised model popular for learning disentangled factors of variation [29,38,47].

We adopt a conditional version of InfoGAN –denoted by cInfoGAN– which is a conditional GAN augmented with an inference mechanism using the InfoGAN formulation. We experiment with two variants of cInfoGAN; a single latent variable model (cInfoGAN) shown in Fig. 7a, where the discriminator D_x is trained to distinguish between real (x) and fake ($\tilde{x}$) images while the discriminator $D_{x,t}$ distinguishes between the positive pair (x, t) and the corresponding negative pair ($\tilde{x}, t$), where $\tilde{x} = G_x(z,t)$ and t is the conditioning vector representing content. With the help of G_z, InfoGAN's mutual information objective is applied on z which represents the unsupervised style.

We also present a double latent variable model of InfoGAN (D-cInfoGAN) shown in Fig. 7b where in addition to inferring $\hat{z}$ we also infer $\hat{c}$ through cycle consistency using the ℓ_1 norm.

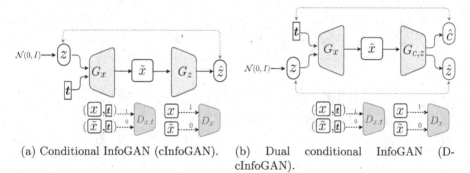

(a) Conditional InfoGAN (cInfoGAN).   (b) Dual conditional InfoGAN (D-cInfoGAN).

Fig. 7. InfoGan baselines.

D.2   cAVAE

Variational Auto-Encoders (VAEs) [28] are latent variable models commonly used for inferring disentangled factors of variation governing the data distribution. Let x be the random variable over the data distribution and z the random variable over the latent space. VAEs are trained by alternating between two phases, an inference phase where an encoder G_z is used to map a sample from the data to the latent space and infer the posterior distribution $q(z|x)$ and a generation phase where a decoder G_x reconstructs the original image using samples of the posterior distribution with likelihood $p(x|z)$.

VAEs maximize the evidence lower bound (ELBO) on the likelihood $p(x)$:

$$\max_G V_{\text{VAE}}(G_x, G_z) \quad = \quad \mathbb{E}_{q(z|x)}[\log p(x|z)] \quad - \quad D_{KL}[q(z|x) \,\|\, p(z)]. \quad (9)$$

Kingma and Welling [28] also introduced a conditional version of VAE (cVAE) where $p(x|z, c)$ is guided by both the latent code z and conditioning factor c. There have also been many attempts in combining VAEs and GANs. Notable efforts are that of Larsen et al. [31,35] and [50].

Conditional Adversarial Variational Autoencoder (cAVAE) is very similar to conditional Variational AutoEncoder (cVAE) but uses an adversarial formulation for the likelihood $p(x|z,c)$. Following the adversarial formulation for reconstruction [32,35], a discriminator D_{cycle} is trained on positive pairs (x,x) and negative pairs $(x,\hat{x})$, where $\hat{x} \sim p(x|t, \hat{z})$ and $\hat{z} \sim q(z|x)$. For the conditional generation we train a discriminator $D_{x,t}$ on positive pairs (x, t) and negative pairs $(\hat{x}, t)$, where t is the conditioning factor. We empirically discover that adding an additional discriminator $D_{x,t,z}$ which also takes advantage of the latent code $\hat{z}$ improves inference. Similar to cInfoGAN, we use two versions of cAVAE: a single latent variable version denoted by cAVAE (Fig. 8a) and a double latent variable version D-cAVAE (Fig. 8b), where in addition to the style posterior $q(z|x)$, we also infer the content posterior $q(c|x)$. Accordingly, to improve inference on the content variable, we add the discriminator $D_{x,t,c}$.

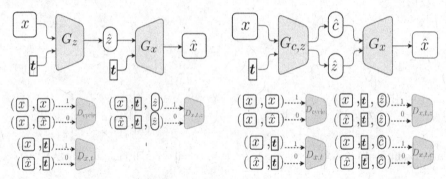

(a) Conditional Adversarial VAE (cAVAE). (b) Dual conditional Adversarial VAE (D-cAVAE).

Fig. 8. Adversarial VAE baselines

D.3   Evaluation Metrics

We explain in detail various evaluation metrics used in our experiments.

Measure of Disentanglement (CIFC). Multiple methods have been proposed to measure the degree of disentanglement between variables [23]. In this work, we propose a measure which evaluates the desired disentanglement characteristics of both the feature generator and the image generator. To have good feature disentanglement, we desire a feature generator (*i.e.*, encoder) that separates the information in an image in two disjoint variables of style and content in such a way that 1) the inferred information is consistent across images. *e.g.*, position and orientation is encoded the same way for all images; and 2) every piece of information is handled by *only* one of the two variables, meaning that the style and content variables do not share features. In order to measure these properties, we propose Cross Image Feature Consistency (CIFC) error where we measure the model's ability to first generate hybrid images of mixed style and content inferred from two different images and then its ability to reconstruct the original images. Figure 9 illustrates this process. As seen in this figure, given two images I_a and I_b, hybrid images I_{ab} and I_{ba} are generated using the pairs $(\hat{c}_a, \hat{z}_b)$ and $(\hat{c}_b, \hat{z}_a)$ respectively. By taking another step of hybrid image generation, I_{aa} and I_{bb} are generated as reconstructions of I_a and I_b respectively. To make the evaluation robust with respect to high frequency image details, we compute the reconstruction error in the feature space. In retrospect, the disentanglement measure is computed as:

$$CIFC = \mathbb{E}_{(I_a, I_b) \sim q_{\text{test}}(x)}[\|\hat{z}_a - \hat{z}_{aa}\| + \|\hat{c}_a - \hat{c}_{aa}\| + \tag{10}$$
$$\|\hat{z}_b - \hat{z}_{bb}\| + \|\hat{c}_b - \hat{c}_{bb}\|],$$

where $q_{\text{test}}(x)$ represents the empirical distribution of the test images.

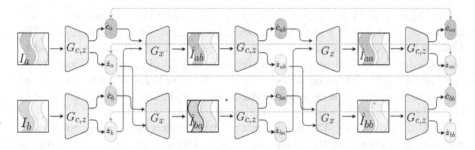

Fig. 9. Cross Image Feature Consistency (CIFC) error. CIFC is computed by first generating hybrid images of mixed style and content across two different images and then reconstructing the original images. For a more robust evaluation, CIFC is measured in the feature space.

FID. The Frechet inception distance (FID) score [22] measures the distance between the real and generated data distributions. An inception model is required for calculating FID, but since the conventional inception model used for FID is pretrained on colored natural images, it is not suitable to be used with LIDC which consists of single channel CT scans. Consequently, we train an inception model on the LIDC dataset to classify benign and malignant nodules. We use InceptionV3 [45] up to layer *"mixed3"* (initialized with pretrained ImageNet weights), and append a global average pooling layer followed by a dense layer.

Inception Score. Inception Score (IS) [41] is another quantitative metric on image generation which is commonly used to measure the diversity of the generated images. We use the same inception model described above to calculate IS. The TensorFlow-GAN library [44] is used to calculate both FID and IS.

E   Related Work

E.1   Connection to Other Conditional GANs in Medical Imaging

While adversarial training has been used extensively in the medical imaging domain, most work uses adversarial training to improve image segmentation and domain adaptation. The methods that use adversarial learning for image generation can be divided into two broad categories; the first group are those which use image-to-image translation as a proxy to image generation. These models use an image mask as the conditioning factor, and the generator generates an image which respects the constraints imposed by the mask [13,13,19,26,37]. Jin et al. [26] condition the generative adversarial network on a 3D mask, for lung nodule generation. In order to embed the nodules within their background context, the GAN is conditioned on a volume of interest whose central part containing the nodule has been erased. A favored approach for generating synthetic fundus retinal images is to use vessel segmentation maps as the conditioning factor. Guibas et al. [19] uses two GANs in sequence to generate fundus images. The first GAN generates vessel masks, and in stage two, a second GAN is trained to generate fundus retinal images from the vessel masks of stage one. Costa et al. [13] first use a U-Net based model to generate vessel segmentation masks from fundus images. An adversarial image-to-image translation model is then used to translate the mask back to the original image.

In Mok and Chung [37] the generator is conditioned on a brain tumor mask and generates brain MRI. To ensure correspondence between the tumour in the generated image and the mask, they further forced the generator to output the tumour boundaries in the generation process. Bissoto et al. [8] uses the semantic segmentation of skin lesions and generate high resolution images. Their model combines the pix2pix framework [25] with multi-scale discriminators to iteratively generate coarse to fine images.

While methods in this category give a lot of control over the generated images, the generator is limited to learning domain information such as low level texture and not higher level information such as shape and composition. Such information is presented in the mask which requires an additional model or an expert has to manually outline the mask which can get tedious for a lot of images.

The second category of methods are those which use high level class information in the form of a vector as the conditioning factor. Hu et al. [24] takes Gleason score vector as input to the conditional GAN to generate synthetic prostate diffusion imaging data corresponding to a particular cancer grade. Baur et al. [6] used a progressively growing model to generate high resolution images of skin lesions.

As mentioned in the introduction one potential pitfall of such methods is that by just using the class label as conditioning factor, it is hard to have control over the nuances of every class. While our proposed model falls within this category, our inference mechanism allows us to overcome this challenge by using the image data itself to discover factors of variation corresponding to various nuances of the content.

E.2   Disentangled Representation Learning

In the literature, disentanglement of style and content is primarily used for domain translation or domain adaptation. Content is defined as domain agnostic information shared between the domains, while style is defined as domain specific information. The goal of disentanglement to preserve as much content as possible and to prevent leakage of style from one domain to another. Gonzalez-Garcia et al. [18] used adversarial disentanglement for image to image translation. In order to prevent exposure of style from domain A to domain B, a Gradient Reversal Layer (GRL) is used to penalize shared information between the generator of domain B and style of domain A. In contrast, our proposed DRAI, uses GRL to minimize the shared information between style and content. In the medical domain, Yang et al. [49] aim to disentangle anatomical information and modality information in order to improve on a downstream liver segmentation task.

Ben-Cohen et al. [7] used adversarial learning to infer content agnostic features as style. Intuitively their method is similar to using GRL to minimize leakage of content information into a style variable. However, while [7] prevents leakage of content into style, it does not prevent the reverse effect which is leakage of style into content and thus does not guarantee disentanglement.

Yang et al. [48] use disentangle learning of modality agnostic and modality specific features in order to facilitate cross-modality liver segmentation. They use a mixture of adversarial training and cycle consistency loss to achieve disentanglement. The cycle-consistency component is used for in-domain reconstruction and the adversarial component is used for cross-domain translation. The two components encourage the disentanglement of the latent space, decomposing it into modality agnostic and modality specific sub-spaces.

To achieve disentanglement between modality information and anatomical structures in cardiac MR images, Chartsias et al. [9] use an autoencoder with two encoders: one for the modality information (style) and another for anatomical structures (content). They further impose constraints on the anatomical encoder such that every encoded pixel of the input image has a categorical distribution. As a result, the output of the anatomical encoder is a set of binary maps corresponding to cardiac substructures.

Disentangled representation learning has also been used for denoising of medical images. In Liao et al. [33], Given artifact affected CT images, metal-artifact reduction (MAR) is performed by disentangling the metal-artifact representations from the underlying CT images.

Sarhan et al. [42] use β-TCAV [10] to learn disentangled representations on an adversarial variation of the VAE. Their proposed model differs fundamentally from our work; its is a single variable model, without a conditional generative process, and does not infer separate style and content information.

Garcia1 et al. [17] used ALI (single variable) on structured MRI to discover regions of the brain that are involved in Autism Spectrum Disorder (ASD).

In contrast to previous work, we use style-content disentanglement to control features for conditional image generation. To the best of our knowledge this is the first time such attempt has been made in the context of medical imaging.

References

1. Abadi, M., et al.: TensorFlow: large-scale machine learning on heterogeneous systems (2015). https://www.tensorflow.org/
2. Agakov, D.B.F.: The IM algorithm: a variational approach to information maximization. Adv. Neural. Inf. Process. Syst. **16**, 201 (2004)
3. The Cancer Imaging Archive. Lung image database consortium - reader annotation and markup - annotation and markup issues/comments (2017). https://wiki.cancerimagingarchive.net/display/public/lidc-idri
4. Armato III, S.G., McLennan, G., Bidaut, L., McNitt-Gray, M.F., et al.: The lung image database consortium (LIDC) and image database resource initiative (IDRI): a completed reference database of lung nodules on CT scans. Med. Phys. **38**(2), 915–931 (2011)

5. Barratt, S., Sharma, R.: A note on the inception score. arXiv preprint arXiv:1801.01973 (2018)
6. Baur, C., Albarqouni, S., Navab, N.: Generating highly realistic images of skin lesions with GANs. In: Stoyanov, D., et al. (eds.) CARE/CLIP/OR 2.0/ISIC - 2018. LNCS, vol. 11041, pp. 260–267. Springer, Cham (2018). https://doi.org/10.1007/978-3-030-01201-4_28
7. Ben-Cohen, A., Mechrez, R., Yedidia, N., Greenspan, H.: Improving CNN training using disentanglement for liver lesion classification in CT. In: 2019 41st Annual International Conference of the IEEE Engineering in Medicine and Biology Society (EMBC), pp. 886–889. IEEE (2019)
8. Bissoto, A., Perez, F., Valle, E., Avila, S.: Skin lesion synthesis with generative adversarial networks. In: Stoyanov, D., et al. (eds.) CARE/CLIP/OR 2.0/ISIC - 2018. LNCS, vol. 11041, pp. 294–302. Springer, Cham (2018). https://doi.org/10.1007/978-3-030-01201-4_32
9. Chartsias, A., et al.: Disentangled representation learning in cardiac image analysis. Med. Image Anal. **58**, 101535 (2019)
10. Chen, R.T.Q., Li, X., Grosse, R.B., Duvenaud, D.K.: Isolating sources of disentanglement in variational autoencoders. In: Bengio, S., Wallach, H., Larochelle, H., Grauman, K., Cesa-Bianchi, N., Garnett, R. (eds.) Advances in Neural Information Processing Systems, vol. 31, pp. 2610–2620. Curran Associates Inc. (2018)
11. Chen, T., Kornblith, S., Norouzi, M., Hinton, G.: A simple framework for contrastive learning of visual representations. arXiv preprint arXiv:2002.05709 (2020)
12. Chen, X., Duan, Y., Houthooft, R., Schulman, J., Sutskever, I., Abbeel, P.: InfoGAN: interpretable representation learning by information maximizing generative adversarial nets. In: NIPS (2016)
13. Costa, P., et al.: Towards adversarial retinal image synthesis. arXiv preprint arXiv:1701.08974 (2017)
14. Donahue, J., Krähenbühl, P., Darrell, T.: Adversarial feature learning. In: ICLR (2017)
15. Dumoulin, V., et al.: Adversarially learned inference. In: ICLR (2017)
16. Ganin, Y., et al.: Domain-adversarial training of neural networks. CoRR, abs/1505.07818 (2015)
17. Garcia1, M., Orgogozo, J.-M., Clare, K., Luck, M.: Towards autism detection on brain structural MRI scans using deep unsupervised learning models. In: Proceedings of Medical Imaging meets NeurIPS Workshop (2019)
18. Gonzalez-Garcia, A., van de Weijer, J., Bengio, Y.: Image-to-image translation for cross-domain disentanglement. In: NIPS (2018)
19. Guibas, J.T., Virdi, T.S., Li, P.S.: Synthetic medical images from dual generative adversarial networks. arXiv preprint arXiv:1709.01872 (2017)
20. Havaei, M., Mao, X., Wang, Y., Lao, Q.: Conditional generation of medical images via disentangled adversarial inference. Med. Image Anal. 102106 (2021)
21. He, K., Fan, H., Wu, Y., Xie, S., Girshick, R.: Momentum contrast for unsupervised visual representation learning. In: Proceedings of the IEEE/CVF Conference on Computer Vision and Pattern Recognition, pp. 9729–9738 (2020)
22. Heusel, M., Ramsauer, H., Unterthiner, T., Nessler, B., Hochreiter, S.: GANs trained by a two time-scale update rule converge to a local nash equilibrium. In: NIPS (2017)
23. Higgins, I., et al.: Beta-VAE: learning basic visual concepts with a constrained variational framework. In: ICLR (2017)

24. Hu, X., Chung, A.G., Fieguth, P., Khalvati, F., Haider, M.A., Wong, A.: Prostate-GAN: mitigating data bias via prostate diffusion imaging synthesis with generative adversarial networks. arXiv preprint arXiv:1811.05817 (2018)

25. Isola, P., Zhu, J.-Y., Zhou, T., Efros, A.A.: Image-to-image translation with conditional adversarial networks. In: Proceedings of the IEEE Conference on Computer Vision and Pattern Recognition, pp. 1125–1134 (2017)

26. Jin, D., Xu, Z., Tang, Y., Harrison, A.P., Mollura, D.J.: CT-realistic lung nodule simulation from 3D conditional generative adversarial networks for robust lung segmentation. In: Frangi, A.F., Schnabel, J.A., Davatzikos, C., Alberola-López, C., Fichtinger, G. (eds.) MICCAI 2018. LNCS, vol. 11071, pp. 732–740. Springer, Cham (2018). https://doi.org/10.1007/978-3-030-00934-2_81

27. Kingma, D.P., Ba, J.: Adam: a method for stochastic optimization. In: ICLR (2015)

28. Kingma, D.P., Welling, M.: Auto-encoding variational bayes. In: ICLR (2014)

29. Kurutach, T., Tamar, A., Yang, G., Russell, S.J., Abbeel, P.: Learning plannable representations with causal InfoGAN. In: Advances in Neural Information Processing Systems, pp. 8733–8744 (2018)

30. Lao, Q., Havaei, M., Pesaranghader, A., Dutil, F., Di Jorio, L., Fevens, T.: Dual adversarial inference for text-to-image synthesis. In: Proceedings of the IEEE International Conference on Computer Vision, pp. 7567–7576 (2019)

31. Larsen, A.B.L., Kaae Sønderby, S., Larochelle, H., Winther, O.: Autoencoding beyond pixels using a learned similarity metric. In: ICML (2016)

32. Li, C., et al.: ALICE: towards understanding adversarial learning for joint distribution matching. In: NIPS (2017)

33. Liao, H., Lin, W.-A., Zhou, S.K., Luo, J.: ADN: artifact disentanglement network for unsupervised metal artifact reduction. IEEE Trans. Med. Imaging 39(3), 634–643 (2019)

34. Mao, X., Li, Q., Xie, H., Lau, R.Y.K., Wang, Z., Smolley, S.P.: Least squares generative adversarial networks. In: Proceedings of the IEEE International Conference on Computer Vision, pp. 2794–2802 (2017)

35. Mescheder, L., Nowozin, S., Geiger, A.: Adversarial variational bayes: unifying variational autoencoders and generative adversarial networks. In: International Conference on Machine Learning, pp. 2391–2400. PMLR (2017)

36. Mirza, M., Osindero, S.: Conditional generative adversarial nets. arXiv preprint arXiv:1411.1784 (2014)

37. Mok, T.C.W., Chung, A.C.S.: Learning data augmentation for brain tumor segmentation with coarse-to-fine generative adversarial networks. In: Crimi, A., Bakas, S., Kuijf, H., Keyvan, F., Reyes, M., van Walsum, T. (eds.) BrainLes 2018. LNCS, vol. 11383, pp. 70–80. Springer, Cham (2019). https://doi.org/10.1007/978-3-030-11723-8_7

38. Ojha, U., Singh, K.K., Hsieh, C.-J., Lee, Y.J.: Elastic-InfoGAN: unsupervised disentangled representation learning in imbalanced data. arXiv preprint arXiv:1910.01112 (2019)

39. van den Oord, A., Li, Y., Vinyals, O.: Representation learning with contrastive predictive coding. arXiv preprint arXiv:1807.03748 (2018)

40. Reed, S., Akata, Z., Yan, X., Logeswaran, L., Schiele, B., Lee, H.: Generative adversarial text to image synthesis. In: ICML (2016)

41. Salimans, T., Goodfellow, I., Zaremba, W., Cheung, V., Radford, A., Chen, X.: Improved techniques for training GANs. In: NIPS (2016)

42. Sarhan, M.H., Eslami, A., Navab, N., Albarqouni, S.: Learning interpretable disentangled representations using adversarial VAEs. In: Wang, Q., et al. (eds.) DART/MIL3ID -2019. LNCS, vol. 11795, pp. 37–44. Springer, Cham (2019). https://doi.org/10.1007/978-3-030-33391-1_5

43. Shen, S., Han, S.X., Aberle, D.R., Bui, A.A.T., Hsu, W.: An interpretable deep hierarchical semantic convolutional neural network for lung nodule malignancy classification. Expert Syst. Appl. **128**, 84–95 (2019)

44. Shor, J.: TensorFlow-GAN (TF-GAN): a lightweight library for generative adversarial networks (2017). https://github.com/tensorflow/gan

45. Szegedy, C., Vanhoucke, V., Ioffe, S., Shlens, J., Wojna, Z.: Rethinking the inception architecture for computer vision. In: 2016 IEEE Conference on Computer Vision and Pattern Recognition (CVPR), pp. 2818–2826 (2016)

46. Tschandl, P., Rosendahl, C., Kittler, H.: The HAM10000 dataset, a large collection of multi-source dermatoscopic images of common pigmented skin lesions. Sci. Data **5**, 1–9 (2018)

47. Wang, N., et al.: Unsupervised classification of street architectures based on InfoGAN (2019)

48. Yang, J., Dvornek, N.C., Zhang, F., Chapiro, J., Lin, M.D., Duncan, J.S.: Unsupervised domain adaptation via disentangled representations: application to cross-modality liver segmentation. In: Shen, D., et al. (eds.) MICCAI 2019. LNCS, vol. 11765, pp. 255–263. Springer, Cham (2019). https://doi.org/10.1007/978-3-030-32245-8_29

49. Yang, J., et al.: Domain-agnostic learning with anatomy-consistent embedding for cross-modality liver segmentation. In: Proceedings of the IEEE International Conference on Computer Vision Workshops (2019)

50. Yu, X., Zhang, X., Cao, Y., Xia, M.: VAEGAN: a collaborative filtering framework based on adversarial variational autoencoders. In: IJCAI, pp. 4206–4212 (2019)

51. Zhu, J.-Y., Park, T., Isola, P., Efros, A.A.: Unpaired image-to-image translation using cycle-consistent adversarial networks. In: Proceedings of the IEEE International Conference on Computer Vision, pp. 2223–2232 (2017)

CT-SGAN: Computed Tomography Synthesis GAN

Ahmad Pesaranghader[1,2(✉)], Yiping Wang[3], and Mohammad Havaei[4]

[1] McGill University, Montreal, Canada
pesarana@mila.quebec
[2] Quebec AI Institute (Mila), Montreal, Canada
[3] University of Waterloo, Waterloo, Canada
[4] Imagia Inc., Montreal, Canada

Abstract. Diversity in data is critical for the successful training of deep learning models. Leveraged by a recurrent generative adversarial network, we propose the CT-SGAN model that generates large-scale 3D synthetic CT-scan volumes ($\geq 224 \times 224 \times 224$) when trained on a small dataset of chest CT-scans. CT-SGAN offers an attractive solution to two major challenges facing machine learning in medical imaging: a small number of given i.i.d. training data, and the restrictions around the sharing of patient data preventing to rapidly obtain larger and more diverse datasets. We evaluate the fidelity of the generated images qualitatively and quantitatively using various metrics including Fréchet Inception Distance and Inception Score. We further show that CT-SGAN can significantly improve lung nodule detection accuracy by pre-training a classifier on a vast amount of synthetic data.

Keywords: Computerized tomography · Data augmentation · Deep learning · Generative adversarial networks · Wasserstein distance

1 Introduction

Recently, deep learning has achieved significant success in several applications including computer vision, natural language processing, and reinforcement learning. However, large amounts of training samples, which sufficiently cover the population diversity, are often necessary to develop high-accuracy machine learning and deep learning models. Unfortunately, data availability in the medical image domain is quite limited due to several reasons such as significant image acquisition costs, protections on sensitive patient information, limited numbers of disease cases, difficulties in data labeling, and variations in locations, scales, and appearances of abnormalities. Despite the efforts made towards constructing large medical image datasets, options are limited beyond using simple automatic methods, huge amounts of radiologist labor, or mining from radiologist reports

Supported by Imagia Inc.

S. Engelhardt et al. (Eds.): DGM4MICCAI 2021/DALI 2021, LNCS 13003, pp. 67–79, 2021.
https://doi.org/10.1007/978-3-030-88210-5_6

[9]. Therefore, it is still a challenge to generate effective and sufficient medical data samples with no or limited involvement of experts.

The production of synthetic training samples is one tempting alternative. However, in practice, it is less appealing due to weaknesses in reproducing high-fidelity data [9]. The advancement in generative models such as the generative adversarial networks (GANs) [5], however, is creating an opportunity in producing real-looking additional (training) data. This possibility has been enhanced with refinements on fully convolutional [19] and conditional GANs [15]. For example, Isola et al. extend the conditional GAN (cGAN) concept to predict pixels from known pixels [8]. Within medical imaging, Nie et al. use a GAN to simulate CT slices from MRI data [16]. For lung nodules, Chuquicusma et al. train a simple GAN to generate simulated images from random noise vectors, but do not condition based on surrounding context [3]. Despite recent efforts to generate large-scale CT-cans with the help of GANs [18,22], to the best of our knowledge, all of these generative models have one or more important pieces missing which are: (1) the (non-progressive) generation of the whole large-scale 3D volumes from scratch with a small number of training samples, (2) the generation of CT-scans without working with a sub-region of a volume or translating from one domain/modality to another domain/modality, and (3) examining their generative model with real-life medical imaging problems such as nodule detection. The absence of these missing pieces can be due to large amounts of GPU memory needed to deal with 3D convolutions/deconvolutions [14,17,21]. This limitation makes even the most well-known GANs for the generation of high-resolution images [11,25] impractical once they are applied to the 3D volumes. On the other hand, the generation of large-scale CT-scan volumes is of significant importance as in these scans, fine parenchymal details such as small airway walls, vasculature, and lesion texture, would be better visible which in turn lead to more accurate prediction models.

In this work, we propose a novel method to generate large-scale 3D synthetic CT-scans ($\geq 224 \times 224 \times 224$) by training a recurrent generative adversarial network with the help of a small dataset of 900 real chest CT-scans. As shown in Fig. 1, we demonstrate the value of a recurrent generative model with which the volumes of CT-scans would be generated gradually through the generation of their sub-component slices and slabs (i.e., series of consecutive slices). By doing so, we can subvert the challenging task of generating large-scale 3D images to one with notably less GPU memory requirement. Our proposed 3D CT-scan generation model, named CT-SGAN, offers a potential solution to two major challenges facing machine learning in medical imaging, namely a small number of i.i.d. training samples, and limited access to patient data. We evaluate the fidelity of generated images qualitatively and quantitatively using Fréchet Inception Distance and Inception Score. We further show training on the synthetic images generated by CT-SGAN significantly improves a downstream lung nodule detection task across various nodule sizes. The contributions of this work are twofold:

– We propose a generative model capable of synthesizing 3D images of lung CT. The proposed CT-SGAN leverages recurrent neural networks to learn slice sequences and thus has a very small memory footprint.

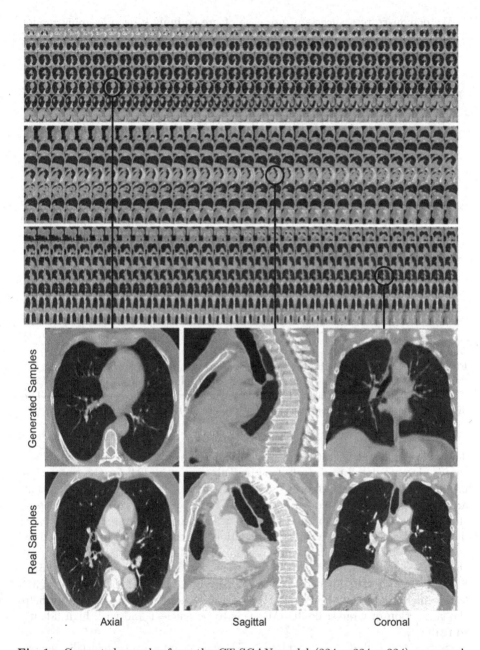

Fig. 1. Generated samples from the CT-SGAN model ($224 \times 224 \times 224$) compared with samples of the same positions from a real CT-scan. The upper part of the figure also demonstrates CT-SGAN, with the help of its BiLSTM network as well as slab discriminator, is capable to learn anatomical consistency across slices within every 900 real training CT-scan volumes reflecting the same behavior in all three perspectives of axial, sagittal, and coronal at generation time. See Appendix for more samples.

– We demonstrate a successful use case of pre-training a deep learning model on 3D synthetic CT data to improve lung nodule detection.

2   Methods

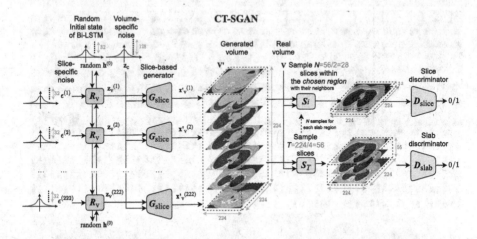

Fig. 2. CT-SGAN has 4 sub-networks: the recurrent neural network (BiLSTM), R_v; the slice generator, G_{slice}; the slice and the slab discriminators, D_{slice} and D_{slab}. G_{slice} generates a volume by getting zs from R_v concatenated with the constant noise. The discriminators verify the consistency and the quality of the generated slices with respect to original CT-scan volumes.

The structure of the CT-SGAN model is depicted in Fig. 2. The model architecture is closely related to a work done by Tulyakov et al. [24] for the generation of videos. In our work, however, instead of considering frames of a video that are meant to be generated sequentially in terms of time, we have a series of slices (i.e., slabs from a volume) that are needed to be generated consecutively in terms of space. Some major modifications in the network designs such as the critical change in z size, consideration of gradient penalty, sampling from the slab regions, consideration of Bidirectional Long/Short-Term Memory (BiL-STM) network for R_v instead of Gated Recurrent Units, and employment of 3D slices differentiate CT-SGAN from their study. These changes specifically help the generation of large-scale 3D CT-scans to be efficient and with high fidelity to real data.

Succinctly, CT-SGAN generates a CT-scan volume by sequentially generating a single slice each of size 224×224 ($\times 3$); from these 3D slices, which lead to more fine-grain quality, the collection of center slices form a volume when piled on top of each other (the first and the last 1D slices in the volumes are the neighbors of their associate 3D slices; hence 222 3D slices in total). At each time step, a slice generative network, G_{slice}, maps a random vector

z to a slice. The random vector consists of two parts where the first is sampled from a constant patient-specific subspace and the second is sampled from a variable slice-specific subspace. Patient-specific noise can also cover the overall color range and contrast within a volume (e.g., scanner-specific information). Since contents of CT-scans slices in one volume usually remains the same, we model the patient/scanner-specific space using a constant Gaussian distribution over all slices for a particular volume. On the other hand, sampling from the slice-specific space is achieved through a BiLSTM network (R_v) where the network is fed with random inputs (ϵ_i and h_0), and its parameters are learned during training. Despite lacking supervision regarding the decomposition of general (patient/scanner-specific) content and slice-specific properties in CT-scan volumes, we noticed that CT-SGAN can learn to disentangle these two factors through an adversarial training scheme as the absence of either noise hurt the overall quality. Both discriminators D_{slice} and D_{slab} play the judge role, providing criticisms to G_{slice} and R_v. The slice discriminator D_{slice} is specialized in criticizing G_{slice} based on individual CT-scan slices. It is trained to determine if a single slice is sampled from a real CT-scan volume, v, or from v' with respect to the slice position. On the other hand, D_{slab} provides criticisms to G_{slice} based on the generated slab. D_{slab} takes a fixed-length slab size, say T (center) slices, and decides if a slab was sampled from a real CT-scan volume or from v'. The adversarial learning problem writes as:

$$\max_{G_{\text{slice}},R_v} \min_{D_{\text{slice}},D_{\text{slab}}} \mathcal{F}_{\text{volume}}\left(D_{\text{slice}}, D_{\text{slab}}, G_{\text{slice}}, R_v\right) \tag{1}$$

where the vanilla Jensen-Shannon divergence objective function $\mathcal{F}_{\text{volume}}$ is (N is the number of slices sampled from a slab region):

$$
\begin{aligned}
&\sum_{i=1}^{N} \big[\mathbb{E}_v\left[-\log D_{slice}\left(S_i(v)\right)\right] \\
&+\mathbb{E}_{v'}\left[-\log\left(1 - D_{slice}\left(S_i(v')\right)\right)\right]\big] \\
&+\mathbb{E}_v\left[-\log D_{slab}\left(S_T(v)\right)\right] \\
&+\mathbb{E}_{v'}\left[-\log\left(1 - D_{slab}\left(S_T(v')\right)\right)\right]
\end{aligned}
\tag{2}
$$

We train CT-SGAN using the alternating gradient update algorithm [4]. Specifically, in one step, we update D_{slice} and D_{slab} while fixing G_{slice} and R_v. In the alternating step, we update G_{slice} and R_v while fixing D_{slice} and D_{slab}.

3   Datasets and Experimental Design

The evaluation of the generated CT-scans was designed to be done under two scrutinies: (1) qualitative inspection of the generated volumes from CT-SGAN where the diverse variation and consistency across all three views of axial, coronal, and sagittal were met (2) quantitative demonstration that synthetic data from CT-SGAN are valuable to build deep learning models for which limited training data are available. We evaluate the efficacy of data augmentation by three nodule detection classifier experiments (i) training with only real dataset (ii) training with only 10,000 synthetic volumes (iii) training with 10,000 synthetic volumes as a pretraining step and then continue to train on the real dataset (i.e., fine-tuning).

3.1   Dataset Preparation

The dataset preprocessed and prepared for the training and evaluation contained 1200 volumes of CT-scans; the first half (clear of any nodules) was from the National Lung Screening Trial (NLST) study [23] and the second half (containing nodules) was from the Lung Image Database Consortium (LIDC) reference database [2] (i.e., the dataset covers at least two distributional domains), where both are publicly available. This combined dataset was divided into 3 stratified non-overlapping splits of training, validation, and test. Explicitly, of this combined dataset (referred as real dataset hereafter), 900 CT-scans were used for CT-SGAN training (as well as the nodule detection classifier when needed), 150 scans for validation of the nodule detection classifier, and the nodule detection results were reported on the remaining 150 CT-scans. The CT-SGAN model was trained only on the training split; however, since CT-SGAN generates samples unconditioned on the presence of the nodules, the nodules from LIDC were removed in advance (i.e., CT-SGAN generates nodule-free CT-scans). Regarding nodule detection experiment, as the real CT-scans came from two resources (i.e., LIDC scanners and NLST scanners), one of which contained nodules, the training, validation and test dataset of the nodule detection classifier was created by following Fig. 3(a) to create an unbiased (source- and device-agnostic) dataset. For this purpose, a *nodule injector* and a *nodule eraser* were adopted based on a lung nodule simulation cGAN [10].

The nodule injector was trained on the LIDC data, which contains nodules information such as location and radius. At training time the inputs were the masked central nodules regions and the volumes of interest containing the nodules, while at inference mode the input was only the masked central nodules regions, see Appendix. The outputs were the simulated nodules when provided with masked regions. In a similar fashion to lung nodule simulation cGAN [10], the nodule eraser was trained to remove nodules on nodule-injected NLST samples. At training time, the inputs of the eraser were the volumes of interest with and without a nodule, and the model learned how to replace the nodules with healthy tissues at inference time, see Appendix.

To mitigate the device noise in the real dataset, the LIDC data was divided evenly and the nodule eraser was applied to remove the nodules from half of the LIDC scans. Similarly, the NLST data was split evenly and the nodule injector was applied to insert the nodules into half of the NLST scans.

To obtain the synthetic volumes for augmentation, the nodule eraser was first applied to the LIDC data to ensure the CT-SGAN training set was nodule-free. We argue the synthetic scans were nodule-free as the nodules in training data were removed. The trained nodule injector was employed to randomly insert nodules inside the lungs of the synthetic volumes, and the number of nodules to be inserted was determined by the nodule amount distribution of the LIDC dataset. The 10,000 synthetic data augmentation were created by injecting nodules into half of the volumes and leave the rest untouched.

4   Results and Discussion

4.1   Qualitative Evaluation

The visual qualitative evaluation of the generated volumes was studied based on three criteria: (1) Anatomical consistency of the generated slices, (2) Fidelity of generated slices to the real ones, and (3) diverse generation of CT-scans. Regarding the first two, Fig. 1 shows these requirements were met as in thousand of generated CT-scans we rarely noticed any anomalies. For the high quality of the slices and slabs, we observed consideration of 3D slices and the inclusion of both patient- and slice-specific noises played important roles. As to the diversity in the generated CT-scans, i.e. to avoid mode collapse, and also to ensure stability in training, the discriminators' losses contained the gradient penalty discussed in [13] introducing variations in the generated volumes. While CT-SGAN was preferably trained with Wasserstein loss [1], we did not notice a drastic change when the vanilla Jensen-Shannon loss was employed. Also, even though artifacts could appear in the generated CT-scans, the presence of them was partially related to the noise in real CT-scans produced by scanners.

4.2   Quantitative Evaluation

3D-SqueezeNet [7] was used as the nodule detection classifier. Input volumes were normalized between 0 and 1, and the classifier predicts the existence of nodules as a binary label (0 or 1). Adam optimizer [12] was used with learning rate = 0.0001, $\beta_1 = 0.9$, $\beta_2 = 0.999$. 3 different seeds were used to initialize the classifier and choose the location and number of nodules to inject into the synthetic volumes. Moreover, 6 different sizes of simulated nodules were compared, see Appendix for the distribution of nodule radius in LIDC dataset. Figure 3(b) summarized the nodule detection classification results. We observe that with a larger number of synthetic training data (10,000 generated CT-scans), the trained classifiers have better performance when compared with the classifier trained with 900 real volumes; the accuracy improvement is significant for the nodule sizes of 14 and 16. Also, nodule classification accuracy increases even further by pre-training the classifier on synthetic data and then fine-tuning on the 900 real volumes.

Fréchet Inception Distance (FID) [6] and Inception Score (IS) [20] were also computed, as shown in Table 1. FID measures the disparity between the distribution of generated synthetic data and real volumes. IS indicates how diverse a set of generated CT volumes is. To compute FID, we randomly selected two different scans in the data source (real or synthetic) and computed the FID slice by slice in order. Similarly, IS was also computed slice-by-slice. The average IS for the synthetic data is 2.3703 while the average IS for the real data is 2.2129. We believe that the IS for real volumes is lower due to the limited real volumes (only 900 scans). The FID between synthetic data and real data is 145.187. As a point of reference, the FID between two splits of the real data is 130.729. As the generation of small-size volumes (e.g., $128 \times 128 \times 128$) with a vanilla 3D GAN, and then resizing the generated scans (at the cost of losing details and diversity) determined the baseline in our model comparison Table 1 also provides that.

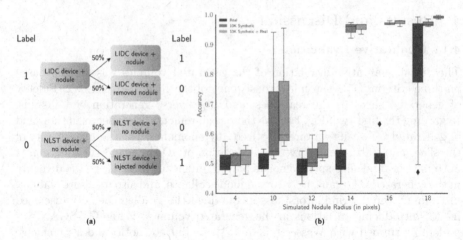

(a)   (b)

Fig. 3. **(a)** The real dataset consists of both LIDC and NLST data. To overcome the effect of device noise on the nodule detection classifier, a nodule *injector* and a nodule *eraser* were trained to mix the nodule information and device information to alleviate the device bias. **(b)** Results for classifiers trained on real and synthetic images. Accuracy is provided for the balanced binary test set. We observe that (pre-)training the classifier on a large amount of synthetic data will notably improve the nodule detection performance.·

Table 1. Inception Score and Fréchet Inception Distance for the real scans and the synthetic scans generated by vanilla 3D GAN and CT-SGAN.

Data Source	IS (↑)	FID (↓)
Real Data	2.21 ± 0.21	130.72 ± 31.05
3D GAN Synthetic Data (128^3)	2.16 ± 0.26	206.34 ± 59.12
CT-SGAN Synthetic Data (224^3)	2.37 ± 0.19	145.18 ± 25.97

5   Conclusions

We introduced CT-SGAN, a novel deep learning architecture that can generate authentic-looking CT-scans. We quantitatively demonstrated the value of data augmentation using CT-SGAN for the nodule detection task. By pretraining the nodule detection classifier on a vast amount of synthetic volumes and fine-tuning on the real data, the performance of the classifier improved notably. For future work, we aim to generate a larger size of CT-scans with the proposed model, as well as extend CT-SGAN to conditional CT-SGAN to avoid external algorithms for the inclusion or exclusion of nodules.

Appendix

A   Sample Synthetic CT-scans from CT-SGAN

See Figs. 4 and 5.

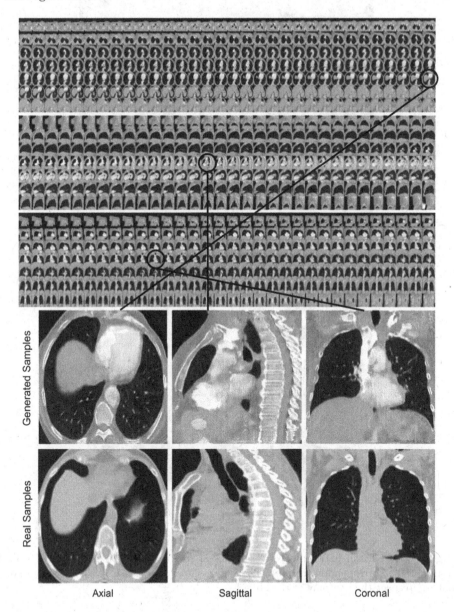

Fig. 4. Middle-region CT-SGAN samples compared with real samples of the same positions.

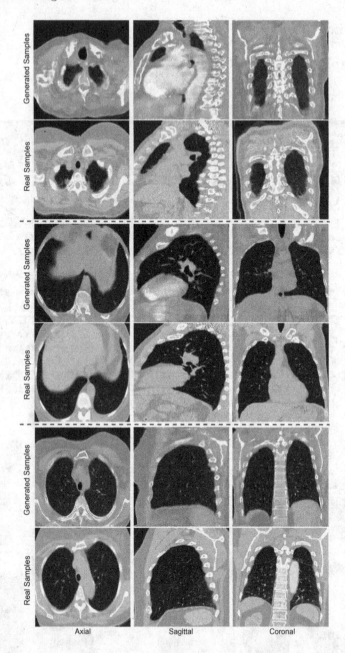

Fig. 5. Comparison between several samples of real and CT-SGAN generated CT, from various corresponding regions. The figure shows how the CT-SGAN generated images, respect the anatomy of the lung.

B   Nodule Injector and Eraser

See Figs. 6, 7 and 8.

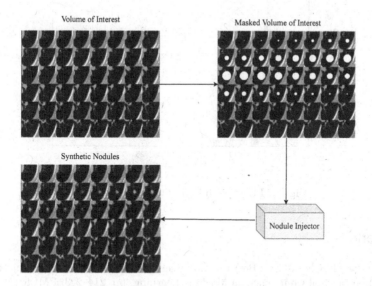

Fig. 6. Illustration of the Nodule Injector during inference mode. Volume of interests (VOIs) were selected as a sub-volume, and a mask was applied to the centre of VOIs as input for the Nodule Injector. The output was the injected synthetic nodules, and the injected VOIs will be pasted back to the CT-scans.

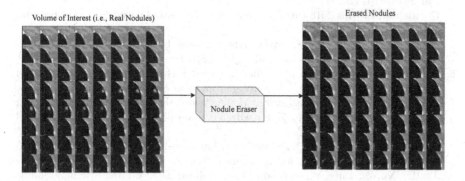

Fig. 7. Illustration of the Nodule Eraser during inference mode. Volume of interests (VOIs) were selected as a sub-volume as input for the Nodule Eraser. The output was the sub-volume without the real nodules, and the erased VOIs will be pasted back to the CT-scans.

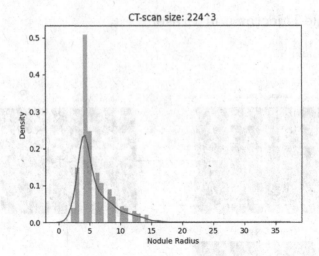

Fig. 8. LIDC Nodule Radius Distribution.

References

1. Arjovsky, M., Chintala, S., Bottou, L.: Wasserstein generative adversarial networks. In: International Conference on Machine Learning, pp. 214–223. PMLR (2017)
2. Armato, S.G., III., et al.: The lung image database consortium (LIDC) and image database resource initiative (IDRI): a completed reference database of lung nodules on CT scans. Med. Phys. **38**(2), 915–931 (2011)
3. Chuquicusma, M.J., Hussein, S., Burt, J., Bagci, U.: How to fool radiologists with generative adversarial networks? A visual turing test for lung cancer diagnosis. In: 2018 IEEE 15th International Symposium on Biomedical Imaging (ISBI 2018), pp. 240–244. IEEE (2018)
4. Goodfellow, I.: Nips 2016 tutorial: generative adversarial networks. arXiv preprint arXiv:1701.00160 (2016)
5. Goodfellow, I., et al.: Generative adversarial nets. In: Advances in Neural Information Processing Systems, pp. 2672–2680 (2014)
6. Heusel, M., Ramsauer, H., Unterthiner, T., Nessler, B., Hochreiter, S.: Gans trained by a two time-scale update rule converge to a local Nash equilibrium (2018)
7. Iandola, F.N., Han, S., Moskewicz, M.W., Ashraf, K., Dally, W.J., Keutzer, K.: SqueezeNet: AlexNet-level accuracy with 50x fewer parameters and <0.5mb model size (2016)
8. Isola, P., Zhu, J.Y., Zhou, T., Efros, A.A.: Image-to-image translation with conditional adversarial networks. arXiv preprint (2017)
9. Jin, D., Xu, Z., Tang, Y., Harrison, A.P., Mollura, D.J.: CT-realistic lung nodule simulation from 3D conditional generative adversarial networks for robust lung segmentation. arXiv preprint arXiv:1806.04051 (2018)
10. Jin, D., Xu, Z., Tang, Y., Harrison, A.P., Mollura, D.J.: CT-realistic lung nodule simulation from 3D conditional generative adversarial networks for robust lung segmentation. In: Frangi, A.F., Schnabel, J.A., Davatzikos, C., Alberola-López, C., Fichtinger, G. (eds.) MICCAI 2018. LNCS, vol. 11071, pp. 732–740. Springer, Cham (2018). https://doi.org/10.1007/978-3-030-00934-2_81

11. Karras, T., Aila, T., Laine, S., Lehtinen, J.: Progressive growing of GANs for improved quality, stability, and variation. arXiv preprint arXiv:1710.10196 (2017)
12. Kingma, D.P., Ba, J.: Adam: a method for stochastic optimization (2017)
13. Mescheder, L., Geiger, A., Nowozin, S.: Which training methods for GANs do actually converge? arXiv preprint arXiv:1801.04406 (2018)
14. Mirsky, Y., Mahler, T., Shelef, I., Elovici, Y.: CT-GAN: malicious tampering of 3D medical imagery using deep learning. arXiv preprint arXiv:1901.03597 (2019)
15. Mirza, M., Osindero, S.: Conditional generative adversarial nets. arXiv preprint arXiv:1411.1784 (2014)
16. Nie, D., et al.: Medical image synthesis with context-aware generative adversarial networks. In: Descoteaux, M., Maier-Hein, L., Franz, A., Jannin, P., Collins, D.L., Duchesne, S. (eds.) MICCAI 2017. LNCS, vol. 10435, pp. 417–425. Springer, Cham (2017). https://doi.org/10.1007/978-3-319-66179-7_48
17. van der Ouderaa, T.F., Worrall, D.E., van Ginneken, B.: Chest CT super-resolution and domain-adaptation using memory-efficient 3D reversible GANs. In: Medical Imaging with Deep Learning (2019)
18. Park, H.Y., et al.: Realistic high-resolution body computed tomography image synthesis by using progressive growing generative adversarial network: visual turing test. JMIR Med. Inform. 9(3), e23328 (2021)
19. Radford, A., Metz, L., Chintala, S.: Unsupervised representation learning with deep convolutional generative adversarial networks. arXiv preprint arXiv:1511.06434 (2015)
20. Salimans, T., Goodfellow, I., Zaremba, W., Cheung, V., Radford, A., Chen, X.: Improved techniques for training GANs (2016)
21. Shin, H.-C., et al.: Medical image synthesis for data augmentation and anonymization using generative adversarial networks. In: Gooya, A., Goksel, O., Oguz, I., Burgos, N. (eds.) SASHIMI 2018. LNCS, vol. 11037, pp. 1–11. Springer, Cham (2018). https://doi.org/10.1007/978-3-030-00536-8_1
22. Sun, L., Chen, J., Xu, Y., Gong, M., Yu, K., Batmanghelich, K.: Hierarchical amortized training for memory-efficient high resolution 3D GAN. arXiv preprint arXiv:2008.01910 (2020)
23. National Lung Screening Trial Research Team: The national lung screening trial: overview and study design. Radiology 258(1), 243–253 (2011)
24. Tulyakov, S., Liu, M.Y., Yang, X., Kautz, J.: MoCoGAN: decomposing motion and content for video generation. In: Proceedings of the IEEE Conference on Computer Vision and Pattern Recognition, pp. 1526–1535 (2018)
25. Zhang, H., et al.: StackGAN: text to photo-realistic image synthesis with stacked generative adversarial networks. In: Proceedings of the IEEE International Conference on Computer Vision, pp. 5907–5915 (2017)

Applications and Evaluation

Hierarchical Probabilistic Ultrasound Image Inpainting via Variational Inference

Alex Ling Yu Hung$^{(\boxtimes)}$ ⓘ, Zhiqing Sun, Wanwen Chen ⓘ, and John Galeotti ⓘ

Carnegie Mellon University, Pittsburgh, PA 15213, USA
{lingyuh,zhiqings,wanwenc,jgaleotti}@andrew.cmu.edu

Abstract. Image inpainting is a well established problem in computer vision which aims to fill in missing regions in images. In medical applications, it can be combined with other tools to remove artifacts or fill in occluded regions, allowing better understanding of the images from doctors or downstream algorithms. However, current methods that solve the problem usually pay no attention to the underlying pixel-intensity distributions in the missing input regions, and most approaches deterministically provide only one result per input region instead of several plausible results. We estimate the intensity distributions within each masked region using a novel Variational Autoencoder (VAE) based hierarchical probabilistic network. Our approach then generates a diverse set of inpainted images, all of which appear visually appropriate.

Keywords: Image inpainting · Variational inference · Artifacts removal · Needle tracking · Probabilistic graphical models

1 Introduction

Image inpainting is a fundamental problem whose goal is to fill in the missing patches in images with reasonable contents. In medical image analysis, it could be used as a tool to remove artifacts or fill in unclear regions. Because neural networks are very sensitive to small perturbation in noises and pixel values [1], correctly removing artifacts and occlusions in the images is important in medical domain to perform computer-aided diagnosis. When humans perform such task, different experts would agree on the semantic contents to fill in, but disagree on the fine details of the contents (e.g. the distribution of speckle noises in ultrasound images). The results should be plausible as long as 1) the filled-in content is semantically correct; 2) the general continuity of the image is kept; 3) the inpainted content is visually realistic, i.e. has the same noise distribution as a real image. However, the subjectivity of this problems are rarely discussed in literature, and most works focus on inpainting images deterministically. Besides,

Electronic supplementary material The online version of this chapter (https://doi.org/10.1007/978-3-030-88210-5_7) contains supplementary material, which is available to authorized users.

S. Engelhardt et al. (Eds.): DGM4MICCAI 2021/DALI 2021, LNCS 13003, pp. 83–92, 2021.
https://doi.org/10.1007/978-3-030-88210-5_7

due to the noisy nature of ultrasound images, recent approaches with traditional loss functions (e.g. L2 loss) would smooth the image, making the results unrealistic. Therefore, we propose a variational inpainting method with more complicated training objectives to solve the issue.

Unlike traditional diffusion-based or patch-based methods [2,3], the majority of recent works have been done on natural images where they mainly focus on developing neural networks to either suppress the piece-meal artifact near the boundaries of the masks [9,28], or recover fine detailed textures in the masked region [14,16,23,25,28]. Besides, a few works have been working on generating multiple reasonable results instead of one [7,29]. In medical image analysis, prior semantic information was incorporated into the network to guide inpainting in [27] while a pyramid Generative Adversarial Network (GAN) was proposed to remove cross symbols in thyroid ultrasound images by inpainting. Due to the fact that training images only provide one plausible inpainted content, a traditional Variational Autoencoder (VAE) would underestimate the variance in the coded distribution. To synthesize the images by sampling from some distributions to generate multiple outputs instead of directly predict an output deterministically, we propose a hierarchical probabilistic approach, which utilizes the distribution of the masked images to resemble the distribution of the unmask images.

To generate multiple plausible results on ultrasound images, we follow the design of the state-of-the-art NVAE [22] and HPU-Net [13] in terms of the model architecture. A series of hierarchical latent variables are injected into the decoder part of a U-Net [19]. During training, we can get a posterior from the pair of incomplete image and complete image, and during inference, we directly sample from the prior distribution. To generate the correct noise pattern, we use loss functions including deep losses enforced by VGG along with other traditional losses [9]. Our method is capable of generating multiple plausible inpainting results on ultrasound images, and removing artifacts.

2   Methods

We propose a VAE-based method that employs a hierarchical latent space decomposition. Shown in Fig. 1, our method aims to learn the posterior given the complete and incomplete image and the prior given the incomplete images by maximizing the variational lower bound (ELBO). During inference, the method estimates the complete image given the incomplete one and the prior.

2.1   Learning

In learning phase, both incomplete image x and complete image y are observed. With respect to latent variables $\{z_i\}$, we aim to learn n posterior distributions

$$
\begin{aligned}
&q_{\phi_n}(z_n|x, y), \\
&q_{\phi_i}(z_i|x, y, z_{>i}),
\end{aligned}
\tag{1}
$$

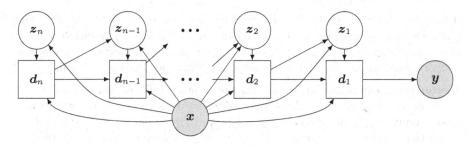

Fig. 1. Probabilistic Graphical Model (PGM) interpretation of our method, where $\{z_i\}$ denotes Gaussian latent variables, while d_i represents a compressed representation of $z_{\geq i}$. x represents the incomplete images and y represents the completed image, where both of them are observed during training and only x is observed during inference.

n prior distributions

$$p_{\theta_n}(z_n|x),$$
$$p_{\theta_i}(z_i|x, z_{>i}), \tag{2}$$

and a final projection function

$$p_\theta(y|x, z_1, \ldots, z_n), \tag{3}$$

such that the ELBO can be written as

$$\mathbb{E}_{q_{\phi_i}}\left[\log p_{\{\theta_i\}}(y|x, \{z_i\}) - \sum_i D_{KL}(q_{\phi_i}(z_i|x, y, z_{>i})\|p_{\theta_i}(z_i|z_{>i}))\right] \tag{4}$$

$D_{KL}(\cdot)$ is the Kullback-Leibler (KL) divergence between two distributions.

2.2 Inference

In the inference stage, only the incomplete image x are observed. Therefore, we can first sample latent variables from the priors:

$$\tilde{z}_n \sim p_{\theta_i}(z_n|x)$$
$$\tilde{z}_i \sim p_{\theta_i}(z_i|x, z_{>i}) \tag{5}$$

And then perform the inference using $\{\tilde{z}_i\}$:

$$\tilde{y} \sim p_\theta(y|x, \tilde{z}_1, \ldots, \tilde{z}_n) \tag{6}$$

2.3 Objectives

To generate correct semantic information and textures, we divide the loss function enforced on the outputs into three parts: reconstruction loss, perceptual loss, and style loss. [9] shows that similar division is effective. The reconstruction loss constrains the outputs to structurally resemble the ground truth, while it only has limited capability in reconstructing the fine details. Perceptual loss and style loss, are used to obtain the fine details in textures. In addition, the KL divergence between the posterior and the prior are also important because the network would learn the correct distribution it samples from by minimizing it.

Reconstruction Loss. Since matching the semantic information is more crucial than the exact same noise, we assign higher weights to the layers with lower resolution in Laplacian pyramid loss [4]. Besides, we assign lower weights to the pixels that are further away from the seen regions. Denote the generated image as $\hat{I}$ and the ground truth image as I, $\hat{I}_k$ and I_k as the k th layer of the respective Laplacian Pyramid, d_k as k th layer of Gaussian pyramid of the distance map which represents the distance to the closest seen pixel, and K as the total number of layers the pyramid. The reconstruction loss L_{rec} is given by:

$$L_{rec} = w^{rec} \sum_{k=1}^{K} \frac{r^k \alpha^{-d_k}}{H_k W_k} ||I_k - \hat{I}_k|| \tag{7}$$

where H_k and W_k are the height and width of the image in the k th layer of the pyramid. α is a positive constant that controls how much the weight decays in the masked region as it gets further away from the unmasked region. The weight parameters of different layers r^k are designed to have the loss give larger weights to low frequency contents and w^{rec} is the weight for the reconstruction loss.

Perceptual Loss and Style Loss. Perceptual and style loss [8] are enforced by the high-level features extracted by a VGG [20] that is pre-trained on ImageNet [6]. The error metric is enforced on the deep features instead of the original images, since deep features consist of some form of semantic information about the images. Perceptual loss L_p retains the spatial correlation of the deep features, so minimizing it would recover the significant content of the missing region. Let $\psi_j(x)$ be the function to extract the output of layer j of VGG from x, and have w_j^p as the weight of the loss at layer j for perceptual loss, and J as the set of layers in VGG.

$$L_p = \sum_{j \in J} w_j^p ||\psi_j(I) - \psi_j(\hat{I})|| \tag{8}$$

Style loss is enforced by the gram matrix, which does not preserve the spatial correlation between pixel. Therefore, style loss only makes sure that the generated style matches the ground truth. Denote the weight of the loss at layer j of VGG for style loss as w_j^s, the style loss L_s could be expressed as:

$$L_s = \sum_{j \in J} w_j^s ||G_j(I) - G_j(\hat{I})|| \tag{9}$$

where $G_j(x)$ is the $C_j \times C_j$ gram matrix of the feature map of layer j of image x. Let C_j, H_j, and W_j be the number of channels, height, and width of the feature map of layer j respectively, then the gram matrix would be:

$$G_j(x)(u, v) = \frac{\sum_{h=1}^{H_j} \sum_{w=1}^{W_j} \psi_j(x)(h, w, u)\psi_j(x)(h, w, v)}{C_j H_j W_j} \tag{10}$$

KL Divergence. Inspired by [22], we use a similar KL divergence representation as theirs, where instead of directly predicting the posterior distribution, we predict the difference between the prior and posterior distributions. Specifically, denote the distribution for the j th variable in z_i in prior as a Normal distribution $p_{\theta_i}(z_i^j|z_{<i}) = \mathcal{N}(\mu_j(z_{<i}), \sigma_j(z_{<i}))$. We define the posterior as $q_{\phi_i}(z_i^j|z_{<i}, x, y) = \mathcal{N}(\mu_j(z_{<i}) + \Delta\mu_j(z_{<i}, x, y), \sigma_j(z_{<i}, x)\Delta\sigma_j(z_{<i}, x, y))$, where $\Delta\mu_j$ and $\Delta\sigma_j$ are the predicted mean difference and variance difference. Therefore, the KL divergence between prior and posterior can be denoted as

$$KL(q_{\phi_i}(z_i^j|z_{<i}, x, y)||, p_{\theta_i}(z_i^j|z_{<i})) = \frac{1}{2}(\frac{\Delta\mu_j^2}{\sigma_j^2} + \Delta\sigma_j^2 - \log\sigma_i^2 - 1) \qquad (11)$$

The total KL divergence loss L_{KL} is the sum over all the KL divergence between all the priors and posteriors by the weight w_i^{KL}.

2.4 Implementation

Our model is implemented in the form of the network in Fig. 2, where the prior and posterior are computed by different U-Net-like [19] network separately and are optimized at the same time by maximizing the ELBO. We utilize dilated convolution [15] in the middle of the network to improve the fine details in the output by increasing the receptive field. In our ConvBlock shown in Fig. 2, batch normalization [11] is used followed by swish activation [18] after convolution. CBAM [24] is put after each swish activation to perform channel-wise and spatial attention. Swish activation and CBAM have been shown effective in image generation tasks [17,22]. Each layer of the network consist of 1 ConvBlock. In addition, we merge the generated image and the original image based on the mask to create the final output.

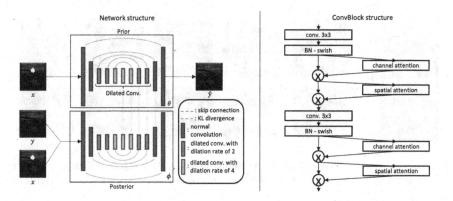

Fig. 2. The structure of the network and ConvBlock. Each layer of the network consists of 1 ConvBlock. Blue, green, and yellow layers in the network indicate dilation rate of 1 (which is normal convolution), 2 and 4 respectively. At the blue layers, max pooling or transpose convolution is used to down- or up-sample. (Color figure online)

We have 6 layers in the encoder while the number of filter are 64, 128, 256, 512, 1024 and 1024. The structure of the decoder is the inverse of the encoder while there are Gaussian latent variables in all the layers of the decoder. The weights for perceptual loss w^p and style loss w^s are set to 0, 0, 0, 0, 0.1 and 0.2, 0.005, 0.002, 0.02, 20 for *block1_conv1*, *block2_conv1*, *block3_conv1*, *block4_conv1*, *block5_conv1* of VGG respectively. These weights are selected based on empirical results. In addition, the weight for reconstruction loss w^{rec} ranges from 100 to 500 depending on different application, while all the weights for KL divergence loss w_i^{KL} are set to 10. The model is optimized by Adam optimizer [12] with a learning rate of 1×10^{-5}.

3 Experiments

Images are captured by UF-760AG Fukuda Denshi on a live-pig, and a blue-gel anthropomorphic tissue phantom with simulated blood vessels (Advanced Medical Technologies, WA), with a linear transducer with 51 mm scanning width. The models in this paper are either trained on 4 Nvidia Tesla V100 GPUs, or on a single Nvidia Titan RTX GPU. The training masks in Sects. 3.1 and 3.2 is based on the mask generation algorithm in [26].

3.1 Inpainting on Live-Pig Images

In this experiment, we train (800 images) and test (180 images) our algorithm on live-pig ultrasound images with 180 images in validation set (1,160 total images). We randomly sample from the latent space to generate multiple plausible results. As shown in Fig. 3, we inpaint in the regions within the green boundaries. Our results have different noise patterns which are visually correct. The differences between random samples are more apparent in the supplementary video.

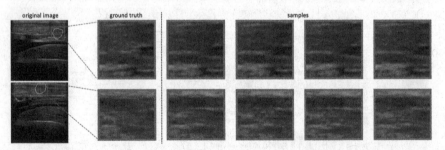

Fig. 3. Demonstration of different random samples. We inpaint within the green boudaries. Different samples have different generated noise patterns and brightness.

Furthermore, we compare our hierarchical probabilistic inpainting algorithm (HPI) against Telea et al.'s method [21] (Telea), Deep Fusion [9] with **both** *our training objective* (DF) and *L2 loss* (DF-L2), Partial Convolution [16] with **both** *our training objectives* (PConv) and *L2 loss* (PConv-L2), and with our network structure *without variational inference* (HPI-woVI). As shown in the top 2 rows of Fig. 4, Telea cannot handle the textures while PConv and PConv-L2 completely fail on pig images. DF and DF-L2 learned the semantic information but results are both blurry. HPI-woVI can correctly learn the semantic information but creates significant stitching artifacts at the boundaries of the mask. HPI is able to generate visually correct images with correct semantic information even though they do not look exactly the same as the ground truth. To evaluate quantitatively, we draw a bounding box around the masked region and extend 5 pixels in each direction to crop out the patch. We then calculate the L1, PSNR, SSIM on these patches. The results are shown in Table 1, where HPI outperforms other methods, even though our goal is not to completely reconstruct the ground truth images. Note that even though DF-L2 has decent results by numbers, it does not generate visually correct textures.

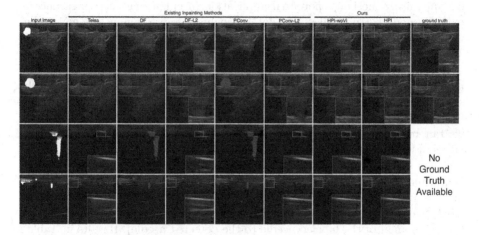

Fig. 4. First 2 rows: pig images. Last 2 rows: artifact removal. Only our approach generates visually correct results without introducing significant new artifacts.

Table 1. Comparison between our methods against others

	Telea	DF	DF-L2	PConv	PConv-L2	HPI-woVI	HPI
L1	0.030	0.056	**0.023**	0.093	0.027	0.026	**0.023**
PSNR	19.99	16.81	22.30	12.75	21.12	21.33	**23.00**
SSIM	0.545	0.373	0.616	0.246	0.578	0.564	**0.635**

3.2 Filling in Artifact Regions After Segmentation

To demonstrate filling in artifact regions, needles were put into the blue-gel phantom to create reverberation artifacts. Such artifacts are generated by sound waves bouncing between the posterior and anterior surfaces of highly reflective objects, resulting in additional echoes in the image that do not exist in reality [30]. We apply a reverberation artifact segmentation algorithm [10] and then inpaint where the artifacts were segmented. We trained the inpainting networks on 1145 images with no reverberation artifacts and test them on reverberation-artifact-corrupted data. Like the previous experiments, we also compare our method with those other methods. Shown in the last 2 rows of Fig. 4, other methods create either completely wrong pixel values in the artifacts region (DF, PConv), checkerboard patterns (HPI-woVI), non-existant anatomic boundaries (DF-L2), blurry patches (Telea), or discontinuity in the image (PConv-L2), while ours generates images with reverberation artifacts correctly removed.

3.3   Needle Tracking

We use our inpainting algorithm to assist a needle tracking algorithm [5] to show our inpainting method's potential application in enhancing the appearance of partially non-visible needles. The dataset includes four 2.5-in. echogenic needles with different amount of pre-bent curvature inserted into the blue-gel phantom. The dataset includes 6 trials of insertion for each kind of needles. There are 384, 364, 408, 373 test images for straight, small-curvature, medium-curvature, and large-curvature needles respectively. We select the inpainted regions following the steps below. In [5], we use an intensity thresholding and a Sobel filter response thresholding to select the candidate points for weighted-RANSAC fitting. First, we ran both thresholding methods in the bounding box of the ground-truth labels of the needle to select the candidate pixels for weighted-RANSAC fitting. We then search along the labeled needle axis and detect the boxes where there is no candidate points and inpaint the regions inside the boxes. The evaluation metrics are: the error of the tip localization (distance between the detected and labeled tips) and the shaft fitting error (the mean absolute error of the pixels on the polynomial of the labeled needle to the detected needle). Results in Table 2 show inpainting lowers the tip and shaft errors in most cases, allowing better needle tracking. The visualization of the needle tracking can be found in the supplementary materials.

Table 2. Comparison between the needle tracking accuracy with and w/o inpainting

Needle type	Tip Error/mm		Shaft Error/mm	
	w inpaint	w/o inpaint	w inpaint	w/o inpaint
Straight needle	**1.52 ± 1.22**	2.56 ± 2.17	**0.65 ± 0.71**	0.77 ± 0.63
Small curvature	**2.50 ± 2.19**	2.92 ± 2.62	**0.56 ± 0.67**	0.76 ± 0.95
Medium curvature	2.89 ± 3.53	**2.87 ± 3.43**	**0.77 ± 1.08**	0.78 ± 1.16
Large curvature	**2.51 ± 2.06**	3.35 ± 3.92	**0.60 ± 0.92**	0.84 ± 1.17

4 Conclusion

We propose a hierarchical probabilistic ultrasound image inpainting algorithm that can be trained via variational inference. We show that the algorithm is able to generate visually correct outputs with some variance in the inpainted regions. This method can be used to fill in where artifacts have been automatically detected and to enhance the appearance of partially non-visible needles for improved needle tracking.

References

1. Akhtar, N., Mian, A.: Threat of adversarial attacks on deep learning in computer vision: a survey. IEEE Access **6**, 14410–14430 (2018)
2. Barnes, C., Shechtman, E., Finkelstein, A., Goldman, D.B.: PatchMatch: a randomized correspondence algorithm for structural image editing. ACM Trans. Graph. **28**(3), 24 (2009)
3. Bertalmio, M., Vese, L., Sapiro, G., Osher, S.: Simultaneous structure and texture image inpainting. IEEE Trans. Image Process. **12**(8), 882–889 (2003)
4. Bojanowski, P., Joulin, A., Lopez-Paz, D., Szlam, A.: Optimizing the latent space of generative networks. arXiv preprint arXiv:1707.05776 (2017)
5. Chen, W., Mehta, K.N., Bhanushali, B.D., Galeotti, J.: Ultrasound-based tracking of partially in-plane, curved needles. In: 2021 IEEE 18th International Symposium on Biomedical Imaging (ISBI), pp. 939–943. IEEE (2021)
6. Deng, J., Dong, W., Socher, R., Li, L.J., Li, K., Fei-Fei, L.: ImageNet: a large-scale hierarchical image database. In: 2009 IEEE Conference on Computer Vision and Pattern Recognition, pp. 248–255. IEEE (2009)
7. Dupont, E., Suresha, S.: Probabilistic semantic inpainting with pixel constrained CNNs. In: The 22nd International Conference on Artificial Intelligence and Statistics, pp. 2261–2270. PMLR (2019)
8. Gatys, L.A., Ecker, A.S., Bethge, M.: Image style transfer using convolutional neural networks. In: Proceedings of the IEEE Conference on Computer Vision and Pattern Recognition, pp. 2414–2423 (2016)
9. Hong, X., Xiong, P., Ji, R., Fan, H.: Deep fusion network for image completion. In: Proceedings of the 27th ACM International Conference on Multimedia, pp. 2033–2042 (2019)
10. Hung, A.L.Y., Chen, E., Galeotti, J.: Weakly-and semi-supervised probabilistic segmentation and quantification of ultrasound needle-reverberation artifacts to allow better AI understanding of tissue beneath needles. arXiv preprint arXiv:2011.11958 (2020)
11. Ioffe, S., Szegedy, C.: Batch normalization: accelerating deep network training by reducing internal covariate shift. arXiv preprint arXiv:1502.03167 (2015)
12. Kingma, D.P., Ba, J.: Adam: a method for stochastic optimization. arXiv preprint arXiv:1412.6980 (2014)
13. Kohl, S.A., et al.: A hierarchical probabilistic U-Net for modeling multi-scale ambiguities. arXiv preprint arXiv:1905.13077 (2019)
14. Liao, L., Hu, R., Xiao, J., Wang, Z.: Edge-aware context encoder for image inpainting. In: 2018 IEEE International Conference on Acoustics, Speech and Signal Processing (ICASSP), pp. 3156–3160. IEEE (2018)

15. Lin, G., Wu, Q., Qiu, L., Huang, X.: Image super-resolution using a dilated convolutional neural network. Neurocomputing **275**, 1219–1230 (2018)

16. Liu, G., Reda, F.A., Shih, K.J., Wang, T.C., Tao, A., Catanzaro, B.: Image inpainting for irregular holes using partial convolutions. In: Proceedings of the European Conference on Computer Vision (ECCV), pp. 85–100 (2018)

17. Ma, B., Wang, X., Zhang, H., Li, F., Dan, J.: CBAM-GAN: generative adversarial networks based on convolutional block attention module. In: Sun, X., Pan, Z., Bertino, E. (eds.) ICAIS 2019. LNCS, vol. 11632, pp. 227–236. Springer, Cham (2019). https://doi.org/10.1007/978-3-030-24274-9_20

18. Ramachandran, P., Zoph, B., Le, Q.V.: Searching for activation functions. arXiv preprint arXiv:1710.05941 (2017)

19. Ronneberger, O., Fischer, P., Brox, T.: U-Net: convolutional networks for biomedical image segmentation. In: Navab, N., Hornegger, J., Wells, W.M., Frangi, A.F. (eds.) MICCAI 2015. LNCS, vol. 9351, pp. 234–241. Springer, Cham (2015). https://doi.org/10.1007/978-3-319-24574-4_28

20. Simonyan, K., Zisserman, A.: Very deep convolutional networks for large-scale image recognition. arXiv preprint arXiv:1409.1556 (2014)

21. Telea, A.: An image inpainting technique based on the fast marching method. J. Graph. Tools **9**(1), 23–34 (2004)

22. Vahdat, A., Kautz, J.: NVAE: a deep hierarchical variational autoencoder. arXiv preprint arXiv:2007.03898 (2020)

23. Wang, N., Li, J., Zhang, L., Du, B.: Musical: multi-scale image contextual attention learning for inpainting. In: IJCAI, pp. 3748–3754 (2019)

24. Woo, S., Park, J., Lee, J.Y., Kweon, I.S.: CBAM: convolutional block attention module. In: Proceedings of the European conference on computer vision (ECCV), pp. 3–19 (2018)

25. Yu, J., Lin, Z., Yang, J., Shen, X., Lu, X., Huang, T.S.: Generative image inpainting with contextual attention. In: Proceedings of the IEEE Conference on Computer Vision and Pattern Recognition, pp. 5505–5514 (2018)

26. Yu, J., Lin, Z., Yang, J., Shen, X., Lu, X., Huang, T.S.: Free-form image inpainting with gated convolution. In: Proceedings of the IEEE International Conference on Computer Vision, pp. 4471–4480 (2019)

27. Yu, R., et al.: Generative adversarial network using multi-modal guidance for ultrasound images inpainting. In: Yang, H., Pasupa, K., Leung, A.C.-S., Kwok, J.T., Chan, J.H., King, I. (eds.) ICONIP 2020. LNCS, vol. 12532, pp. 338–349. Springer, Cham (2020). https://doi.org/10.1007/978-3-030-63830-6_29

28. Zeng, Y., Fu, J., Chao, H., Guo, B.: Learning pyramid-context encoder network for high-quality image inpainting. In: Proceedings of the IEEE Conference on Computer Vision and Pattern Recognition, pp. 1486–1494 (2019)

29. Zheng, C., Cham, T.J., Cai, J.: Pluralistic image completion. In: Proceedings of the IEEE Conference on Computer Vision and Pattern Recognition, pp. 1438–1447 (2019)

30. Ziskin, M., Thickman, D., Goldenberg, N., Lapayowker, M., Becker, J.: The comet tail artifact. J. Ultrasound Med. **1**(1), 1–7 (1982)

CaCL: Class-Aware Codebook Learning for Weakly Supervised Segmentation on Diffuse Image Patterns

Ruining Deng[1], Quan Liu[1], Shunxing Bao[1], Aadarsh Jha[1], Catie Chang[1], Bryan A. Millis[2], Matthew J. Tyska[2], and Yuankai Huo[1(✉)]

[1] Vanderbilt University, Nashville, TN 37215, USA
yuankai.huo@vanderbilt.edu
[2] Vanderbilt University Medical Center, Nashville, TN 37215, USA

Abstract. Weakly supervised learning has been rapidly advanced in biomedical image analysis to achieve pixel-wise labels (segmentation) from image-wise annotations (classification), as biomedical images naturally contain image-wise labels in many scenarios. The current weakly supervised learning algorithms from the computer vision community are largely designed for focal objects (e.g., dogs and cats). However, such algorithms are not optimized for diffuse patterns in biomedical imaging (e.g., stains and fluorescence in microscopy imaging). In this paper, we propose a novel class-aware codebook learning (CaCL) algorithm to perform weakly supervised learning for diffuse image patterns. Specifically, the CaCL algorithm is deployed to segment protein expressed brush border regions from histological images of human duodenum. Our contribution is three-fold: (1) we approach the weakly supervised segmentation from a novel codebook learning perspective; (2) the CaCL algorithm segments diffuse image patterns rather than focal objects; and (3) the proposed algorithm is implemented in a multi-task framework based on Vector Quantised-Variational AutoEncoder (VQ-VAE) via joint image reconstruction, classification, feature embedding, and segmentation. The experimental results show that our method achieved superior performance compared with baseline weakly supervised algorithms. The code is available at https://github.com/ddrrnn123/CaCL.

Keywords: Weakly supervised learning · Segmentation · AutoEncoder

1 Introduction

Mapping the location of 19,628 human protein?coding genes plays a critical role as a "census" of proteins, which further increases our knowledge of human biology and enables new insights into principles of life. For instance, the Human Protein Atlas (HPA) project[1] has applied >25,000 antibodies to characterize

---

[1] https://www.proteinatlas.org.

© Springer Nature Switzerland AG 2021
S. Engelhardt et al. (Eds.): DGM4MICCAI 2021/DALI 2021, LNCS 13003, pp. 93–102, 2021.
https://doi.org/10.1007/978-3-030-88210-5_8

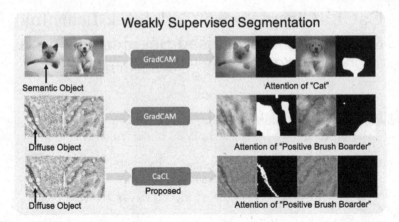

Fig. 1. The performances of object segmentation. This figure shows the performances of object segmentation using different attention-based weakly supervised learning methods. The former method, GradCAM, is designed for focal objects rather than diffuse objects. Our proposed method, CaCL, can obtain better results on diffuse objects.

the tissue-level spatial expression by collecting 10 million immunohistochemistry (IHC) images. The IHC images indicate the location and distribution of protein expression. For example, understanding the area ratio between IHC stained regions and cell body regions at the brush border of the human duodenum reveals the specificity of gene expressions.

The color deconvolution algorithm [13] is regarded as the de facto standard approach to segment IHC stained histopathology images. However, the manual tuning of IHC staining parameters (e.g., segmentation threshold) to deal with heterogeneous image qualities and attributes is labor-intensive. Moreover, color deconvolution cannot understand the semantic information of a figure.

Recent weakly supervised learning techniques have played a critical role in image segmentation with the benefits of only needing image-wise annotation [15]. Zhou et al., [18] proposed Class Activation Mapping (CAM) for Convolutional Neural Network(CNN) with a global average pool to allow CNNs to visualize object localization. Then, Selvaraju et al., [14] developed Gradient-weighted Class Activation Mapping (GradCAM) and Guided GradCAM (G-GradCAM) for better visual explanations with localization information. Later on, Fong et al., [6,7] introduced a framework for learning meta-predictors. However, the current weakly supervised learning algorithms from the computer vision community are mostly designed for focal objects and may display attention with any image, which are not optimized for diffuse patterns in biomedical imaging [4](Fig. 1).

Meanwhile, there have been several weakly supervised learning approaches in histology [12]. Belharbi et al., [3] proposed an active learning framework to jointly perform supervised image-level classification and active learning for segmentation. Xu et al., [16] proposed a weakly supervised learning framework for histopathology image segmentation, using multiple instance learning (MIL)-

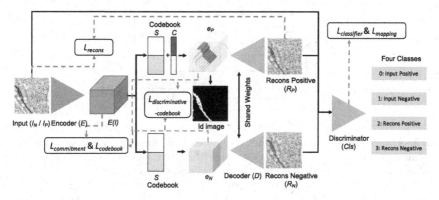

Fig. 2. The backbone of our method. Our method includes CaCL embedding, GAN based reconstruction and classification, and weakly supervised learning segmentation.

based label enrichment and fully supervised training with image-level labels. These methods achieved superior performances. With the development of deep learning technology, more unsupervised segmentation models were proposed for medical image analysis [1,2,10,17]. However, most of the attention-based methods only obtained attention maps for partial classification tasks rather than segmentation tasks. Herein, we provided a weakly supervised learning model to achieve robust segmentation images directly from attention maps.

A new generative model, Vector Quantised-Variational AutoEncoder (VQ-VAE) [9], was proposed to encode an image from an infinite continuous feature space to a finite discrete feature space using a codebook with a fixed number of codes. Inspired by VQ-VAE, we propose a novel class-aware codebook learning (CaCL) algorithm to segment diffuse patterns in medical imaging. The central idea is to split the original codebook into two separate codebooks. One codebook encodes the discriminative class patterns (codebook C), while the other encodes the common image patterns between two groups of images (codebook S). Then, the pixels that used in the codebook $S + C$ during the encoding process are used as an attention to perform weakly supervised segmentation. Briefly, the innovations of the proposed approach is in three-fold: (1) We approach the weakly supervised segmentation from a novel codebook learning perspective; (2) We introduce the CaCL algorithm to segment diffuse image patterns rather than focal objects; (3) The proposed algorithm is implemented in a multi-task framework based on Vector Quantised-Variational AutoEncoder (VQ-VAE) via joint image reconstruction, classification, feature embedding, and segmentation.

2   Methods

The entire framework of the proposed CaCL method is presented in Fig. 2. The CaCL algorithm consists of three sections: (1) a class-aware codebook for feature embedding; (2) generative adversarial image reconstruction and classification; and (3) weakly supervised segmentation from diffuse patterns.

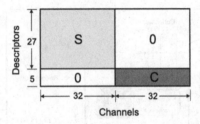

Fig. 3. The design of the class-aware codebook. This figure shows the design of the class-aware codebook. One encodes class discriminative features (codebook C), while another encodes the shared features among two classes (codebook S).

2.1   Class-Aware Codebook Based Feature Encoding

In this study, we design a class-aware codebook inspired by VQ-VAE2 [11]. With the VQ-VAE framework, three steps are used to process an input image. First, the encoder E is used to convert a RGB image into a feature map. Second, the feature map is coded by the codebook from an infinite solution space to a fixed number of codes for each pixel. For example, if the codebook contains 32 codes, each pixel can only be one of the 32 types of features. Last, the coded feature maps were decoded to the input image resolution as a encoder-decoder design.

As opposed to VQ-VAE, which only used one codebook to encode all inputs, we propose to use two codebooks in CaCL. One encodes class discriminative features (codebook C), while another encodes the shared features among two classes (codebook S), as shown in Fig. 3. In this study, the images with positive protein expression patterns (dark brown at the brush broader) are defined as I_P, while the images without protein expression patterns are defined as I_N. Then, for each input image, we will first retrieve one raw feature map from E. Second, two coded feature maps will be obtained by using codebook S only and codebook $S + C$, respectively. Two images will be reconstructed using the same decoder D. One image only has common diffuse patterns across positive and negative images (Recons Negative R_N in Fig. 2), while another image contains both common diffuse patterns and class discriminative patterns (Recons Positive R_P in Fig. 2).

2.2   Loss Definition

Commitment Loss and Codebook Loss: Herein, we implement the commitment loss and codebook loss in VQ-VAE2 that retains the reconstruction features close to the chosen codebook vectors.

$$\mathcal{L}_{commitment}(I, D(e)) = ||sg[e] - E(I)||_2^2 \tag{1}$$

$$\mathcal{L}_{codebook}(I, D(e)) = ||sg[E(I)] - e||_2^2 \tag{2}$$

where e is the coded feature map for the input I, E is the encoder function, and D is the decoder function. The operator sg refers to a non-gradient operation that stops the gradients from flowing into its argument. It uses the exponential moving average updates for the codebook with a decay parameter.

Reconstructive Loss: The reconstructive loss is applied to supervise the quality of reconstruction images R_P and R_N. Each input image I will go through the combined codebook C and S and the single codebook S, which obtain both R_P and R_N. R_P is calculated through the mean-square-error as the reconstruction loss with I_P, and R_N is compared with images I_N, respectively.

$$\mathcal{L}_{recons}(I, R_P, R_N) = (1 - M)||I - R_P||_2^2 + M||I - R_N||_2^2$$

$$\text{Where } M = \begin{cases} 1, & I = I_N \\ 0, & I = I_P \end{cases} \tag{3}$$

Discriminative-Codebook Loss: To encourage the model to use codebook C, we introduce a new discriminative-codebook loss to calculate the mean-square-error of the quantized feature maps e_N and e_P in the non-zero channels from codebook C. Briefly, if the image is negative, we force the feature maps to be identical from two code books. If the image is positive, we force the feature maps to be different from two code books by using $\mathcal{L}_{discriminative-codebook}$.

$$\mathcal{L}_{discriminative-codebook}(I, e_N, e_P) = K||e_N - e_P||_2^2$$

$$\text{Where } K = \begin{cases} 1, & I = I_N \\ -1, & I = I_P \end{cases} \tag{4}$$

Hybrid Discriminator Loss: The hybrid discriminator loss performs both: (1) real/fake; and (2) positive/negative classification tasks on reconstructed images. The implementation of the discriminator and the generator are followed by a generative adversarial network (GAN) design [19]. We create two image pools P_{data} to separately store all fake positive and fake negative images to train the discriminator. We use one resnet18 [8], named as Cls, as the discriminator (classifier).

$$\mathcal{L}_{classifier}(I_P, I_N, R_P, R_N) = T_{R_P} log(Cls(X \sim P_{data(R_P)}))$$
$$+ T_{R_N} log(Cls(X \sim P_{data(R_N)})) \tag{5}$$
$$+ T_{I_P} log(Cls(I_P)) + T_{I_N} log(Cls(I_N))$$

$$\mathcal{L}_{mapping}(R_P, R_N) = T_{I_P} log(Cls(R_P)) + T_{I_N} log(Cls(R_N)) \tag{6}$$

where $T_{I_P}, T_{I_N}, T_{R_P}, T_{R_N}$ are the targets of I_P, I_N, R_P, R_N, respectively.

The aforementioned loss functions are aggregated into $\mathcal{L}_{combine}$ with weights λ. Since the discriminators typically converge faster than generators, we perform back-propagation at different frequencies. During the training, the classification loss ($\mathcal{L}_{classifier}$) is updated in every ten epochs, while $\mathcal{L}_{combine}$ is updated in each epoch.

$$\mathcal{L}_{combine} = \lambda_{mapping}\mathcal{L}_{mapping} + \lambda_{commitment}\mathcal{L}_{commitment} + \lambda_{recons}\mathcal{L}_{recons}$$
$$+ \lambda_{discriminative-codebook}\mathcal{L}_{discriminative-codebook} \tag{7}$$

2.3   Training Strategy

The class consistency is normalized by computing commitment loss and reconstructive loss. For all positive images (I_p), only reconstructed positive images (R_p) are calculated in reconstructive loss. The raw features $E(I_p)$ from positive input images (I_p) are computed in commitment loss with positive coded features (e_p). The same principles are implemented for all negative inputs (I_n).

To train the codebooks, all the vectors in both codebook S and C are updated after quantizing the coded positive features (e_p). In contrast, only the vectors in codebook S are updated after quantizing the coded negative features (e_n). Meanwhile, we use mean-square-error to reduce the difference between encoded features $(e_p$ and $e_n)$ from the negative inputs (I_n), while simultaneously amplifying the distinctions between e_p and e_n from the positive inputs in the discriminative-codebook loss $\mathcal{L}_{discriminative-codebook}$. Such a process guides the codebook S and codebook C to assemble distinctive features in different classes, independently.

Next, a classifier is used to identify four types of images, which are Input Positive (I_p), Input Negative (I_n), Reconstructive Positive (R_p), and Reconstructive Negative(R_n). Meanwhile, we employ a discriminator to reconcile the differences between the input images (I_p, I_n) and the reconstructive images (R_p, R_n). Ideally, only R_p from I_p contain the positive patterns from the codebook C.

2.4   Weakly Supervised Learning Segmentation

Ideally, after training the model, only pixels using class-specific codebook C should contribute to the differences between the two classes. Therefore, we simply mark those pixels as 1, and mark the remaining pixels as 0. The outcome mask is used as our weakly supervised segmentation results.

3   Data and Experiments

This research study was conducted retrospectively using human subject data made available in open access by the Human Protein Atlas (https://www.proteinatlas.org). Ethical approval was not required as confirmed by the license attached with the open access data. 42 high resolution duodenum histological micro-array images were obtained from the Human Protein Atlas. 27 images contained high brush border protein expression, while the remaining ones did not. The protein expression is specified as the IHC staining pattern (dark brown). Patches in an 8×8 grid without overlapping from each high-resolution image were extracted. Due to the GPU memory limitation, we downsample these patches with 375×375 pixels to image patches with 128×128 pixels. We randomly selected 1480 patches for training, 200 patches for validation, and 200 for testing. Half of the testing images were from the brush border area to evaluate the performance of our method.

The design of the class-aware codebook is in Fig. 3. The number of descriptors in codebook S is 27, while the number of descriptors in codebook C is 5.

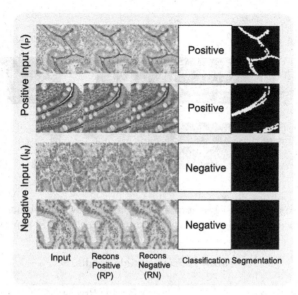

Fig. 4. CaCL at the testing stage. This figure shows the example outcomes from the proposed CaCL framework, which include reconstructed images, classification results and segmentation results.

Each descriptor has 64 channels, where 32 non-overlapping channels are from each codebook S and C, respectively. The remaining locations of the codebook are set to 0. The decay in each codebook update is 0.98. For all the experiments, we use the Adam solver for optimization with a batch size of one. The learning rate of the classification loss is 0.0001, while the learning rate of the combined loss is 0.0003. The size of the image pool is 64. The weights λ of commitment loss, reconstructive loss, discriminative-codebook loss, and discriminator loss are empirically set to 0.25, 100, 50, and 1, respectively. These parameters were determined by fine-tuning process to obtain superior performances in both segmentation metrics and reconstructive visualizations.

The color deconvolution was employed as the current standard IHC stain segmentation method. CAM and GradCAM were utilized as the benchmarks of attention based weakly supervised learning. All experiments were completed on the same workstation, with NVIDIA Quadro P5000 GPU.

4   Results

In Fig. 4, the example input I_P and I_N images, and the corresponding reconstructed R_P and R_N, are presented. Figure 5 shows the qualitative weakly segmentation results, while Table 1 presents the quantitative results. The Dice Similarity Coefficient (DSC), Positive Predictive Value (Precision), Sensitivity (Recall), and Binary Cross-Entropy (BCE) are used as evaluation metrics. For

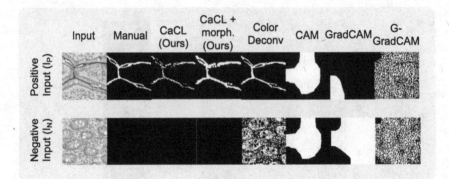

Fig. 5. Pixel-wise attention segmentation. This figure shows the results of brush border segmentation using pixel-wise attention from different weakly supervised learning methods.

Table 1. Segmentation results.

Method	Dice	Recall	Precision	BCE
Color deconv. [13]	0.347	0.400	0.363	7.066
CAM [18]	0.065	0.038	0.260	15.813
GradCAM [14]	0.061	0.035	0.298	19.586
G-GradCAM [14]	0.030	0.018	0.099	14.273
CaCL (Ours)	0.623	**0.787**	0.574	**1.079**
CaCL+morph. (Ours)	**0.703**	0.712	**0.723**	1.258

each I_N image, if all the pixels inside the predicted segmentation masks are 0, then DSC, Precision, and Recall are computed as 1. Otherwise, those metrics are 0, according to [5]. All the results of baseline models in Table 1 are the best performances by iterating all the intensity values as thresholds. A simple morphological dilation operation with radius 1 is also tested in Table 1. As a result, our method achieved the best quantitative performance.

5   Discussion

In this study, we presented a new weakly supervised learning method with a class-aware codebook. The proposed CaCL approach achieved diffuse pattern segmentation without pixel-wise annotation. Our proposed method combines with the classification task and the segmentation task as a whole with "pixel-wise attention" from image-wise weak labels, while previous CAM based attention is more coerce.

The codebook-based reconstruction uses the region-level features from pixel-wise feature maps, which inhibit positive features. The purpose of the dilation enhancement is to decrease this impact from neighbor pixels, which achieve better segmentation results in Table 1.

The goal of our method is to achieve both focal pattern segmentation and controlling the expression of positive patterns with the realistic reconstructive images by class-aware codebooks. In our experiment, simply using the standard reconstruction loss without the discriminator loss $\mathcal{L}_{mapping}$ generates numerous unreasonable noise pixels as fake patterns on the reconstructive images, which can cheat in the classifier $\mathcal{L}_{classifier}$ and fail to control the pattern expression. In Fig. 5, our design can receive segmentation results while achieving consistent expression control in IHC stained histopathology images.

At current stage, there are still major limitations. One obvious limitation is the size of our dataset. The number of the training and testing images is limited due to the limitation of resources and extensive labor costs, as well as time needed to achieve pixel-wise manual annotations. More training data would lead to better segmentation performance. In the future, one promising improvement of the proposed method would be to extend the current binary classification and segmentation approach to multi-class scenarios.

References

1. Baumgartner, C.F., Koch, L.M., Tezcan, K.C., Ang, J.X., Konukoglu, E.: Visual feature attribution using Wasserstein GANs. In: Proceedings of the IEEE Conference on Computer Vision and Pattern Recognition, pp. 8309–8319 (2018)
2. Baur, C., Wiestler, B., Albarqouni, S., Navab, N.: Deep autoencoding models for unsupervised anomaly segmentation in brain MR images. In: Crimi, A., Bakas, S., Kuijf, H., Keyvan, F., Reyes, M., van Walsum, T. (eds.) BrainLes 2018. LNCS, vol. 11383, pp. 161–169. Springer, Cham (2019). https://doi.org/10.1007/978-3-030-11723-8_16
3. Belharbi, S., Ben Ayed, I., McCaffrey, L., Granger, E.: Deep active learning for joint classification & segmentation with weak annotator. In: Proceedings of the IEEE/CVF Winter Conference on Applications of Computer Vision, pp. 3338–3347 (2021)
4. Chan, L., Hosseini, M.S., Plataniotis, K.N.: A comprehensive analysis of weakly-supervised semantic segmentation in different image domains. Int. J. Comput. Vision 129(2), 361–384 (2021). https://doi.org/10.1007/s11263-020-01373-4
5. Chicco, D., Jurman, G.: The advantages of the Matthews correlation coefficient (mcc) over F1 score and accuracy in binary classification evaluation. BMC Genom. 21(1), 6 (2020). https://doi.org/10.1186/s12864-019-6413-7
6. Fong, R., Patrick, M., Vedaldi, A.: Understanding deep networks via extremal perturbations and smooth masks. In: Proceedings of the IEEE International Conference on Computer Vision, pp. 2950–2958 (2019)
7. Fong, R.C., Vedaldi, A.: Interpretable explanations of black boxes by meaningful perturbation. In: Proceedings of the IEEE International Conference on Computer Vision, pp. 3429–3437 (2017)
8. He, K., Zhang, X., Ren, S., Sun, J.: Deep residual learning for image recognition. In: Proceedings of the IEEE conference on computer vision and pattern recognition, pp. 770–778 (2016)

9. Oord, A.v.d., Vinyals, O., Kavukcuoglu, K.: Neural discrete representation learning. arXiv preprint arXiv:1711.00937 (2017)
10. Pawlowski, N., et al.: Unsupervised lesion detection in brain CT using Bayesian convolutional autoencoders. Medical Imaging with Deep Learning (2018)
11. Razavi, A., van den Oord, A., Vinyals, O.: Generating diverse high-fidelity images with VQ-VAE-2. In: Advances in Neural Information Processing Systems, pp. 14866–14876 (2019)
12. Rony, J., Belharbi, S., Dolz, J., Ayed, I.B., McCaffrey, L., Granger, E.: Deep weakly-supervised learning methods for classification and localization in histology images: a survey. arXiv preprint arXiv:1909.03354 (2019)
13. Ruifrok, A.C., Johnston, D.A., et al.: Quantification of histochemical staining by color deconvolution. Anal. Quant. Cytol. Histol. **23**(4), 291–299 (2001)
14. Selvaraju, R.R., Cogswell, M., Das, A., Vedantam, R., Parikh, D., Batra, D.: Grad-CAM: visual explanations from deep networks via gradient-based localization. In: Proceedings of the IEEE International Conference on Computer Vision, pp. 618–626 (2017)
15. Wang, Y., et al.: Weakly supervised universal fracture detection in pelvic x-rays. In: Shen, D., et al. (eds.) MICCAI 2019. LNCS, vol. 11769, pp. 459–467. Springer, Cham (2019). https://doi.org/10.1007/978-3-030-32226-7_51
16. Xu, G., et al.: Camel: a weakly supervised learning framework for histopathology image segmentation. In: Proceedings of the IEEE/CVF International Conference on Computer Vision, pp. 10682–10691 (2019)
17. You, S., Tezcan, K.C., Chen, X., Konukoglu, E.: Unsupervised lesion detection via image restoration with a normative prior. In: International Conference on Medical Imaging with Deep Learning, pp. 540–556. PMLR (2019)
18. Zhou, B., Khosla, A., Lapedriza, A., Oliva, A., Torralba, A.: Learning deep features for discriminative localization. In: Proceedings of the IEEE Conference on Computer Vision and Pattern Recognition, pp. 2921–2929 (2016)
19. Zhu, J.Y., Park, T., Isola, P., Efros, A.A.: Unpaired image-to-image translation using cycle-consistent adversarial networks. In: Proceedings of the IEEE International Conference on Computer Vision, pp. 2223–2232 (2017)

BrainNetGAN: Data Augmentation of Brain Connectivity Using Generative Adversarial Network for Dementia Classification

Chao Li[1,2], Yiran Wei[1], Xi Chen[3(✉)], and Carola-Bibiane Schönlieb[4]

[1] Department of Clinical Neurosciences, University of Cambridge, Cambridge, UK
[2] Shanghai General Hospital, Shanghai Jiao Tong University, Shanghai, China
[3] Department of Computer Science, University of Bath, Bath, UK
xc841@bath.ac.uk
[4] Department of Applied Mathematics and Theoretical Physics,
University of Cambridge, Cambridge, UK

Abstract. Alzheimer's disease (AD) is the most common age-related dementia, which significantly affects an individual's daily life and impact socioeconomics. It remains a challenge to identify the individuals at risk of dementia for precise management. Brain MRI offers a non-invasive biomarker to detect brain aging. Previous evidence shows that the structural brain network generated from the diffusion MRI promises to classify dementia accurately based on deep learning models. However, the limited availability of diffusion MRI challenges the model training of deep learning. We propose the BrainNetGAN, a variant of the generative adversarial network, to efficiently augment the structural brain networks for dementia classifying tasks. The BrainNetGAN model is trained to generate fake brain connectivity matrices, which are expected to reflect the latent distribution and topological features of the real brain network data. Numerical results show that the BrainNetGAN outperforms the benchmarking algorithms in augmenting the brain networks for AD classification tasks.

Keywords: Data augmentation · Generative adversarial network · Brain connectivity · Classification

1 Introduction

Alzheimer's disease (AD) is the most common age-related dementia that significantly impacts the cognitive performance and the socioeconomic status of patients [13]. The structural brain network, constructed from the diffusion MRI (dMRI), is an emerging technique to quantify the complex brain white matter structure and incorporate the prior knowledge of brain anatomy. Specifically,

---

C. Li and Y. Wei—Equal Contribution.

© Springer Nature Switzerland AG 2021
S. Engelhardt et al. (Eds.): DGM4MICCAI 2021/DALI 2021, LNCS 13003, pp. 103–111, 2021.
https://doi.org/10.1007/978-3-030-88210-5_9

tractography performed on dMRI can quantify *the connectivity strength* between the separated anatomical brain regions defined on the brain atlas. Numerous studies have reported the usefulness of the structural brain network in characterising a broad spectrum of neurological diseases [8, 19], including AD [2].

In parallel, machine learning approaches based on the structural network have shown promise in the classification tasks to distinguish AD patients from healthy controls (CN). Particularly, state-of-the-art deep learning models have been widely used in predicting dementia using end-to-end training schemes [1]. A deep learning model, however, often requires a large amount of training data as well as balanced class labels for reasonable classification performance, which both are not feasible in clinical practice.

To mitigate this challenge, recent studies employed data augmentation techniques to increase the training sample sizes. For the augmentation of traditional images, image rotation and flipping schemes are generally effective. Nevertheless, a brain network matrix cannot be simply rotated or flipped, as it could completely change the order of brain regions and introduce artifacts to the predictive model. Therefore, data augmentation approaches tailored for brain networks are desired.

1.1 Related Work

Other data augmentation approaches have been developed to synthesize brain network matrices and adjust the imbalanced classes of the training data. Among these approaches, oversampling techniques, such as Synthetic Minority Oversampling Technique (SMOTE) [5], and Adaptive Synthetic Sampling (ADASYN) [10], are widely used. Unlike the naive augmentation methods that randomly replicate the minority samples, SMOTE generates synthetic samples by linearly interpolating two neighboring real samples. The neighborhood is identified by a standard K-nearest neighbor (KNN) approach. Developed based on the SMOTE, ADASYN produces fake samples according to the density of the class label distribution. More fake samples of minority classes are generated than the majority classes. ADASYN also adopts the KNN to cluster the samples and adjust the boundary of multiple minority classes. In recent years, ADASYN has been used to synthesize brain networks for the classification of AD [17]. However, the KNN-assisted oversampling techniques may not effectively capture the topological property of the high dimensional brain networks, leading to significant information loss in constructing the synthetic samples.

Generative adversarial networks (GAN) [7] is a generative model invented recently for data synthesis. It is well known in computer vision applications, where synthetic images can be indistinguishable from real images. Previous research shows that data augmentation using GAN has been successful in image classification tasks [3]. However, it is unclear whether the GAN can retain the topological properties of brain networks, which would require a tailored GAN architecture.

1.2 BrainNetGAN

We propose a new variant of GAN with specialized architecture for conditional brain network synthesis in this work. Inspired by [3], we hypothesize that data augmentation could improve the performance of dementia classification tasks where brain network matrices are used as the input. The proposed GAN variant, BrainNetGAN, can generate fake brain network matrices for both AD and CN. The main contributions of this work include:

- To our best knowledge, this is the first GAN variant developed specialized for synthesizing brain network matrices in dementia classification tasks.
- A specialized 2D convolutional kernel was applied to learn the topological property of brain networks [11].
- By adopting the architecture of the Wasserstein GAN with gradient penalty [9], fast and stable training of GAN was enabled on generating network matrices.
- A graph neural network (GNN) was specially adopted as the classifier to effectively evaluate the ability of algorithms in learning the topological property of brain networks.

Our experiments show that the proposed method outperforms the baseline and other benchmarking techniques, suggesting the advantage of using GAN for data augmentation in brain networks.

2   Methods

2.1   Structural Brain Networks

Adjacency matrices of the structural brain network were generated using the following steps (Fig. 1A). Firstly, dMRI was pre-processed in the FSL (FMRIB software library). Tractography was then performed on the processed dMRI using the Diffusion Toolkit [14]. The grey matter regions of dMRI were divided into 90 brain regions using the Automated Anatomical Labelling (AAL) atlas, after a nonlinear registration to the standard space using Advanced Normalization Tools [4,18]. The structural brain networks were constructed by counting the number of tracts between each pair of brain regions to produce an adjacency matrix, as the inputs of the following models. The tract counts of the brain network were normalized to between 0 and 1.

2.2   Data Augmentation Using BrainNetGAN

The proposed BrainNetGAN consists of three components: a generator (G), a discriminator (D), and a classifier (C). The G network is a feed-forward deep neural network (DNN) with four hidden layers. G takes both the random vector z sampled from a standard Gaussian distribution and one-hot coded class labels c of the brain networks (i.e., AD or CN). The output of the G network is a 1×4005 vector that is then reshaped to a 90×90 brain network matrix (with diagonal values equal to zero) as shown in Fig. 1C.

Both the D and C networks are convolutional neural networks (CNN) consisting of three convolutional layers with specialized kernels developed by [11]

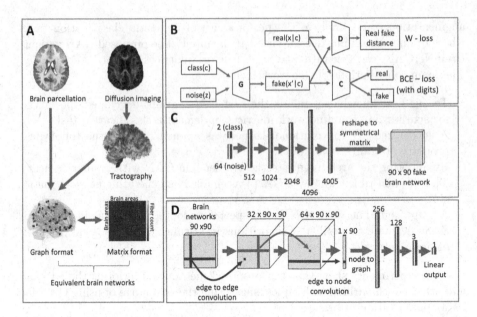

Fig. 1. Architecture of the BrainNetGAN and its elements. **A**. Preparation of brain network matrices from dMRI, see Sect. 2.1 for details. **B**. Design of BrainNetGAN. The framework consists of a generator G, a discriminator D and a classifier C. BCE-loss: binary cross entropy loss, W-loss: Wasserstein loss. **C**. The G network is a feed-forward DNN with ReLU activation functions. The numbers in **C** indicate the layer-wise input/output information. **D**. The D and C networks both contain special convolution kernels for the adjacency matrix of networks, see Sect. 2.2 for details.

and five fully connected layers (Fig. 1D). Unlike standard convolutional kernels that only consider local neighbors of an element in a matrix, the adopted kernels take the entire row and column of the element for convolution operation, simulating both edge-edge and edge-node convolutions (e.g. a row of adjacency matrix represents all edges connecting to one node in the graph). The input of both D and C networks is a 90×90 matrix, either synthetic (fake) or real brain network matrices. Although the architecture of both the D and the C networks are identical, the loss functions are different. Specifically, a cross-entropy loss is adopted for the C network to perform binary classification (AD/CN), while the Wasserstein loss is employed in the D network to evaluate the difference between the real and fake network matrices.

The objective function of the BrainNetGAN consists of two components:

$$L_D = E_{x \sim P_r}[D(x)] - E_{\tilde{x} \sim P_g}[D(\tilde{x})] + \lambda E_{\hat{x} \sim P_g}[(\|\nabla_{\hat{x}} D(\hat{x})\|_2 - 1)^2] \quad (1)$$

$$L_C = E_{\tilde{x} \sim P_g}[\log(C(\tilde{x}))] + E_{x \sim P_r}[\log(C(x))] \quad (2)$$

Equation 1 represents the Wasserstein loss with gradient penalty. The first two components denote the Earth-Mover distance $W(P_r, P_g)$ between real and fake distributions that the generator tries to minimize. P_r denotes the distribution of real samples, D is the discriminator, $D(x)$ represents the outcome of D given input x, P_g denotes the distribution of fake samples defined by $\tilde{x} = G(z|c)$, $z \sim p(z)$. $G(z|c)$ describes the generative process using a Gaussian noise z conditioned on label c. The third component of Eq. 1 represents gradient penalty that enforces the 1-Lipschitz continuity [9] by penalizing the gradient norm of the random samples $\hat{x} \sim P_{\hat{x}}$. λ denotes the gradient penalty coefficient. Equation 2 is the loss of classifier C which represents the log-likelihood of the correct class that the generator G and the classifier C both try to maximize.

2.3 Data Acquisition and Experimental Setup

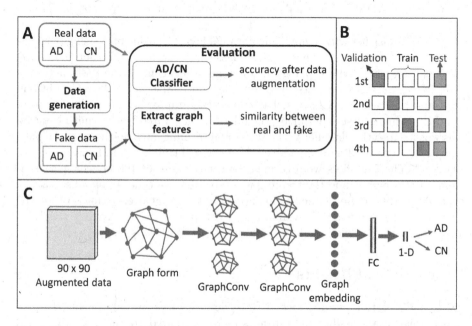

Fig. 2. Components of the experiments. **A.** Overall experimental design. **A.Left**: Real data was first used to generate the same amount of fake data using different data generating methods. **A.Right**: The fake samples were used in AD/CN classification tasks to quantitatively evaluate the performance of data augmentation methods. The similarity between real and fake network matrices was also calculated using the graph features extracted from the GNN. **B.** An 4-fold cross-validation was used in the network training. **C.** A graph convolutional neural network was adopted as the classifier to classify AD from CN, and at the same time, evaluate (through the extracted graph features) the performance of the data augmentation methods in capturing the topological property of the brain networks.

In this study, diffusion MRIs of 110 AD and 110 age-matched CN were obtained from the Alzheimer's Disease Neuroimaging Initiative (ADNI) (adni.loni.usc.edu).

The generated data from BrainNetGAN and baseline methods was evaluated by both classifying AD/CN and similarity metrics of graph features (Fig. 2A). In assessing the similarity metrics, all real data was used to generate fake data. For the classification task, real data was shuffled and split into the training, validation and testing sets with a ratio of 3 : 1 : 1 to perform an 4-fold cross-validation (Fig. 2B). For each training set, three different methods produced the same sample size as the training real data.

For BrainNetGAN, the learning rate (lr) of G, D, and C were the same during hyper-parameters tuning, the value of lr ranged between 0.00001 and 0.001. In an empirical study, 0.0005 was the large lr to guarantee the fastest convergence. The input dimension of noise z was empirically tested from 64 to 256, and 64 was used in this work as it produced the most reliable convergence results. The SMOTE and ADASYN methods were both implemented using *imblearn library* [12].

A GNN classifier with graph convolutional layers implemented using pytorch-geometric [6] was used as the AD/CN classifier to evaluate whether the generated data captured the topological differences between AD and CN networks (Fig. 2C). The GNN consists of graph convolutional layers called GraphConv [15], and fully connected layers that embed the graphs. Two experiments were conducted: in the first experiment, the fake data generated by the multiple methods was fed into the classifier without the real training data. In the second experiment, the generated data was mixed with the real training data for training the classifier. The classifiers were trained for a maximum of 100 epochs with random initial weights, and the training was repeated five times to avoid statistical bias. An early stopping scheme was applied to prevent over-fitting so that the training is stopped when the validation loss stops decreasing. Average accuracy, precision, and recalls of validation and test are reported.

3   Numerical Results

Data augmentation performance of BrainNetGAN was evaluated using the proposed dementia classifier, and results were compared to those in the baseline dataset (no augmented data) and augmented data from different methods (Table 2). The results show that the model with data generated by BrainNetGAN achieved higher performance in the classification compared to other models. Notably, the higher recall and precision of the model augmented by BrainNetGAN imply that the distribution of the augmented data by BrainNetGAN is less biased than other methods.

Moreover, the results on the testing set showed that the classification accuracy was improved from 0.79 to 0.83, when the fake samples generated by BrainNetGAN were added to the real samples, which doubled the sample size of the training dataset.

In addition, we verify the similarity between the graph features of real and fake networks generated by different methods. Kullback-Leibler divergence and Maximum mean discrepancy were used to compare three data augmentation

Table 1. The similarity between graph features of real and fake brain networks

	Kullback-Leibler Divergence		Maximum Mean Discrepancy	
	CN	AD	CN	AD
Edge weight				
SMOTE	0.510	0.482	**0.101**	**0.119**
ADASYN	0.523	0.533	0.120	0.132
BrainNetGAN	**0.473**	**0.460**	0.110	0.123
Node strength				
SMOTE	0.603	0.613	0.045	0.054
ADASYN	0.611	0.587	0.056	0.059
BrainNetGAN	**0.545**	**0.559**	**0.025**	**0.037**
Local efficiency				
SMOTE	0.428	0.398	0.035	0.037
ADASYN	0.523	0.410	0.042	0.039
BrainNetGAN	**0.247**	**0.214**	**0.009**	**0.014**
Global efficiency				
SMOTE	0.539	0.484	0.042	0.064
ADASYN	0.514	0.543	0.056	0.064
BrainNetGAN	**0.277**	**0.243**	**0.013**	**0.020**

Table 2. Evaluate data augmentation performance using graph neural networks

	Validation			Test		
	Accuracy	Precision	Recall	Accuracy	Precision	Recall
Baseline	0.819 ± 0.083	0.790 ± 0.061	0.710 ± 0.059	0.791 ± 0..027	0.658 ± 0.083	0.668 ± 0.063
Substitute real training data for generated fake data in AD/CN classification task						
SMOTE	0.819 ± 0.069	0.709 ± 0.049	0.790 ± 0.068	0.767 ± 0.023	0.725 ± 0.096	0.748 ± 0.094
ADASYN	0.793 ± 0.075	0.712 ± 0.040	0.705 ± 0.087	0.742 ± 0.105	0.702 ± 0.115	0.702 ± 0.098
BrainNetGAN	**0.831 ± 0.072**	**0.772 ± 0.086**	**0.822 ± 0.062**	**0.812 ± 0.068**	**0.795 ± 0.104**	**0.803 ± 0.102**
Combine real training data with generated fake data in AD/CN classification task						
Baseline ± SMOTE	0.820 ± 0.092	0.801 ± 0.051	0.820 ± 0.055	0.802 ± 0.025	0.798 ± 0.085	0.812 ± 0.119
Baseline ± ADASYN	0.778 ± 0.093	0.745 ± 0.065	0.712 ± 0.070	0.788 ± 0.124	0.718 ± 0.094	0.711 ± 0.132
Baseline ± BrainNetGAN	**0.852 ± 0.085**	**0.805 ± 0.091**	**0.825 ± 0.043**	**0.829 ± 0.058**	**0.809 ± 0.142**	**0.825 ± 0.113**

methods (Table 1). Both metrics evaluate the distance between the distribution of real and fake data, where a lower value indicates high similarity. Edge weight, node strength, local efficiency, and global efficiency were computed using the Brain Connectivity Toolbox [16]. We compared the similarities of those graph features calculated from the brain networks augmented by the three methods.

As shown in the Table 1, the data generating performance of the two methods were consistent between AD and CN. The BrainNetGAN outperformed other methods in the comparison of most topological features, indicating that the topological properties of the fake brain networks generated by BrainNetGAN approximate the real data.

4   Discussion and Conclusion

We propose the BrainNetGAN to perform data augmentation of brain network matrices for dementia classification. Numerical results demonstrated that the BrainNetGAN outperformed the benchmarking methods and generated high-quality fake samples which effectively improved the classification performance. Future work can focus on improving the brain network edge performance in fake data generation and more deliberated analysis of different types of entries in the connectivity matrix (therefore to generate better fake samples). To conclude, GAN based brain network augmentation is a promising technique that can provide clinical values in training deep learning models for the classification of neuropsychiatric diseases.

References

1. Ahmed, M.R., Zhang, Y., Feng, Z., Lo, B., Inan, O.T., Liao, H.: Neuroimaging and machine learning for dementia diagnosis: recent advancements and future prospects. IEEE Rev. Biomed. Eng. **12**, 19–33 (2018)
2. Ajilore, O., Lamar, M., Kumar, A.: Association of brain network efficiency with aging, depression, and cognition. Am. J. Geriatr. Psychiatry **22**(2), 102–110 (2014)
3. Antoniou, A., Storkey, A., Edwards, H.: Data augmentation generative adversarial networks. arXiv preprint arXiv:1711.04340 (2017)
4. Avants, B.B., Tustison, N., Song, G.: Advanced normalization tools (ants). Insight j **2**(365), 1–35 (2009)
5. Chawla, N.V., Bowyer, K.W., Hall, L.O., Kegelmeyer, W.P.: Smote: synthetic minority over-sampling technique. J. Artif. Intell. Res. **16**, 321–357 (2002)
6. Fey, M., Lenssen, J.E.: Fast graph representation learning with PyTorch geometric. In: ICLR Workshop on Representation Learning on Graphs and Manifolds (2019)
7. Goodfellow, I.J., et al.: Generative adversarial networks. arXiv preprint arXiv:1406.2661 (2014)
8. Griffa, A., Baumann, P.S., Thiran, J.P., Hagmann, P.: Structural connectomics in brain diseases. Neuroimage **80**, 515–526 (2013)
9. Gulrajani, I., Ahmed, F., Arjovsky, M., Dumoulin, V., Courville, A.: Improved training of Wasserstein GANs. arXiv preprint arXiv:1704.00028 (2017)
10. He, H., Bai, Y., Garcia, E.A., Li, S.: ADASYN: adaptive synthetic sampling approach for imbalanced learning. In: 2008 IEEE International Joint Conference on Neural Networks (IEEE World Congress On Computational Intelligence), pp. 1322–1328. IEEE (2008)
11. Kawahara, J., et al.: BrainNetCNN: convolutional neural networks for brain networks; towards predicting neurodevelopment. Neuroimage **146**, 1038–1049 (2017)
12. Lemaître, G., Nogueira, F., Aridas, C.K.: Imbalanced-learn: a python toolbox to tackle the curse of imbalanced datasets in machine learning. J. Mach. Learn. Res. **18**(17), 1–5 (2017). http://jmlr.org/papers/v18/16-365
13. Mattson, M.P.: Pathways towards and away from Alzheimer's disease. Nature **430**(7000), 631–639 (2004)
14. Mori, S., Crain, B.J., Chacko, V.P., Van Zijl, P.C.: Three-dimensional tracking of axonal projections in the brain by magnetic resonance imaging. Ann. Neurol.: Off. J. Am. Neurol. Assoc. Child Neurol. Soc. **45**(2), 265–269 (1999)

15. Morris, C., Ritzert, M., Fey, M., Hamilton, W.L., Lenssen, J.E., Rattan, G., Grohe, M.: Weisfeiler and leman go neural: higher-order graph neural networks. In: Proceedings of the AAAI Conference on Artificial Intelligence, vol. 33, pp. 4602–4609 (2019)

16. Rubinov, M., Sporns, O.: Complex network measures of brain connectivity: uses and interpretations. Neuroimage **52**(3), 1059–1069 (2010)

17. Song, T.A., et al.: Graph convolutional neural networks for Alzheimer's disease classification. In: 2019 IEEE 16th International Symposium on Biomedical Imaging (ISBI 2019), pp. 414–417. IEEE (2019)

18. Tzourio-Mazoyer, N., et al.: Automated anatomical labeling of activations in SPM using a macroscopic anatomical parcellation of the MNI MRI single-subject brain. Neuroimage **15**(1), 273–289 (2002)

19. Wei, Y., Li, C., Price, S.: Quantifying structural connectivity in brain tumor patients. medRxiv (2021)

Evaluating GANs in Medical Imaging

Lorenzo Tronchin[1(✉)], Rosa Sicilia[1], Ermanno Cordelli[1], Sara Ramella[2], and Paolo Soda[1]

[1] Unit of Computer Systems & Bioinformatics, Department of Engineering,
Università Campus Bio-Medico di Roma, Rome, Italy
`l.tronchin@unicampus.it`
[2] Radiotherapy Unit, Università Campus Bio-Medico di Roma, Rome, Italy

Abstract. Generative Adversarial Networks (GANs) have recently gained large interest in computer vision being used in many tasks, but their evaluation is still an open issue. This is especially true in medical imaging where GAN application is at its infancy, and where the use of scores based on models trained on datasets far away from the medical domain, e.g. the Inception score, can lead to misleading results. To overcome such limitations we propose a framework to evaluate images generated by GANs in terms of fidelity and structural similarity with the real ones. On the one hand, we measure the distance between the probability densities of the real and generated samples by exploiting feature representations given by a Convolutional Neural Network (CNN) trained as a discriminator. On the other hand, we compute domain-independent metrics catching the image high-level quality. We also introduce a visual layer explaining the CNN. We extensively evaluate the proposed approach with 4 state-of-the-art GANs over a real-world medical dataset of CT lung images.

Keywords: GAN · Evaluation · CNN · Explainability

1 Introduction

Recent years have witnessed a rising interest towards Generative Adversarial Networks (GANs) in Medical Image Analysis (MIA), which are able to synthesise images with an unprecedented level of realism. Although their use in this field is in an early stage, it already accounts for a wide set of applications from medical image synthesis to segmentation or classification [11]. Besides these opportunities, the way the GANs should be evaluated is still an open and critical issue. While the interested readers can find in [3] a comprehensive review of GAN evaluation metrics, categorised into *qualitative* and *quantitative* scores, there is no general consensus on which should be employed in fair comparisons. The most famous qualitative measure, widely used in the medical domain [4,5,8,11], is based on human visual judgement. Its validity is still questionable [11] and raises the need of exploring quantitative measures suited for real-world medical tasks. The literature employing quantitative measures is dominated by those

S. Engelhardt et al. (Eds.): DGM4MICCAI 2021/DALI 2021, LNCS 13003, pp. 112–121, 2021.
https://doi.org/10.1007/978-3-030-88210-5_10

based on the Inception Net [6], such as the Inception Score (IS) [21] or rather the Fréchet Inception Distance (FID) [7], which assume that the Inception Net is able to grasp the underlying properties and information hidden in the data. However, the Inception Net is not trained on a medical dataset and, hence, it exploits domain-specific representations that may lead to misleading results in the medical domain [2]. To avoid this limitation, in MIA several work adopt image quality scores working at pixel level rather than domain-specific measures. Nevertheless, they are also limited since they do not grasp the benefits introduced by GANs, i.e. rendering of coherent image features beyond the pixel-level [11].

These observations suggest that assessing GANs in MIA is a topic deserving further research efforts. In this respect, we hereby introduce two main contributions. First, we present a framework to evaluate images generated by GANs in terms of: (i) discriminability, i.e. the fidelity between generated and real images at feature level measured exploiting a Convolutional Neural Network (CNN), which acts as a general-purpose discriminator revealing the underlying distributions of the data; (ii) structural similarity between generated and real images measured by means of domain-independent metrics catching the image high level quality; these scores permit also to detect risk of failure modes, i.e. overfitting (when the GAN merely memorises the real training) and mode collapse (when a GAN generates only a small subset of samples, also known as a mode). Second, we introduce a visual layer that permits us to explain the CNN results.

We exploit this framework to compare 4 state-of-the-art GANs tested on a real-world medical dataset. Indeed, determining which architecture can harvest all the relevant features held in the available dataset is a first and essential step within a more general setting, aiming to combine a GAN and a CNN in the MIA application described in Sect. 3.

2   Methods

The proposed framework evaluates n GANs used to synthesise medical images. It is divided into two steps: the first measures sample discriminability, whereas the second carries out the structural evaluation comparing directly a set of synthetic samples with the real ones. As general procedure, we first divide the real image set R into two disjoint portions, i.e. Real Training set (RTr) and Real Test set (RTs), so that $RTr \cap RTs = \emptyset$. Each GAN$_i$ learns the distribution of RTr to generate a Synthetic Training set (STr_i) and a Synthetic Test set (STs_i).

Discriminability Evaluation. The first step of our assessment analyses the underlying information hidden in the synthetic images grasped by a specific representation. The generative ability of a GAN is measured by a classifier discriminating the synthetic samples from the real ones. Such an analysis, inspired by the Classifier Two Sample Test [13], determines whether two samples are drawn from the same distribution exploiting a binary classifier as a proxy. The intuition behind this evaluation relies on creating a representation specifically tailored for distinguishing real and synthetic images. The GANs distributions are assessed in a features space built ad-hoc for the fidelity-evaluation task: if the synthetic

Table 1. Evaluation metrics for GAN assessment. Rows 1–2: CNN-based evaluation; Rows 3–4: structural data evaluation.

METRIC	DEFINITION
MAHALANOBIS DISTANCE	$d_M^2 = (\mathbf{x} - \mu_Y)^T \sum_Y^{-1}(\mathbf{x} - \mu_Y)$
FRÉCHET DISTANCE	$d_F^2 = \mid \mu_X - \mu_Y \mid^2 + tr(\sum_X + \sum_Y - 2(\sum_X \sum_Y)^{1/2})$
MS-SSIM	$MS\text{-}SSIM(\tilde{x}, y) = L_M(\tilde{x}, y)^{\alpha_M} \prod_{j=1}^{M} C_j(\tilde{x}, y)^{\beta_j} S_j(\tilde{x}, y)^{\gamma_j}$
TEMPLATE MATCHING	$CCR(p, q) = \dfrac{\sum_{\hat{p},\hat{q}}(\tilde{x}(\hat{p},\hat{q}) \cdot \Gamma(p+\hat{p}, q+\hat{q}))}{\sqrt{\sum_{\hat{p},\hat{q}} \tilde{x}(\hat{p},\hat{q})^2 \cdot \sum_{\hat{p},\hat{q}} \Gamma(p+\hat{p}, q+\hat{q})^2}}$

representation can fool the classifier in a space designed to separate synthetic from real samples, the generator has learned to synthesise useful features from real data support. Note that the use of an unsupervised model instead of a supervised one would build a not-specific representation space. Accordingly, the dichotomisers are n CNNs (described in Sect. 4), trained to distinguish between real and synthetic samples, acting as independent discriminators that can be compared across the different GANs. Note that each CNN is trained with the images synthesised by GAN_i (STr_i) with RTr and are tested on the independent set generated by GAN_i (STs_i) and RTs. The recall on each class can be regarded as a proxy measure of the discriminability: the lower the recall of the CNN on an $STs_i \cup RTs$, the better the generation of GAN_i, i.e. the synthetic samples distribution lies closer to the real samples. To further prove this intuition and to *explain* the CNNs behaviour, we leverage the embedding given by the last dense layer of each CNN which is regarded as a continuous feature space. To simplify the notation, X and Y indicate the representations of synthetic and real samples in such a feature space, respectively. With X and Y we compute the Mahalanobis distance [16] and the Fréchet Inception Distance (FID) [3], i.e. two measures estimating the distance of the probability distributions of samples. The first computes the distance between each feature vector $\mathbf{x}$ of synthetic samples from the distribution of real samples. Its formal definition is reported in Table 1, where μ_Y and $\sum_Y$ denote the mean and covariance matrix of real samples representation Y. The FID measure [3] is currently used for GANs evaluation in combination with the Inception Net: we revisit it for MIA by substituting the Inception Net representation with those returned by the CNNs discriminators. The formal definition in Table 1 considers the distribution distance between X and Y, where μ and $\sum$ denote the mean and covariance matrix of each set.

We also add to this quantitative evaluation a visual layer providing a graphical representation of the distribution proximity in a low-dimensional feature space. To this end we work as follows: (i) we compute a common reference of real images as the average representation returned by the CNNs; (ii) we apply the PCA on this dataset, keeping the first 2 principal components; (iii) in this 2D space we project samples in $STs_i \cup RTs$; (iv) we plot this representation, adding an ellipsoid centred in the mean of each distribution, where the orthogonal axes have length equal to one standard deviation and direction given by covariance

Table 2. GAN loss functions. The upper panel lists the losses of unsupervised GANs, whereas the lower panel reports the supervised loss functions.

GAN		DISCRIMINATOR LOSS	GENERATOR LOSS
Unsupervised	DCGAN	$-\mathbb{E}_{x \sim p_{data}}[\log D(x)] - \mathbb{E}_{z \sim p_z}[\log(1 - D(G(z)))]$	$-\mathbb{E}_{z \sim p_z}[\log(D(G(z)))]$
	LSGAN	$-\mathbb{E}_{x \sim p_{data}}[(D(x) - 1)^2] + \mathbb{E}_{z \sim p_z}[(D(G(z)))^2]$	$-\frac{1}{2}\mathbb{E}_{z \sim p_z}[(D(G(z)))^2]$
	WGAN-GP	$-\mathbb{E}_{x \sim p_{data}}[D(x)] + \mathbb{E}_{z \sim p_z}[D(G(z))] + \lambda\mathbb{E}_{\hat{x} \sim p_{\hat{x}}}[(\|\nabla_{\hat{x}} D(\hat{x})\|_2 - 1)^2]$	$-\mathbb{E}_{z \sim p_z}[D(G(z))]$
Supervised	ACGAN	$L_{source} + L_{class}$	$L_{source} - L_{class}$
		where $L_{source} = \mathbb{E}[\log P(S = real \mid p_{data})] + \mathbb{E}[\log P(S = fake \mid p_g)]$; $L_{class} = \mathbb{E}[\log P(C = c \mid p_{data})] + \mathbb{E}[\log P(C = c \mid p_g)]$	

matrix eigenvectors. Straightforwardly, the larger the fidelity between generated and real images, the closer the corresponding ellipses.

Structural Evaluation. Here we evaluate the synthesis result against the ground truth with quantitative scores indicating the similarity between artificial and real images. The rationale is based on two assumptions: (i) the exhaustive comparison between synthetic and real allows to catch *overfitting*; hence, we search for a high one-to-one similarity between generated and real distribution; (ii) comparing the intra-group similarity of both synthetic and real images we search for low samples variability assessing whether the GAN is subjected to *mode collapse*. According to [3], we exploit measures with well-defined bounds and low sensitivity to image distortions, e.g. translations, rotations, etc. On these premises, we compare images generated by each GAN with the real ones using 2 measures. One is the Multiscale Structural Similarity Index (MS-SSIM), a variant of the SSIM score working at multiple scales [23] and ranging from 0 (low similarity) to 1 (high similarity). Its definition is in Table 1, where $\tilde{x}$ and y denote the synthetic and the real images, and where we omit to detail the other symbols for space reason, referring the reader to [23]. As translation invariant measure we employ the Template Matching (TM), i.e. a method for searching the location of a template image in a larger one. Each synthetic image is regarded as a template image $\tilde{x}$, which is convoluted in 2D over Γ, an image composing in a 2D lattice several real samples y. TM compares $\tilde{x}$ and Γ by computing for each location $(p, q) \in \Gamma$ the normalised cross-correlation (Table 1). The result is a grayscale heatmap in $[0, 1]$, where each pixel denotes how much its neighbourhood matches with the template: the larger the heatmap, the better the match.

Table 3. Hyperparameters of the used GANs. Batch Normalisation (BN) resulted in sample oscillation and training instability for all GANs except the WGAN-GP. Abbreviations: Kernel (K), LeakyReLU (LReLU), Label Smoothing (LS).

GAN	α_G	α_D	β_1	β_2	K G, D	z size	Activation G, D	BN	LS	Training parameters	
DCGAN	0.0001	0.0001	0.5	0.99	3, 5	100	LReLU(0.2), LReLU(0.2)	×	✓	Training epochs, batch size	1000, 128
LSGAN	0.0001	0.0001	0.5	0.99	3, 5	1024	ReLU, LReLU(0.2)	×	×	Optimiser	Adam
WGAN-GP	0.0002	0.0001	0	0.90	3, 3	128	LReLU(0.2), LReLU(0.2)	G	×	$D : G$ update ratio	2:1
AC-GAN	0.0002	0.0002	0.5	0.99	5, 3	110	ReLU, LReLU(0.2)	×	×	Random noise (z)	Spherical Gaussian distribution

We acknowledge that such metrics have been widely discussed in the literature [3,7,23]: but here they are originally employed to interpret the different aspects of GANs generation.

2.1 Competing GANs

The goal of GANs is to learn the real data distribution p_{data} from a set of samples to generate new samples. To this end, a generative network G maps the input z, that is random noise , to the image space $p_g = G(z)$ and aims to achieve $p_g = p_{data}$. Then a discriminator network D aims to discern between real and generated images [9]. The generator aims to synthesise realistic images $G(z)$ that D will consider real with high probability.

We hereby considered for comparison 3 unsupervised architectures and 1 supervised architecture, whose losses are in Table 2. The first is **DCGAN**, the most famous employed in MIA, which has improved the standard GANs [9] introducing convolutional layers to GAN architecture. To improve the gradient information we deploy the DCGAN with non-saturating loss [9]. **LSGAN** solves the problem of the vanishing gradient when updating the generator weights using the synthetic samples that are on the right side (i.e. the real data side) of the decision boundary but are still far from real data distribution [17]. It employs a least-squares loss function for the discriminator penalising the samples lying far away on the right side of the decision boundary, even if they are correctly classified. **WGAN-GP** extends the Wasserstein GAN introduced in [1], which uses the Earth-Mover or Wasserstein distance to train a GAN and relies on a hard clipping of discriminator weights ensuring that are K-Lipschitz for some K. WGAN-GP provides an alternative and soft way to enforce the Lipschitz constraint besides weight clipping, penalising with a factor λ the discriminator if the gradient norm for interpolated samples between real and synthetic data points deviates from 1 [10]. The Auxiliary Classifier **AC-GAN** leverages side information as samples class, adding more structure to the generator latent space and resulting in higher quality samples [18]. Here D provides the probability that a sample is real along with its class probability, and G is fed with the noise z along with the class label information to generate synthetic samples $p_g = G(z, c)$.

3 Materials

We use a radiomic dataset with CT scans of 123 patients with Locally Advanced stage III non-small cell lung cancer (protocol ID: NCT03583723). They belong to two classes: *adaptive* patients, who showed a tumour volume reduction during the chemo-radiotherapy treatment, and *non-adaptive* patients, who on the contrary didn't show any reduction [20]. We considered 2690 2D slices of patients' CT scans, where the region-of-interests are given by the clinical target volumes manually delineated by expert radiation oncologists.

Table 4. Explainable classifier-based evaluation results. Best scores are in bold.

	ADAPTIVE				NON-ADAPTIVE			
	DCGAN	LSGAN	WGAN-GP	AC-GAN	DCGAN	LSGAN	WGAN-GP	AC-GAN
RECALL	.82 ± .13	.93 ± .04	.88 ± .09	.80 ± .09	**.66 ± .13**	.91 ± .03	.87 ± .09	.87 ± .05
SPECIFICITY	**.78 ± .12**	.94 ± .02	.81 ± .10	.87 ± .07	**.71 ± .13**	.94 ± .03	.90 ± .08	.79 ± .09
MAHALANOBIS DISTANCE	20.45 ± 1.12	23.63 ± 0.46	19.51 ± 1.02	**19.22 ± 0.33**	**16.96 ± 0.37**	18.32 ± 0.51	18.03 ± 0.87	20.38 ± 1.27
FRÉCHET DISTANCE	**19.20 ± 14.12**	140.19 ± 32.03	27.67 ± 9.18	26.44 ± 8.27	**9.44 ± 4.23**	114.10 ± 36.48	42.82 ± 19.12	63.71 ± 21.40

4   Experimental Results

To ensure a fair comparison among the GANs, their architectures correspond to those presented in the original paper [10, 17–19], adapting the depth of the networks and the feature maps of each layer for grayscale 80×80 inputs. Furthermore, pixel intensities of CT images were rescaled to $[-1, 1]$. Table 3 summarises the GAN hyperparameters and their training parameters. To generate labelled lung lesion and exploit the methodology presented in Sect. 2, we include 99 cases in RTr and 24 cases in RTs. This results in 2182 training slices (1024 adaptive, 1158 non-adaptive), and 508 evaluation slices (225 adaptive, 283 non-adaptive). We trained the unsupervised GANs once per class, whilst the supervised AC-GAN is trained directly on RTr including both classes.

Let us now briefly describe the CNN working as a general-purpose discriminator: its architecture is inspired to the VGG Net [22] and accepts input images of 80×80, whose pixel intensities range in $[0, 1]$. Our CNN has 4 consecutive blocks, each containing a stack of two convolutional layers and a max-pooling layer, then implements LeakyReLU [15] as activation (slope = 0.2), a 128 neurons dense layer after the last convolutional layer and a final dense layer with one neuron and the sigmoid activation function. Weights initialisation uses Glorot Uniform and loss is measured with binary cross-entropy. Each convolutional block, with a kernel of 3, doubles the number of filters in each block. We train each CNN_i for 20 epoch using Adam [12] ($\alpha = 0.001$). We allowed a moderate degree of overfitting on the training set since our CNN acts as a discriminator: high-specialised networks allow to better understand the difference between generated and real samples coherently with GANs theory [10]. The proposed method was deployed using Python 3.5.6 and Tensorflow 2.2.0, whilst all the experiments were performed with a NVIDIA TESLA V100 16GB.

As discussed in Sect. 1, measures as the IS or FID cannot be used in MIA. To prove this observation, we fine-tuned the representation learnt from the Inception Net to classify our CT ROIs as adaptive or non-adaptive. We got a low AUC score (61%) at the patient level, confirming the impossibility of applying scores that leverage the ImageNet representation. Turning now the attention to our discriminability evaluation, the upper panel of Table 4 shows the recall and the specificity returned by each CNN_i, averaged over 50 runs. As presented in Sect. 2, a lower performance indicates that the GAN is able to generate samples hard to be distinguished from the real ones. These results show that DCGAN is the best architecture for the task at hand, except for the recall on the adaptive class. This finding was also confirmed using other machine learning classifiers

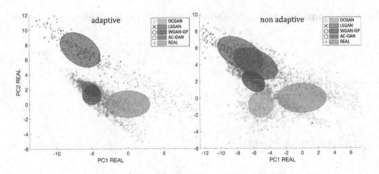

Fig. 1. Test sample distribution in the 2D PCA space of real samples.

(Decision Tree, Random Forest, Support Vector Machine with a linear kernel, k-Nearest Neighbours and AdaBoost) trained as discriminators on the feature representations extracted from the CNNs, whose results are omitted for brevity. The high fidelity of DCGAN generation we obtained also agrees with [14], where the authors report that, with proper training, it can successfully learn the real distribution. Moreover, analysing the distribution distance using both the Mahalanobis and Fréchet distances (lower panel of Table 4), we observe that: (i) the DCGAN *STs* shows the lowest distance from *RTs* on both classes with the Fréchet distance; (ii) with the Mahalanobis distance DCGAN is outperformed only on the adaptive class by AC-GAN. This last finding can be explained recalling that, whilst Fréchet computes the distance among the whole distributions of samples, Mahalanobis considers the distance of each synthetic sample from the reference real distribution. Hence, if the synthetic samples have outliers very close to the real distribution, the average Mahalanobis distance would be lower than the average Fréchet distance. As final confirmation of this considerations, Fig. 1 presents the information extracted from the visual layer of the CNN to explain the results as described in Sect. 2. Indeed, the DCGAN ellipsoid is the closest one to that of the real samples for non-adaptive class, while it has a high overlap with AC-GAN for the adaptive class, a result that agrees with the metrics shown in Table 4.

Whilst the assessment of synthetic images *fidelity* reported so far neglects the *memory-GAN* phenomenon, the proposed structural evaluation favours models synthesising *diverse* samples, overcoming this issue. Figure 2 shows the distributions of the MS-SSIM maximum value for the adaptive and non-adaptive class when we compare: (i) each synthetic image against all synthetic ones to estimate the intra-group variability of the generated distribution (denoted as Synthetic-vs-Synthetic (S_S)), (ii) each synthetic image against all real samples to catch the inter-group variability (denoted as Synthetic-vs-Real (S_R)), (iii) each real sample opposed to all real ones to provide a baseline of the real samples similarity (denoted as Real-vs-Real (R_R)). In S_R distributions (yellow), we note that all GAN distributions have a median value near 0.8, indicating a high correlation. Even if we cannot present an interpretation of which GAN wins, these results

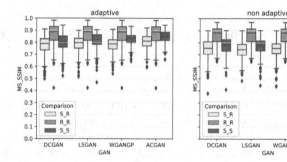

Fig. 2. MS-SSIM maximum value distribution across GANs. Abbreviations: Synthetic-vs-Synthetic (S_S), Synthetic-vs-Real (S_R), Real-vs-Real (R_R).

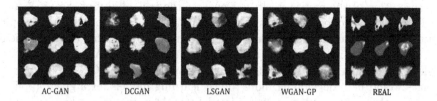

Fig. 3. Comparison between GANs images for adaptive class.

suggest that GANs succeeded as they synthesise samples with high structural similarity to the real ones. The lower structural similarity of S_R distribution with respect to the S_S (blue) and R_R (green) distributions highlights that GANs do not simply memorise the real dataset. These results are also confirmed analysing the S_S and the baseline R_R distributions: S_S is shifted downwards with respect to R_R, indicating a lower correlation within the synthetic dataset despite a real dataset characterised by high similarity. Indeed, the 2690 slices were extracted from a small pool of patients. These considerations further prove that all GANs do not suffer from overfitting and mode collapse. The structural analysis allows extending the previous findings but considering a diverse aspect than fidelity: (i) the representation learned for *non-adaptive* seems to be less prone to overfitting with respect to the *adaptive* class considering the lowest S_R median value; (ii) the overlap of the real and AC-GAN ellipsoids can be related to a small degree of overfitting on *adaptive class*; (iii) all GANs capture the real dataset support generating more diverse features than the feature retained in the real images. The study carried out using TM confirms the results obtained (the corresponding figures are omitted for brevity). Besides these insights, the structural analysis has a limitation: since there is no absolute threshold to assess the appropriate similarity between the distributions, it can't determine which GAN is the most effective in the generation task, but it can only identify eventual overfitting or mode collapse.

Finally, Fig. 3 shows synthetic and real images for the adaptive class, proving that the visual assessment of CT slices is a challenging task. Indeed, the human eye fails in extracting structural and general features needed to: (i) determine which GAN is the most suitable for the generation task; (ii) assess overfitting.

5   Conclusions

In this work, we study how to assess GANs in MIA since visual inspection can be prone to error, further to be a burdensome task and measures relying on a pre-trained Inception Net cannot be used. Since a single score cannot detail the whole generation process, we introduce a framework that measures the sample discriminability via an explainable CNN and structural information of synthetic data. We provide empirical evidence that such an approach is a feasible method to assess the fidelity of synthetic samples with respect to real ones, providing an interpretation of the metric itself that ensures the trustability of the process.

Overall, the proposed framework results suitable to be adapted to MIA applications since it does not rely on any prior information, providing the data scientist with an effective method to choose the more effective GAN for the clinical task at hand. As a next step, we aim to show that using the DCGAN's synthetic images would help developing more robust learning models for a given application, e.g. the radiomic dataset considered here. Also, future research includes: testing other GANs, including other image datasets, investigating a clinical-based metric in order to evaluate and explain GANs.

Acknowledgements. This work was partially founded by Universita' Campus Bio-Medico di Roma under the programme "University Strategic Projects".

References

1. Arjovsky, M., Chintala, S., Bottou, L.: Wasserstein GAN (2017). arXiv:1701.07875
2. Barratt, S., Sharma, R.: A note on the Inception Score (2018). arXiv:1801.01973
3. Borji, A.: Pros and Cons of GAN evaluation measures. Comput. Vis. Image Underst. **179**, 41–65 (2019)
4. Calimeri, F., et al.: Biomedical data augmentation using generative adversarial neural networks. In: International Conference on Artificial Neural Networks, pp. 626–634 (2017)
5. Chuquicusma, M.J.M., et al.: How to fool radiologists with generative adversarial networks? A visual turing test for lung cancer diagnosis (2018)
6. Deng, J., et al.: Imagenet: A large-scale hierarchical image database. In: 2009 IEEE Conference on Computer Vision and Pattern Recognition, pp. 248–255 (2009)
7. Dowson, D., Landau, B.: The Fréchet distance between multivariate normal distributions. J. Multivar. Anal. **12**(3), 450–455 (1982)
8. Frid-Adar, M., et al.: GAN-based synthetic medical image augmentation for increased CNN performance in liver lesion classification. Neurocomputing **321**, 321–331 (2018)
9. Goodfellow, I.J., et al.: Generative Adversarial Nets (2014). arXiv:1406.2661
10. Gulrajani, I., Ahmed, F., Arjovsky, M., Dumoulin, V., Courville, A.C.: Improved training of wasserstein GANs. In: Advances in Neural Information Processing Systems, pp. 5767–5777 (2017)

11. Kazeminia, S., et al.: GANs for medical image analysis. Artificial Intelligence in Medicine, p. 101938 (2020)
12. Kingma, D., Ba, J.: Adam: A Method for Stochastic Optimization (December 2014). arXiv:1412.6980
13. Lopez-Paz, D., Oquab, M.: Revisiting Classifier Two-Sample Tests (2018)
14. Lucic, M., et al.: Are GANs created equal? A large-scale study (2018)
15. Maas, A.L., Hannun, A.Y., Ng, A.Y.: Rectifier nonlinearities improve neural network acoustic models. In: Proceedings of the ICML, vol. 30, p. 3 (2013)
16. Mahalanobis, P.C.: On the generalized distance in statistics (1936)
17. Mao, X., Li, Q., Xie, H., Lau, R.Y., Wang, Z., Paul Smolley, S.: Least squares generative adversarial networks. In: Proceedings of the IEEE International Conference on Computer Vision, pp. 2794–2802 (2017)
18. Odena, A., et al.: Conditional image synthesis with auxiliary classifier GANs. In: International Conference on Machine Learning, pp. 2642–2651 (2017)
19. Radford, A., Metz, L., Chintala, S.: Unsupervised representation learning with Deep Convolutional Generative Adversarial networks. arXiv:1511.06434 (2015)
20. Ramella, S., et al.: A radiomic approach for adaptive radiotherapy in non-small cell lung cancer patients. PLOS ONE **13**(11), 1–14 (2018). https://doi.org/10.1371/journal.pone.0207455
21. Salimans, T., Goodfellow, I., Zaremba, W., Cheung, V., Radford, A., Chen, X.: Improved techniques for training GANs. In: Advances in Neural Information Processing Systems, pp. 2234–2242 (2016)
22. Simonyan, K., Zisserman, A.: Very deep convolutional networks for large-scale image recognition (2015)
23. Wang, Z., Simoncelli, E.P., Bovik, A.C.: Multiscale structural similarity for image quality assessment. In: The Thirty-Seventh Asilomar Conference on Signals, Systems & Computers, 2003, vol. 2, pp. 1398–1402 (2003)

AdaptOR Challenge

Improved Heatmap-Based Landmark Detection

Huifeng Yao[1]([✉]), Ziyu Guo[1], Yatao Zhang[1], and Xiaomeng Li[2]

[1] Shandong University, Shandong, China
[2] Hong Kong University of Science and Technology, Hongkong, China

Abstract. Mitral valve repair is a very difficult operation, often requiring experienced surgeons. The doctor will insert a prosthetic ring to aid in the restoration of heart function. The location of the prosthesis' sutures is critical. Obtaining and studying them during the procedure is a valuable learning experience for new surgeons. This paper proposes a landmark detection network for detecting sutures in endoscopic pictures, which solves the problem of a variable number of suture points in the images. Because there are two datasets, one from the simulated domain and the other from real intraoperative data, this work uses cycleGAN to interconvert the images from the two domains to obtain a larger dataset and a better score on real intraoperative data. This paper performed the tests using a simulated dataset of 2708 photos and a real dataset of 2376 images. The mean sensitivity on the simulated dataset is about $75.64 \pm 4.48\%$ and the precision is about $73.62 \pm 9.99\%$. The mean sensitivity on the real dataset is about $50.23 \pm 3.76\%$ and the precision is about $62.76 \pm 4.93\%$. The data is from the AdaptOR MICCAI Challenge 2021, which can be found at https://zenodo.org/record/4646979#.YO1zLUxCQ2x.

Keywords: Heatmap · Landmark detection · CycleGAN

1 Introduction

In mitral valve repair, the surgeon repairs part of the damaged mitral valve to allow the valve to fully close and stop leaking. The surgeon may tighten or reinforce the ring around a valve by implanting an artificial ring. The surgeon may place approximately 12 to 15 sutures on the mitral annulus [1]. We need to know how sutures are placed because analyzing the pattern and distances between them can help us improve the quality of this surgery. Furthermore, the position of the sutures may aid the medico in learning how to perform this surgery by reconstructing it in a 3D virtual environment.

Deep learning methods have been widely used in the field of medical images. This task belongs to the landmark detection task in computer vision. In general, people mainly use the heatmap-based [6] method, coordinate regression method,

---

Huifeng Yao and Ziyu Guo—contributed equally to this work and should be considered joint first-authors.

S. Engelhardt et al. (Eds.): DGM4MICCAI 2021/DALI 2021, LNCS 13003, pp. 125–133, 2021.
https://doi.org/10.1007/978-3-030-88210-5_11

and patch-based method. Payer et al. [6] used the SpatialConfiguration-Net which combines the local appearance of landmarks with their spatial configuration. Because the coordinate regression method is too difficult to converge and the patch-based method is difficult to distinguish adjacent points, we choose the heatmap-based method.

Many state-of-the-art heatmap-based deep learning methods focus on detecting fixed key points which are not suitable for our task. Stern et al. [9] proposed a heatmap-based method to detecting a varying number of key points. Inspired by that, we present an improved heatmap-based method that can deal with a varying number of sutures and get better performance than that.

The data set is mainly split into two endoscopic sets. One is simulation data set and the other is real data set. Inspired by Engelhardt et al. [2], we also implement the image to image translation to get more real data. We use the cycleGAN [11] network to do this task.

The work proposed a network to detect a varying number of landmarks and used the cycleGAN to translate images from two different domains. And we are participating within the scope of the AdaptOR challenge.

2   Materials and Methods

2.1   Data Set

Our data set comes from the AdaptOR challenge [8]. The data set is mainly split into two endoscopic sets:

(1) Sim-Domain is the image acquired during simulating mitral valve repair on a surgical simulator. More information on the simulator can be found in [3] and [4]. The simulator dataset used for training consists of 2708 frames, which were extracted from 10 surgeries. We divide it into 5 fold. To prevent data leakage, dataset splitting was always carried out on the level of the surgeries.

(2) Intraop-Domain is the Intraoperative endoscopic data from real minimally invasive mitral valve repair. Since the intraoperative dataset consists of 2376 frames extracted from 4 simulated surgeries, we split it into 4 fold with each surgery comprising one fold.

The Label of this data set is stored in the format of a JSON file. In addition, the data splitting is shown in Table 1.

Table 1. Data set.

Domain	Split	Number of frames				
		$f1$	$f2$	$f3$	$f4$	$f5$
Sim	Train	2246	2144	1960	2174	2308
	Validation	462	564	748	534	400
Intraop	Train	1582	1852	2004	1690	–
	Validation	794	524	372	686	–

2.2   Outline of the Proposed Method

We have a lot of simulated data, but we don't have enough real data. So the first step is the image to image translation. We use a cycleGAN to convert simulated data to real data in order to obtain more real data, which will help our model score higher on the real dataset. The second step is to generate the heatmap. Unlike other tasks about landmark detection, which use one channel for each landmark, we do not have fixed points in this task. So we generate all the points in one channel. And each of them is a 2D Gaussian kernel. We do some augmentation for both the original image and heatmap. Then the enhanced images would be the input of the U-net-based [7] network. The corresponding heatmap would be the label of the image. Then, we use the Otsu [5] to get the thresholding image. We also Use the open operation to remove the noise in the image and make the binarized area smoother. Finally, we use the cutting method to separate very close points and the centroid of each region is taken as the final result. All of these are shown in Fig. 1.

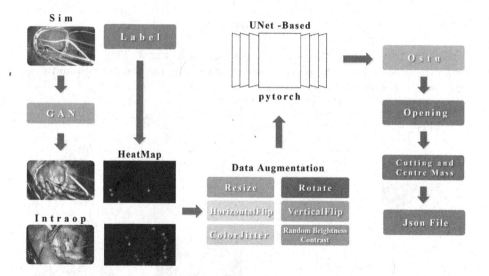

Fig. 1. Outline

2.3   Pre-processing

Image to Image GAN. In this task, our datasets come from two domains, one is the simulation domain and the other is the Intraop domain. The data set of the Intraop domain is smaller than the data set of the simulation domain. We decided to transform the simulation domain data into Intraop domain data to get a higher score on the Intraop domain. We introduced cycleGAN to solve this problem.

The cycleGAN has two mapping functions, as shown in Fig. 2, one is G and the other is F. G transforms the image of X domain into the image of Y domain, and F transforms the image of Y domain into the image of X domain. Two discriminators identify the real domain image and the generated image.

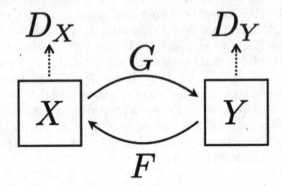

Fig. 2. CycleGAN

Applied to this task, the overall flow is shown in Fig. 3. This diagram only shows the process from the simulation domain to the Intraop domain and vice versa, which is not shown here.

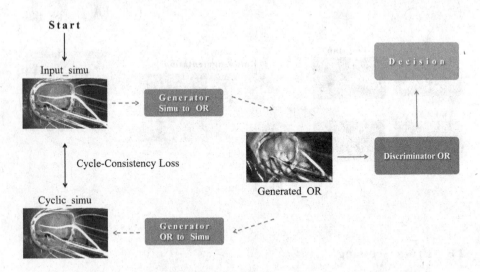

Fig. 3. CycleGAN in this task

Heatmap. Unlike the traditional landmark detection method, we do not generate a heatmap for each point but generate all the points onto the same heatmap.

Because in other tasks, the number of feature points is fixed, while in our task, the number of feature points varies with the image, ranging roughly from 0 to 15.

Each of our points is a 2D Gaussian kernel, and a variable number of points make up this heatmap, which will be used as the model's label. The heatmap is shown in Fig. 4.

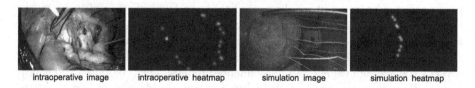

intraoperative image   intraoperative heatmap   simulation image   simulation heatmap

Fig. 4. Heatmap

Data Augmentation. During training the images are randomly augmented using Albumentations functions: horizontally and vertically with a probability of 50%, rotation of $\pm40°$, ColorJitter with a probability of 50%, RandomBrightnessContrast with a probability of 50%.

2.4   Point Detection

This work uses a U-Net-based architecture with a depth of 5. After each 3×3-convolution, batch normalization is applied. The first convolutional layer has 16 filter maps, while the bottleneck layer has 512 filter maps. We choose the Resnext [10] network as our encoder. We don't have an activation function after the final 1×1-convolutional layer while training, but we apply the sigmoid function when we predict the heatmap. The loss function is dice loss.

The input images are RGB images with 3 channels. One channel output is the heatmap. The heatmap becomes the real output point after a series of subsequent operations.

2.5   Post-processing

Otsu. The maximum between-class variance method is a nonparametric and unsupervised method of automatic threshold selection for picture segmentation. According to the gray characteristics of the image, the image is divided into background and objects. Among them, the greater the variance between the background and objects shows that the difference between the two parts of the image is also greater. This method calculates the relationship between the average gray level between background pixels and foreground pixels and their proportion in the whole image, so as to obtain the global threshold when the image segmentation effect is the best, and finally segment the image according to this value.

Opening. We discovered that the network's predicted images were connected together in blocks that should have been separated after Otsu. The open operation is used to separate them. This also smoothes the edges of the segmented blocks and removes some of the noise.

Centre Mass and Cutting. After the opening process, we identify the centroid of each segmented block in the output image, and these are regarded the final predicted points, but we discovered that the form of some of these blocks compared to the circle generated by the point in the heatmap image is somewhat irregular. Therefore, we assess whether a segmentation block should be clipped depending on whether the area of each segmented block in an output image exceeds the average value of all its segmented blocks. Then, based on the height and width of the segmented blocks' bounding box, decide the cutting direction. The cutting point is the centroid of the segmented blocks that need to be sliced. Cutting is done in the x-axis direction if the bounding box's height is higher than its width. If the bounding box's height is less than its width, the cutting is done using Cut in the y-axis direction. We recalculate the centroid of the partitioned block after cutting as the output points and save them in JSON files.

The example of post-processing is shown in Fig. 5.

Fig. 5. Example of post-processing. (a) input, (b) predict, (c) Otsu, (d) opening, (e) Centre mass and Cutting

2.6 Evaluation

A point detection is considered successful if the centres of mass of ground truth and prediction are less than 6 pixels apart. On an image of size 512×288, this radius roughly corresponds to the thickness of a suture when it enters the tissue. Every matched point from the produced mask is considered a true positive (TP). Predicted points that could not be matched to any ground truth point are defined as false positives (FP) and all ground truth points without a corresponding point in the produced mask are false negatives (FN). Precision and sensitivity are computed over all landmarks. And F1-score presents the harmonic mean of precision and sensitivity.

$$Precision = \frac{TP}{TP + FP} \tag{1}$$

$$Sensitivity = \frac{TP}{TP + FN} \tag{2}$$

$$F1 = \frac{2 * Precision * Sensitivity}{Precision + Sensitivity} \tag{3}$$

3   Results

There are some visual examples in Fig. 6(a) and Fig. 6(b).

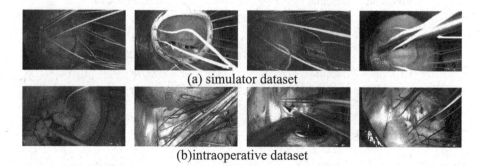

(a) simulator dataset

(b)intraoperative dataset

Fig. 6. Example results of two domain. The green circles are true positives (TP). The red circles show false positives (FP). The yellow circles represent false negatives (FN).

The result of the simulation domain is shown in Table 2, and the result of the intraop domain is shown in Table 3.

The baseline results come from this paper [9]. We can not calculate the standard deviation of the baseline F1 score since the baseline does not give experimental data for F1 score.

As shown in Tables 2 and 3, while our precision is lower than the baseline, our sensitivity is much higher. As a result, when comparing F1 scores, our method outperforms the baseline on both the simulation and intraop domains. Because the images in the intraop domain have more interference factors and less data, the recognition effect of the two methods in the intraop domain is slightly inferior to that of the simulation domain.

Table 2. Simu result.

Cross-validation result on Simu data							
Metric	Model	$f1$	$f2$	$f3$	$f4$	$f5$	$\mu \pm \sigma$
Precision	Baseline	–	–	–	–	–	81.50 ± 5.77
	Ours	84.37	54.79	76.84	74.18	77.89	73.62 ± 9.99
Sensitivity	Baseline	–	–	–	–	–	61.60 ± 6.11
	Ours	79.63	72.25	68.64	80.20	77.48	$\mathbf{75.64 \pm 4.48}$
F1 score	Baseline	–	–	–	–	–	69.78
	Ours	81.94	62.33	72.51	77.07	77.69	$\mathbf{74.31 \pm 6.69}$

Table 3. Intra result.

Cross-validation result on Intra data						
Metric	Model	$f1$	$f2$	$f3$	$f4$	$\mu \pm \sigma$
Precision	Baseline	–	–	–	–	66.68 ± 4.67
	Ours	62.24	67.35	54.92	66.54	62.76 ± 4.93
Sensitivity	Baseline	–	–	–	–	24.45 ± 5.06
	Ours	51.81	54.44	44.22	50.45	**50.23 ± 3.76**
F1 score	Baseline	–	–	–	–	35.78
	Ours	56.56	60.22	48.99	57.38	**55.79 ± 4.15**

4   Conclusions

We present a novel method for predicting multiple key points in endoscopic images in this paper. Our method differs from traditional key point detection methods, which have a fixed number of prediction key points. Our method can detect multiple key points at the same time, significantly reducing detection time and model calculation. We also introduce cycleGAN, which can interconvert images from two domains to create a larger dataset. Our results outperform the baseline as well as other related methods after many repeated and rigorous experiments.

References

1. Carpentier, A., Adams, D.H., Filsoufi, F.: Carpentier's Reconstructive Valve Surgery E-Book. Elsevier Health Sciences, Amsterdam (2011)
2. Engelhardt, S., De Simone, R., Full, P.M., Karck, M., Wolf, I.: Improving surgical training phantoms by hyperrealism: deep unpaired image-to-image translation from real surgeries. In: Bildverarbeitung für die Medizin 2019. I, pp. 282–282. Springer, Wiesbaden (2019). https://doi.org/10.1007/978-3-658-25326-4_62
3. Engelhardt, S., Sauerzapf, S., Brčić, A., Karck, M., Wolf, I., De Simone, R.: Replicated mitral valve models from real patients offer training opportunities for minimally invasive mitral valve repair. Interact. Cardiovasc. Thorac. Surg. **29**(1), 43–50 (2019)
4. Engelhardt, S., Sauerzapf, S., Preim, B., Karck, M., Wolf, I., De Simone, R.: Flexible and comprehensive patient-specific mitral valve silicone models with chordae tendineae made from 3d-printable molds. Int. J. Comput. Assist. Radiol. Surg. **14**(7), 1177–1186 (2019)
5. Otsu, N.: A threshold selection method from gray-level histograms. IEEE Trans. Syst. Man Cybern. **9**(1), 62–66 (1979)
6. Payer, C., Štern, D., Bischof, H., Urschler, M.: Integrating spatial configuration into heatmap regression based CNNs for landmark localization. Med. Image Anal. **54**, 207–219 (2019)
7. Ronneberger, O., Fischer, P., Brox, T.: U-Net: convolutional networks for biomedical image segmentation. In: Navab, N., Hornegger, J., Wells, W.M., Frangi, A.F. (eds.) MICCAI 2015. LNCS, vol. 9351, pp. 234–241. Springer, Cham (2015). https://doi.org/10.1007/978-3-319-24574-4_28

8. Sharan, L., et al.: Mutually improved endoscopic image synthesis and landmark detection in unpaired image-to-image translation. arXiv preprint arXiv:2107.06941 (2021)

9. Stern, A., et al.: Heatmap-based 2d landmark detection with a varying number of landmarks. Bildverarbeitung für die Medizin 2021. Informatik aktuell. Springer Vieweg, Wiesbaden (2021)

10. Xie, S., Girshick, R., Dollár, P., Tu, Z., He, K.: Aggregated residual transformations for deep neural networks. In: Proceedings of the IEEE Conference on Computer Vision and Pattern Recognition, pp. 1492–1500 (2017)

11. Zhu, J.Y., Park, T., Isola, P., Efros, A.A.: Unpaired image-to-image translation using cycle-consistent adversarial networks. In: Proceedings of the IEEE International Conference on Computer Vision, pp. 2223–2232 (2017)

Cross-Domain Landmarks Detection in Mitral Regurgitation

Jiacheng Wang, Haojie Wang, Ruochen Mu, and Liansheng Wang$^{(\boxtimes)}$

Department of Computer Science at School of Informatics, Xiamen University,
Xiamen, China
{jiachengw,haojiew,ruochenmu}@stu.xmu.edu.cn,
lswang@xmu.edu.cn

Abstract. Mitral regurgitation (MR) is a frequent indication for valve surgery. One of its treatments, mitral valve performed with endoscopic video recordings, is a complex minimally invasive procedure which is facing the problem of data availability and data privacy. Therefore, the simulation cases are widely used to form surgery training and planning. However, the cross-domain gap may affect the performance significantly as Deep Learning methods rely heavily on data. We propose to develop an algorithm to reduce the domain gap between simulation and intra-operative cases. The task is to learn the distance and location information of the points and predict a series of 2D landmarks' location, the coordinates of the landmark were both marked on real and simulate dataset by the AdaptOR Challenge organizer. Our work has merged the data from both domains by using a relation heatmap generation algorithm, which can generate a relation key point heatmap based on the distance measurement of landmarks and explicitly represent the geometric relation between landmarks. The MSE loss function is used to minimize the error between the ground-truth and predicted heatmaps. We test our methods on a challenge dataset, in which the model has achieved a good F1 score of 66.19%.

Keywords: Landmark detection · Mitral regurgitation · Deep learning

1 Introduction

Mitral regurgitation (MR) is the second most frequent indication for valve surgery in Europe and may occur for organic or functional causes [1]. Previous study has reported that the endoscopic video recordings are playing a more and more important role in one of treatments of MR, mitral valve repair by minimally invasive procedures. For example, mitral valve repair, which is aimed at re-storing the function of the mitral valve by inserting an annuloplasty ring, the blue and white mattress sutures anchor the ring. The points where the sutures enter or exit the tissue are our major research objectives. However, the endoscopic evaluations are facing the problem that is the lack of dataset, so the

S. Engelhardt et al. (Eds.): DGM4MICCAI 2021/DALI 2021, LNCS 13003, pp. 134–141, 2021.
https://doi.org/10.1007/978-3-030-88210-5_12

physical mitral valve simulator is widely used as a method of surgical training and simulation [4]. Both the simulation and intraoperative cases are marked by the 2D landmarks. The landmarks are referred for doctors to perform the mitral valve repair surgical operations. In this work, the challenge proposes to reduce the cross-domain gap between simulation and reality cases and use the data from both domains to predicted the coordinates of the points in a new validation frame.

The goal is to predict the landmarks coordinates by using these labels with the method of what is called relation heatmap generation algorithm. We convert both intraop-domain (intraoperative cases) and sim-domain (simulation cases) points images to the heatmap mask, and merge them as the training dataset, then use the backbone network, HRNet [8], to find the similar location or distribution features.

Our system is trained with 33872 annotated intraoperation landmarks and 23938 annotated simulation landmarks in total. An intro-domain dataset which have 524 images was our validation dataset, the evaluation results show that our approach can achieve an accurate and automatic estimate which is equivalent of the challenge's request. The work in this paper has been accomplished and will participate in the AdaptOR Challenge [3].

2   Method

We first introduce our heatmap generation algorithm to produce relation heatmap that best represents the correspondences of points where these sutures enter or exit the tissue before the ring is sewed. Then, the inference details are described in the end.

2.1   Generating Heatmap of Key Points for Training

Because our label is the coordinate of the point, we convert the points to the heatmap mask with the help of Gaussian-shaped kernel. The value of the center of the area is 1 and the values of nearby pixels gradually decay until it reaches its boundary (we set this hype-parameters which control the size of boundary named radius).

Most networks such as ResNet follow the design of LeNet-5. The rule is to: (i) gradually reduce the spatial size of the feature maps, (ii) connect the convolutions from high resolution to low resolution in series, and (iii) lead to a low-resolution representation, which is further processed for classification. However, high-resolution representations are needed for position-sensitive tasks such as semantic segmentation. HRNet [8] is able to maintain high-resolution representations through the whole process, so we choose HRNet to be our backbone in this task. Its overall architecture has been illustrated in Fig. 1, which starts from a high resolution convolution stream, then gradually add high-to-low resolution convolution streams one by one, and connect the multi-resolution streams in parallel at last.

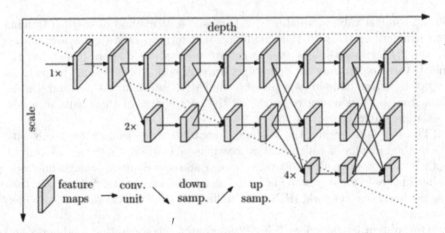

Fig. 1. The architecture of HRNet. The resulting network consists of several (4 in this paper) stages, and the n-th stage contains n streams corresponding to n resolutions. We conduct repeated multi-resolution fusions by exchanging the information across the parallel streams over and over.

What's more, We have also take some extensive attempts to make better use of the two types of data, sim-domain and intraop-domain.

(1) We simply join intraop-domain and sim-domain images into one dataset as training set and take them with their corresponding heatmaps as input to train our model, we take HRNet and Deeplabv3+ [2] as backbone to find the better performing model.

$$\mathcal{L}_{backbone} = \sum_{x_{intraop}} \mathcal{L}_{mse}^{intraop} + \lambda_{sim} \sum_{x_{sim}} \mathcal{L}_{mse}^{sim} \tag{1}$$

λ_{sim} is a hyper-parameters ranging from 0 to 1, and the first Sub-formula is MSE loss same as the second Sub-formula as follows, and $I_{X_{sim}}$ is the output of backbone:

$$\mathcal{L}_{mse}^{sim} = \mathcal{L}_{mse}^{sim}(I_{x_{sim}}, heatmap_{sim}) \tag{2}$$

Having datasets and its labels of two domain, we try to compare the following two methods to investigate whether these methods can make better use of the different datasets and get more robust representations.

(2) We combine two datasets same as first method and we added adversarial training approach [5]. It involves two net-works.One network generate the predictions for the input image, which could be from intraop-domain or sim-domain, while another network acts as a discriminator which takes the feature maps from the backbone network and tries to predict domain of the input. The backbone network tries to fool the discriminator, thus making the features from the two domains have a similar distribution. However, it didn't work well in this task.

The loss of discriminator is:

$$\mathcal{L}_D = \sum_{x_{intraop}} \mathcal{L}_D(I_{x_{intraop}}, 1) + \sum_{x_{sim}} \mathcal{L}_D(I_{x_{sim}}, 0) \tag{3}$$

We construct a fully-convolutional discriminator network D taking I_x as input and that produces domain classification outputs class label 1 (0) for the intraop-domain(sim-domain).

The loss of our backbone is:

$$\mathcal{L}_{backbone} = \sum_{x_{intraop}} \mathcal{L}_{mse}^{intraop} + \lambda_{sim} \sum_{x_{sim}} \mathcal{L}_{mse}^{sim} + \lambda_{adv} \sum_{x_{sim}} \mathcal{L}_D(I_{x_{sim}}, 1) \tag{4}$$

(3) We call this method as Collaborative training. Different from the first methods combining two domain datasets, We split datasets based their domain and use two datasets to train two different model which their backbone are identical. If the input is from the domain that the model's training set are from, this model can perform better than the other model. By using the average of Kullback-Leibler (KL) divergence loss [6], a model learn from one domain can teach the model trained on the other one, which allows that model to learn the knowledge from two domains which are different but share many similarities. The loss of two models is quite similar:

$$\mathcal{L}_{backbone} = \sum_{x_{intraop}} \mathcal{L}_{mse}^{intraop} + \lambda_{KL} \sum_{k} \mathcal{L}_{k->i}^{kl}(\mathcal{I}_{S_k}^{S_k} || \mathcal{I}_{S_i}^{S_k}), \tag{5}$$

$$\mathcal{L}_{k->i}^{kl}(\mathcal{I}_{S_k}^{S_k} || \mathcal{I}_{S_i}^{S_k}) = -\frac{1}{X_{S_k}} \sum \sigma(\mathcal{I}_{S_K}^{S_k}) log(\frac{\sigma(\mathcal{I}_{S_i}^{S_k})}{\sigma(I_{S_k}^{S_k})}), \tag{6}$$

where $\mathcal{X}$, $\mathcal{I}$, $\mathcal{Y}$ are the input image, outputs of networks and the corresponding heatmap respectively. subscript indicates that $\mathcal{I}$ is generated by M_{s_i} while the superscript indicates $\mathcal{I}$ is the feature computed for images from domain S_k, $\sigma()$ indicates the sigmoid function, $|X|$ represents the number of pixels in image X.

2.2   Inference Procedure

When predicting landmark heatmap I, we detect all responses whose value is greater or equal to its connected neighbors(the number of neighbors are controlled by the kernel size of the layer of max pooling). For each peak $\hat{p}_i$, we adopt the keypoint value I_{x_i, y_i} (i.e., prediction probability in the keypoint location ($\hat{x}_i, \hat{y}_i$)) as its detection confidence score, and keep the top 30 confident peaks. In order to prevent the confidence score of certain images are too low or too high, we set Upper bound(4) and lower bound(10), and the specific number of each images is controlled by the confidence score (score ≥ 0.1).

3   Experiments

3.1   Datasets

There is two domain datasets with resolution of 512×288 in this task. One is intraop-domain, 2708 mono frames from 10 simulations (192–374 frames each) with 33872 annotated landmarks, the other is sim-domain, 2376 mono frames from 4 patients (372–794 frames each) with 23938 annotated landmarks. The purpose of this task is getting a well-performing model in intraop-domain with the help of datasets in sim-domain.

Our test set is aicm13, the subset of datasets in intraop-domain, having 524 images. The training set consists of all datasets of sim-domin, having 2708 images. and remaining datasets of intraop-domin, having 1852 images.

3.2   Implementation Details

When we use the Gaussian kernel [7] to smooth each annotated boundary box around its center point to obtain the ground truth of landmark keypoint heatmap Y, the hyper-parameters sigma and radius will affect the generation of heatmap. But in this task, this two hyper-parameters has little affect and we set sigma = 5, radius = 24.

We will describe the specific details of the three methods next. We employ PyTorch deep learning framework in the implementations. All experiments are done on a single NVIDIA 2080TI GPU with 11 GB memory.

(1) Intuitively, sim-domain datasets share many similarities with intraop-domain datasets, we experiments whether combine this extra sim-domain datasets with intraop-domain datasets will help our model perform well.So our loss function is defined as follows:

$$\mathcal{L}_{seg} = \mathcal{L}_{seg}^{intraop} + \lambda_{sim} \cdot \mathcal{L}_{seg}^{sim} \tag{7}$$

when $\lambda_{sim} = 0$, this means we don't use sim-domain datasets, when $0 < \lambda_{sim} \leq 1$, this means we use sim-domain datasets. Our model is trained using Stochastic Gradient Descent optimizer with learning rate 5×10^{-5}, momentum 0.9 and weight decay 10^{-4}. And in the inferences procedure, we keep the top 30 confident peaks points, and select at least 10 points until its confidence scores is lower than threshold (Table 1).

Table 1. Different value of λ_{sim}.

value of λ_{sim}	f1 score
0	0.5230
0.5	0.5516
1	0.5632

As illustrated in the methods section, we try there methods for this task. But the other 2 methods didn't work well. And finally we submitted the model trained by method one.

(2) In the adversarial training approach, Our model, except the adversarial discriminator, is trained using Stochastic Gradient Descent optimizer with learning rate 1×10^{-5}, momentum 0.9 and weight decay 10^{-4}, We use Adam optimizer with learning rate 1×10^{-5} to train the discriminator. And λ_{adv} set value 1.0 which balance the loss of segmentation and loss of discriminator.

(3) In the collaborative training method, these four model are trained with different Stochastic Gradient Descent optimizer with same hyper-parameters that learning rate 1×10^{-5}, momentum 0.9 and weight decay 10^{-4}. And λ_{KL} set value 0.1 which balance the loss of segmentation and Kullback-Leibler (KL) divergence loss.

What's more, after analysing the general number of points, the first inference rules don't generate points well, so we change the rules of generating points: in order to prevent the confidence score of certain images are too low or too high, we set Upper bound(4) and lower bound(10) through experimental verification, and the specific number of each images is controlled by the confidence score(score ≥ 0.1).

3.3   Results

After using new inference rules, the f1 score improve from 0.5632 to 0.6619. However, The f1 score of methods2(adversarial training) is 0.5842 and f1 score of methods3(two model trained with KL loss) is 0.4481 as illustrated in Table 2. And finally we submitted the model trained by method one. We visualize the heatmap of images and the prediction of heatmap in Fig. 2.

Table 2. The results of three methods.

Method	f1 score	Sensitivity	Precision
Method 1	0.6619	0.6010	0.7367
Method 2	0.5842	0.6368	0.5397
Method3	0.4481	0.4884	0.4139

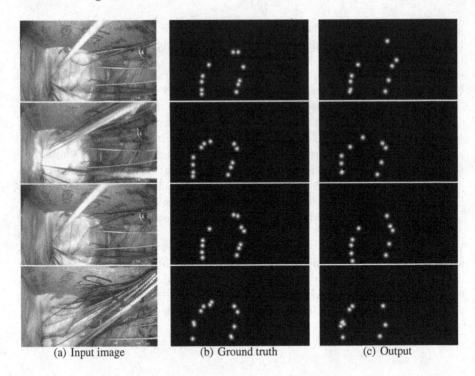

(a) Input image   (b) Ground truth   (c) Output

Fig. 2. Visualization of method 1.

4   Conclusion

In this work, we present an algorithm to achieve the cross-domain landmarks detection in both simulation and intra-operative cases. Specifically, we design a task to learn the distance and location information of the points and predict a series of 2D landmarks' location. We train the network on both domains through a relation heatmap generation algorithm, which can generate a relation key point heatmap based on the distance measurement of landmarks. We test this method on a public challenge dataset and achieve good detection.

References

1. Ancona, R., Pinto, S.C.: Mitral valve incompetence: epidemiology and causes. https://www.escardio.org/Journals/E-Journal-of-Cardiology-Practice/Volume-16/ Mitral-valve-incompetence-epidemiology-and-causes (2018)
2. Chen, L.C., Zhu, Y., Papandreou, G., Schroff, F., Adam, H.: Encoder-decoder with atrous separable convolution for semantic image segmentation. In: Proceedings of the European Conference on Computer Vision (ECCV), pp. 801–818 (2018)
3. Engelhardt, S., et al.: Deep Generative Model Challenge for Domain Adaptation in Surgery 2021 (March 2021). https://doi.org/10.5281/zenodo.4646979

4. Engelhardt, S., Sauerzapf, S., Brčić, A., Karck, M., Wolf, I., De Simone, R.: Replicated mitral valve models from real patients offer training opportunities for minimally invasive mitral valve repair. Interact. Cardiovasc. Thorac. Surg. **29**(1), 43–50 (2019)
5. Goodfellow, I., et al.: Generative adversarial nets. Adv. Neural Inf. Process. Syst. **27** (2014)
6. He, J., Jia, X., Chen, S., Liu, J.: Multi-source domain adaptation with collaborative learning for semantic segmentation. In: Proceedings of the IEEE/CVF Conference on Computer Vision and Pattern Recognition, pp. 11008–11017 (2021)
7. Law, H., Deng, J.: Cornernet: detecting objects as paired keypoints. In: Proceedings of the European Conference on Computer Vision (ECCV), pp. 734–750 (2018)
8. Sun, K., Xiao, B., Liu, D., Wang, J.: Deep high-resolution representation learning for human pose estimation. In: Proceedings of the IEEE/CVF Conference on Computer Vision and Pattern Recognition, pp. 5693–5703 (2019)

DALI 2021

Scalable Semi-supervised Landmark Localization for X-ray Images Using Few-Shot Deep Adaptive Graph

Xiao-Yun Zhou[1(✉)], Bolin Lai[2], Weijian Li[3], Yirui Wang[1], Kang Zheng[1], Fakai Wang[4], Chihung Lin[5], Le Lu[1], Lingyun Huang[2], Mei Han[1], Guotong Xie[2], Jing Xiao[2], Kuo Chang-Fu[5], Adam Harrison[1], and Shun Miao[1]

[1] PAII Inc., Bethesda, MD, USA
[2] Ping An Technology, Shenzhen, China
[3] University of Rochester, Rochester, NY, USA
[4] University of Maryland, College Park, MD, USA
[5] Chang Gung Memorial Hospital, Linkou, Taiwan, Republic of China

Abstract. Landmark localization plays an important role in medical image analysis. Learning based methods, including convolutional neural network (CNN) and graph convolutional network (GCN), have demonstrated the state-of-the-art performance. However, most of these methods are fully-supervised and heavily rely on manual labeling of a large training dataset. In this paper, based on a fully-supervised graph-based method, deep adaptive graph (DAG), we proposed a semi-supervised extension of it, termed few-shot DAG, *i.e.*, five-shot DAG. It first trains a DAG model on the labeled data and then fine-tunes the pre-trained model on the unlabeled data with a teacher-student semi-supervised learning (SSL) mechanism. In addition to the semi-supervised loss, we propose another loss using Jensen–Shannon (JS) divergence to regulate the consistency of the intermediate feature maps. We extensively evaluated our method on pelvis, hand and chest landmark detection tasks. Our experiment results demonstrate consistent and significant improvements over previous methods.

Keywords: Few-shot learning · GCN · Landmark localization · X-ray images · Deep adaptive graph · Few-shot DAG

1 Introduction

Landmark localization is a fundamental tool for a wide spectrum of medical image analysis applications, including image registration, developmental dysplasia diagnosis, and scoliosis assessment. Although many recent improvements

Electronic supplementary material The online version of this chapter (https://doi.org/10.1007/978-3-030-88210-5_13) contains supplementary material, which is available to authorized users.

S. Engelhardt et al. (Eds.): DGM4MICCAI 2021/DALI 2021, LNCS 13003, pp. 145–153, 2021.
https://doi.org/10.1007/978-3-030-88210-5_13

have been proposed [3,18], most of them are still based on fully-supervised learning and rely heavily on the manual labeling of a large amount of training data. Human labeling is prohibitively expensive and requires medical expertise. Thus, it is challenging to obtain large-scale labeled training data in practical applications and decreased scales of training data can impede achieving strong performance. Yet, large-scale unlabeled X-ray images can be efficiently collected from picture archiving and communication systems (PACSs). Hence, a promising strategy is to adopt semi-supervised learning (SSL) scheme, which enables efficient learning from both labeled and unlabeled data.

State-of-the-art landmark detection methods are typically learning-based, *i.e.*, graph convolutional network (GCN) [6,7], heatmap regression [12,16], and coordinate regression [8,15,19,20]. Among these, deep adaptive graph (DAG) [6] employs GCNs to exploit both the visual and structural information to localize landmarks. By incorporating a shape prior, DAG reaches a higher robustness compared to heatmap-based and coordinate regression-based methods [6]. While there are efforts towards SSL for landmark localization [2,13], they are based off of convolutional neural network (CNN)-only approaches, rather than the state-of-the-art GCN-based DAG. Some other prominent SSL successes in medical imaging have also been reported, such as for segmentation [1,10] and abnormality detection [17], but these are also CNN-based. Most other SSL methods are mainly developed for natural image classification tasks [4,5,14], which likewise do not address GCN-based landmark localization.

In this work, we propose few-shot DAG, an effective SSL approch for landmark detection. Few-shot DAG can achieve strong landmark localization performance with only a few training images (*e.g.*, five). The framework of few-shot DAG is illustrated in Fig. 1. We first train a fully-supervised DAG model on the labeled data and then fine-tune the pre-trained DAG model using SSL on the unlabeled data. Inspired by [14], for SSL, dual models are used, *i.e.*, a teacher and a student model. The output of the teacher model is used as the pseudo ground truth (GT) to supervise the training and back-propagation of the student model. The parameters of the teacher model are updated by the exponential moving average (EMA) of the parameters of the student model. In addition to the semi-supervised loss inspired by [14], we further add a Jensen–Shannon (JS) divergence loss on the intermediate feature map, to encourage similar feature distributions between the teacher and student models. The proposed few-shot DAG is validated on pelvis, hand and chest X-ray images with 10, 10, and 20 labeled samples and 5000, 3000, and 5000 unlabeled samples, showing consistent, notable and stable improvements compared with state-of-the-art fully-supervised methods [9] and other semi-supervised methods [4,11].

2   Method

Our work enhances prior efforts at using DAG for landmark detection [6]. We first briefly describe the network structure and training mechanism of DAG in Sect. 2.1 and then introduce the proposed SSL extension of DAG with the JS divergence loss in Sect. 2.2.

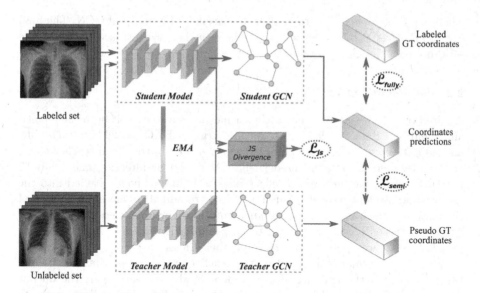

Fig. 1. An illustration of the proposed few-shot DAG framework with the teacher student SSL scheme and JS divergence loss.

2.1   Deep Adaptive Graph

DAG [6] formulates landmark localization as a graph evolution task, where the vertices of the graph represent the landmarks to be localized. The evolution starts from the mean shape generated from the training data, and the evolution policy is modeled by a CNN feature extractor followed by two GCNs. The CNN encodes the input image as a feature map, from which graph features are extracted via bi-linear interpolation at the vertex locations. The graph with features is further processed by cascading a global GCN and local GCNs to respectively estimate the affine transformation and vertex displacements toward the targets.

The DAG is trained via fully-supervised learning using a global GCN and local GCN loss. Specifically, the global GCN outputs an affine transformation that globally transforms the initial graph vertices (*i.e.*, the mean shape). The average L1 distances between the affine transformed vertices and the GT locations are calculated as the global loss:

$$\mathcal{L}_{global} = \left[\mathbb{E} \left(|\mathbf{v}_{global} - \mathbf{v}_{gt}| \right) - m \right]_{+}, \tag{1}$$

where $[x]_{+} := max(0, x)$, $\mathbf{v}_{global}$ and $\mathbf{v}_{gt}$ denote the affine transformed and GT vertices, respectively. m is a hyper-parameter specifying the margin of allowable error. The local GCN iteratively displaces $\mathbf{v}_{global}$ to refine their locations. The average L1 distances between the displaced vertices and the GT are calculated as the local loss:

$$\mathcal{L}_{local} = \mathbb{E} \left(|\mathbf{v}_{local} - \mathbf{v}_{gt}| \right), \tag{2}$$

where $\mathbf{v}_{local}$ denotes the vertices after displacement by the local GCN. The final loss for DAG is $\mathcal{L}_{global} + w_1 \times \mathcal{L}_{local}$, where w_1 is a weight used to adjust the ratio between the global and local loss.

2.2 Few-Shot DAG

Inspired by [14], we adopt a mean teacher mechanism to exploit both the labeled and unlabeled data. In particular, we first train a DAG model using only the labeled dataset, referred to as the *pre-trained model*. In the mean teacher training, the teacher and student models share the same architecture and are both initialized using the *pre-trained model*. The same input images are fed into the teacher and student models. Gaussian noises are added to the input images of the student model as an additional augmentation. For unlabeled images, a consistency loss is enforced between the teacher and student models. In the proposed few-shot DAG, we apply two forms of unlabeled loss.

First, the landmarks detected by the teacher model are used as the pseudo GT to supervise the training of the student model. In particular, the output $\mathbf{v}_{local}$ of the teacher model is used as the pseudo GT for the student model to calculate the global and local losses of Eq. (1) and (2), respectively. While this is helpful, it only applies a sparse consistency constraint on the GCN outputs. As a result, we apply a second loss in the form of JS divergence between the CNN feature maps of the teacher and student model, encouraging a similar distribution between the two. Specifically, the output feature maps of CNN are converted to pseudo-probabilities via a Softmax along the channel dimension. The JS divergence loss is then formulated as:

$$\mathcal{L}_{js} = \frac{1}{2|\Omega|} \sum_{x \in \Omega} (D(\mathbf{a}_S(x), \mathbf{m}(x)) + D(\mathbf{a}_T(x), \mathbf{m}(x)) \tag{3}$$

where $D(.)$ is the Kullback-Leibler divergence, $\mathbf{a}_S$ and $\mathbf{a}_T$ are the student and teacher activation maps, respectively, Ω is their batch, spatial and channel domain, and $\mathbf{m}$ is the mean of $\mathbf{a}_S$ and $\mathbf{a}_T$.

For the labeled data, we use the fully-supervised loss: $\mathcal{L}_{global} + w_1 \times \mathcal{L}_{local}$. For the unlabeled data, we calculate $\mathcal{L}'_{global} + w_1 \times \mathcal{L}'_{local} + w_2 * \mathcal{L}_{js}$, where $\mathcal{L}'_{global} + w_1 \times \mathcal{L}'_{local}$ use the pseudo GT produced by the teacher model and w_2 balances the contribution of the JS divergence loss. The labeled and unlabeled batches are fed with a ratio $1 : R$ (R is 100 in our experiments) to form the semi-supervised training iterations. Finally, the weights of the student model are updated through back-propagation of the loss. The weights of the teacher model are updated iteratively via the EMA of the student model's weights [14]:

$$\theta_T^t = \alpha \theta_T^{t-1} + (1 - \alpha)\theta_S^t, \tag{4}$$

where θ_T and θ_S are the weights of the teacher and student models, respectively, t is the training step, and α is a smoothing coefficient to control the pace of knowledge updates.

Table 1. The mean and std Euclidean error and the failure rate of the proposed method, with comparisons to Payer *et al.*, pseudo label, Π-model and temporal ensemble on the pelvis, hand and chest datasets. Best performance is in bold.

Data	Method	Mean error	Std error	Failure rate
Pelvis	Payer *et al.* [9]	46.29	106.63	12.62%
	Pseudo label [11]	20.50	34.27	2.21%
	Π-Model [4]	58.31	98.41	14.66%
	Temporal ensemble [4]	21.12	42.90	2.07%
	DAG [6]	25.89	44.60	4.29%
	Few-shot DAG	19.63	34.29	**1.27%**
	Few-shot DAG + JS	*18.45*	*30.69*	*1.31%*
Hand	Payer *et al.* [9]	12.29	37.81	1.24%
	Pseudo label [11]	9.27	24.82	0.77%
	Π-Model [4]	17.96	45.07	3.56%
	Temporal ensemble [4]	10.20	22.35	0.81%
	DAG [6]	10.97	27.60	1.51%
	Few-shot DAG	9.07	21.76	0.50%
	Few-shot DAG + JS	*9.07*	*19.67*	*0.47%*
Chest	Payer *et al.* [9]	61.41	131.27	5.75%
	Pseudo label [11]	55.33	57.84	8.32%
	Π-Model [4]	208.38	138.45	64.80%
	Temporal ensemble [4]	52.41	47.54	5.92%
	DAG [6]	58.99	73.55	12.35%
	Few-shot DAG	54.94	55.00	9.37%
	Few-shot DAG + JS	*43.46*	*47.22*	*5.28%*

3   Results

Experimental Setup. The proposed few-shot DAG is validated on three X-ray datasets: pelvis, hand, and chest. All datasets were collected from Chang Gung Memorial Hospital, Linkou, Taiwan, ROC, after de-identification of the patient information. In the pelvis, hand, and chest experiments, 6029, 93, and 1092 images are used for the testing while 10, 6, and 10 images are used for the validation. In the comparison experiment in Table 1, 10, 6, and 20 labeled images are used for the training for the pelvis, hand and chest dataset, respectively. In the ablation study experiment in Table 2, our method is trained using 1/5/50, 1/5/30 and 1/5/10/50 images for the pelvis, hand and chest dataset, respectively. The Euclidean distance between the GT and the predicted landmarks is used as the main evaluation metric. In addition, the failure rate (defined as error larger than 5% of the image width) is supplied as a supplementary evaluation metric. As the failure rate may change along the chosen threshold, we view the Euclidean error as more important and objective.

Table 2. The mean and std Euclidean error and the failure rate of DAG, few-shot DAG and few-shot DAG + JS on different numbers of training examples. Best performance is in bold. - indicates no convergence.

Data	Training samples	Method	Mean error	Std error	Failure rate
Pelvis	1	DAG [6]	–	–	–
	5	DAG [6]	55.53	80.05	14.72%
		Few-shot DAG	34.48	44.17	3.22%
		Few-shot DAG + JS	*27.31*	*39.14*	*3.19%*
	10	DAG [6]	25.89	44.60	4.29%
		Few-shot DAG	19.63	34.29	**1.27%**
		Few-shot DAG + JS	*18.45*	*30.69*	*1.31%*
	50	DAG [6]	15.62	34.40	1.29%
		Few-shot DAG	13.44	30.03	**0.55%**
		Few-shot DAG + JS	*13.37*	*28.26*	*0.58%*
Hand	1	DAG [6]	–	–	–
	5	DAG [6]	24.17	47.05	5.07%
		Few-shot DAG	23.30	36.13	2.52%
		Few-shot DAG + JS	*15.41*	*31.99*	*1.78%*
	10	DAG [6]	10.97	27.60	1.51%
		Few-shot DAG	9.07	21.76	0.50%
		Few-shot DAG + JS	*9.07*	*19.67*	*0.47%*
	30	DAG [6]	8.44	22.00	0.67%
		Few-shot DAG	8.09	**17.51**	**0.40%**
		Few-shot DAG + JS	*7.74*	*17.55*	*0.43%*
Chest	1 & 5	DAG [6]	–	–	–
	10	DAG [6]	133.09	121.57	38.49%
		Few-shot DAG	128.74	102.61	39.73%
		Few-shot DAG + JS	*76.11*	*78.01*	*14.80%*
	20	DAG [6]	58.99	73.55	12.35%
		Few-shot DAG	54.94	55.00	9.37%
		Few-shot DAG + JS	*43.46*	*47.22*	*5.28%*
	50	DAG [6]	27.31	42.33	2.69%
		Few-shot DAG	23.32	28.64	1.07%
		Few-shot DAG + JS	*22.49*	*27.29*	*0.85%*

Comparison with Other Baselines. We compare against the fully-supervised DAG [6] and also Payer *et al.* [9], who introduced a heatmap based method focusing on leveraging the spatial information. Three semi-supervised methods are used for the comparison: (1) pseudo label [11], which trains the SSL model with the pseudo GT generated by the *pre-trained model* in Sect. 2.1; (2) Π-Model [4], which maintains only a student model with the semi-supervised loss between the two input batches (one is with Gaussian noise and one is not); (3) temporal ensemble [4], which updates the pseudo GT after each epoch via the EMA of historically and currently generated pseudo GT. For the proposed method, performance in stages is offered, including few-shot DAG and few-shot

DAG + JS. It is worth mentioning that temporal ensemble consumes much longer training time than other methods.

Detailed results of validating the seven methods on the pelvis, hand and chest data set are presented in Table 1. We can see that for the main evaluation metric, *i.e.,* mean and std Euclidean error, out of the fully-supervised methods, DAG noticeably outperforms Payer *et al.* For the semi-supervised methods, few-shot DAG + JS outperforms Π-Model with large margins while also outperforms pseudo label and temporal ensemble with notable margins. Furthermore, the proposed method shows consistent performance gains from DAG to few-shot DAG, and to few-shot DAG + JS, demonstrating the value of the proposed SSL scheme and JS divergence loss. For the supplementary evaluation metric - failure rate, similar trends can be observed. Even though few-shot DAG + JS does not achieve the lowest failure rate in one experiment, its failure rate is very close to the lowest value, *i.e.,* 1.31% vs. 1.27% on the pelvis.

Ablation Study on the Scalability of Few-Shot DAG. We show that the proposed few-shot DAG can work well for different scales of training data. To illustrate this, we conduct experiments on varied numbers of labeled data. The corresponding results are shown in Table 2. We can see that the fully-supervised DAG cannot converge on extremely few training examples, *i.e.,* 1 for pelvis, 1 for hand, 1 and 5 for chest, resulting in non-converged few-shot DAG as well. On converged experiments with few training examples, for the main evaluation metric, the proposed SSL scheme (DAG vs. few-shot DAG) achieves notable performance improvements on all experiments. The proposed JS divergence loss (few-shot DAG vs. few-shot DAG + JS) performs best on most experiments, except one (hand-50) where the std error is comparable (17.55 vs. 17.51). For the supplementary evaluation metric, the proposed semi-supervised methods, including few-shot DAG and few-shot DAG + JS fail much less than the fully-supervised method. While in most experiments, few-shot DAG + JS fails less than few-shot DAG; only in three experiments (pelvis-10, pelvis -50, hand-50), comparable failure rates are achieved for semi-supervised DAG with or without the JS divergence loss.

4   Conclusion

In this paper, we introduced few-shot DAG, an SSL enhancement of DAG that significantly improves landmark localization. It first trains a DAG model on a few labeled training examples (*e.g.,* five), and then fine-tunes the trained model on a large number of unlabeled training examples using consistency losses tailored for CNN and GCN outputs. Overall, our approach achieves strong landmark localization performances with only a few training examples. As shown in the validation on three datasets, the proposed few-shot DAG consistently out-performs both previous fully-supervised and semi-supervised methods with notable margins, indicating its good performance, robustness and potentially wide application in the future.

References

1. Cui, W., et al.: Semi-supervised brain lesion segmentation with an adapted mean teacher model. In: Chung, A.C.S., Gee, J.C., Yushkevich, P.A., Bao, S. (eds.) IPMI 2019. LNCS, vol. 11492, pp. 554–565. Springer, Cham (2019). https://doi.org/10.1007/978-3-030-20351-1_43

2. Honari, S., Molchanov, P., Tyree, S., Vincent, P., Pal, C., Kautz, J.: Improving landmark localization with semi-supervised learning. In: CVPR, pp. 1546–1555 (2018)

3. Juneja, M., et al.: A review on cephalometric landmark detection techniques. Biomed. Signal Process. Control **66**, 102486 (2021)

4. Laine, S., Aila, T.: Temporal ensembling for semi-supervised learning. arXiv preprint arXiv:1610.02242 (2016)

5. Lee, D.H.: Pseudo-label: the simple and efficient semi-supervised learning method for deep neural networks. In: Workshop on challenges in representation learning, ICML, vol. 3 (2013)

6. Li, W., et al.: Structured landmark detection via topology-adapting deep graph learning. arXiv preprint arXiv:2004.08190 (2020)

7. Lu, Y., et al.: Learning to segment anatomical structures accurately from one exemplar. In: Martel, A.L., et al. (eds.) MICCAI 2020. LNCS, vol. 12261, pp. 678–688. Springer, Cham (2020). https://doi.org/10.1007/978-3-030-59710-8_66

8. Lv, J., Shao, X., Xing, J., Cheng, C., Zhou, X.: A deep regression architecture with two-stage re-initialization for high performance facial landmark detection. In: CVPR, pp. 3317–3326 (2017)

9. Payer, C., Štern, D., Bischof, H., Urschler, M.: Integrating spatial configuration into heatmap regression based cnns for landmark localization. Media **54**, 207–219 (2019)

10. Raju, A., et al.: Co-heterogeneous and adaptive segmentation from multi-source and multi-phase CT imaging data: a study on pathological liver and lesion segmentation. In: Vedaldi, A., Bischof, H., Brox, T., Frahm, J.-M. (eds.) ECCV 2020. LNCS, vol. 12368, pp. 448–465. Springer, Cham (2020). https://doi.org/10.1007/978-3-030-58592-1_27

11. Sohn, K., Zhang, Z., Li, C.L., Zhang, H., Lee, C.Y., Pfister, T.: A simple semi-supervised learning framework for object detection. arXiv preprint arXiv:2005.04757 (2020)

12. Sun, K., Xiao, B., Liu, D., Wang, J.: Deep high-resolution representation learning for human pose estimation. In: CVPR, pp. 5693–5703 (2019)

13. Tang, X., Guo, F., Shen, J., Du, T.: Facial landmark detection by semi-supervised deep learning. Neurocomputing **297**, 22–32 (2018)

14. Tarvainen, A., Valpola, H.: Mean teachers are better role models: weight-averaged consistency targets improve semi-supervised deep learning results. In: NeurIPS, pp. 1195–1204 (2017)

15. Trigeorgis, G., Snape, P., Nicolaou, M.A., Antonakos, E., Zafeiriou, S.: Mnemonic descent method: a recurrent process applied for end-to-end face alignment. In: CVPR, pp. 4177–4187 (2016)

16. Valle, R., Buenaposada, J.M., Valdes, A., Baumela, L.: A deeply-initialized coarse-to-fine ensemble of regression trees for face alignment. In: ECCV, pp. 585–601 (2018)

17. Wang, Y., et al.: Knowledge distillation with adaptive asymmetric label sharpening for semi-supervised fracture detection in chest x-rays. arXiv preprint arXiv:2012.15359 (2020)

18. Wu, Y., Ji, Q.: Facial landmark detection: a literature survey. IJCV **127**(2), 115–142 (2019)
19. Yu, X., Zhou, F., Chandraker, M.: Deep deformation network for object landmark localization. In: Leibe, B., Matas, J., Sebe, N., Welling, M. (eds.) ECCV 2016. LNCS, vol. 9909, pp. 52–70. Springer, Cham (2016). https://doi.org/10.1007/978-3-319-46454-1_4
20. Zhang, Z., Luo, P., Loy, C.C., Tang, X.: Learning deep representation for face alignment with auxiliary attributes. TPAMI **38**(5), 918–930 (2015)

Semi-supervised Surgical Tool Detection Based on Highly Confident Pseudo Labeling and Strong Augmentation Driven Consistency

Wenjing Jiang[1,2], Tong Xia[1], Zhiqiong Wang[2], and Fucang Jia[1,3(✉)]

[1] Shenzhen Institute of Advanced Technology, Chinese Academy of Sciences,
Shenzhen, China
`fc.jia@siat.ac.cn`

[2] College of Medicine and Biomedical Information Engineering, Northeastern
University, Shenyang, China

[3] Pazhou Lab, Guangzhou, China

Abstract. Surgical tool detection in computer-assisted intervention system aims to provide surgeons with specific supportive information. Existing supervised methods heavily rely on the volume of labeled data. However, manually annotating location of tools in surgical videos is quite time-consuming. To overcome this problem, we propose a semi-supervised pipeline for surgical tool detection, using strategies of highly confident pseudo labeling and strong augmentation driven consistency. To evaluate the proposed pipeline, we introduce a surgical tool detection dataset, Cataract Dataset for Tool Detection (CaDTD). Compared to the supervised baseline, our semi-supervised method improves mean average precision (mAP) by 4.3%. In addition, an ablative study was conducted to validate the effectiveness of the two strategies in our tool detection pipeline, and the results show the mAP improvement of 1.9% and 3.9%, respectively. The proposed dataset, CaDTD, is publicly available at https://github.com/evangel-jiang/CaDTD.

Keywords: Surgical video analysis · Tool detection · Semi-supervised learning · Pseudo labeling · Strong augmentation

1 Introduction

In recent years, the development of computer assisted intervention (CAI) system has improved surgeons' work efficiency and surgical safety in modern operating rooms, especially for endoscopic and microscopic surgeries [1]. The CAI system usually consists of multiple sub-components, such as tool presence detection, tool detection, tool segmentation, and workflow recognition, etc. [2]. The semantic recognition and understanding of surgical scenes can provide reliable supports for surgeons to make better intra-operative decisions [3] and perform effective surgical analysis post-operatively [4,5].

© Springer Nature Switzerland AG 2021
S. Engelhardt et al. (Eds.): DGM4MICCAI 2021/DALI 2021, LNCS 13003, pp. 154–162, 2021.
https://doi.org/10.1007/978-3-030-88210-5_14

It is challenging to achieve surgical tool detection for the lack of datasets with annotated tool boundaries. Annotating the location of tools present in surgical videos requires experts to delineate the bounding box of each instance frame by frame, which is a complex and time-consuming work. Only few surgical tool detection datasets [5–7] are publicly available, especially those based on in-vivo surgical videos [6]. Existing studies [6,8] on surgical tool detection are quite limited and mainly based on supervised learning. Jin et al. [6] utilized Faster R-CNN [9] to locate surgical tools. Zhang et al. [8] deployed a modulated anchoring network to achieve surgical tools detection. Limited amount of labeled data in current tool detection datasets makes it difficult for supervised learning to obtain promising results. On the contrary, numerous unlabeled videos are easy to access, which can be utilized by semi-supervised learning.

With the awareness of utilizing unlabeled data, semi-supervised learning has shown promising results on improving model performance in deep learning [10]. One semi-supervised method for surgical tool detection is pseudo labeling via a combination of object detector and tracker [11]. However, it is reliant upon manual post-correction to remove the unlabeled images with wrong pseudo labels. An effective way to generate high-quality pseudo labels without post-correction is self-training and augmentation driven consistency (STAC) regularization [12].

To overcome the challenges above and inspired by [12], we focus on improving the reliability of pseudo labels and the model robustness in semi-supervised surgical tool detection. The contributions of this paper are two-fold: 1) We introduce highly confident pseudo labeling and strong augmentation driven consistency in semi-supervised surgical tool detection without manual post-correction; 2) We generate CaDTD, a cataract surgical tool detection dataset with two optional tool setups, and thus provide a reference for subsequent research.

2 Methodology

2.1 Dataset

To relieve the shortage of datasets for surgical tool detection and to reduce annotation cost, CaDTD, a dataset annotated with the bounding box of surgical tools is proposed in this paper and will be released publicly. As mentioned before, public datasets for surgical tool detection are scanty. One representative dataset is m2cai16-tool-locations [6], which consists of endoscopic surgery videos. To provide a reference for subsequent research on microscopic surgery, we present a new tool detection dataset on cataract surgery. Cataract is one of the main causes for blindness and cataract surgery is one of the most commonly performed surgeries [13]. Therefore, introducing tool detection dataset for cataract surgery is of significant importance.

CaDTD consists of 50 cataract surgical videos of tool-tissue interactions derived from CATARACTS dataset [13]. All these surgeries were performed at Brest University Hospital, and most of them are performed by a renowned expert. All videos were obtained at 30 fps and have the resolution of 1920×1080.

Among the 50 videos, 25 videos are unlabeled and others are labeled with bounding box based on the semantic segmentation information in CaDIS [14]. In order to reduce the annotation workload, the annotated frames are down sampled to a resolution of 720×540 and the adjacent frames are at least 3 s apart.

Table 1. The class IDs and names of surgical tools in CaDTD.

ID	Class Name	ID	Class Name
1	Phacoemulsifier Handpiece*	7	Hydrodissection Cannula*
2	Lens Injector*	8	Viscoelastic Cannula
3	I/A Handpiece*	9	Capsulorhexis Cystotome*
4	Secondary Knife*	10	Bonn Forceps*
5	Micromanipulator	11	Secondary Knife*
6	Capsulorhexis Forceps	12	Primary Knife*

In CaDTD, we define two types of tool setups, setup 1 and setup 2, as shown in Table 1. Setup 1 focuses on the whole tool, including both the head and the handle of tools, while setup 2 focuses only on the head of tools. The classes marked with * indicate tools those were annotated differently on the two tool setups. It is noteworthy that, for the tools present only in few videos or even few frames, we consider them as noise and ignore them in CaDTD. Besides, in this work, we only concern about tools so that the anatomy is ignored as well.

2.2 Methods

The proposed semi-supervised pipeline for surgical tool detection, as shown in Fig. 1, is based on a teacher-student framework which consists of the following steps: a) Train a teacher model using labeled data; b) Generate pseudo labels for unlabeled data via a predicting process on the teacher model; c) Train a student model using both labeled and unlabeled data.

Specifically, the teacher model is trained to predict labels (bounding box of each tool) for unlabeled data, thus the student model trained on the unlabeled data can be supervised with those pseudo labels. In addition, a solution to generate highly confident pseudo labels and a strong augmentation driven consistency regularization strategy are applied in our framework to improve model robustness. The details are described in the following subsections.

Highly Confident Pseudo Labeling. The quality of pseudo labels has a significant impact on the training of student model. Therefore, the pseudo labels for tools (bounding boxes) are not straightforwardly generated from the teacher model in this paper. Different from the existing semi-supervised method for surgical tool detection [11], in our pipeline, manual post-correction for pseudo labels is

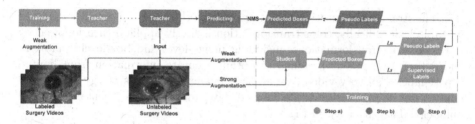

Fig. 1. The proposed semi-supervised pipeline for surgical tool detection. There are two models in the pipeline, a teacher and a student. The modules in orange, blue, and green stand for step a), b), and c), separately. (Color figure online)

optimized by an automatic filtering process with a confidence threshold τ. Since the detector predicts multiple results for each target, the filtering process utilizes a non-maximum suppression (NMS) to remove repetitive bounding boxes. The bounding box with a confidence score lower than τ is therefore dropped out. The threshold τ is a hyper-parameter set to a high value in our work to automatically select the most reliable pseudo labels. With reliable pseudo labels, the unlabeled data together with the labeled data can be utilized to train the student model with sufficient supervised information.

Table 2. The list of strong augmentation operations used in this paper.

Operation Name	Description
Color transformation	Change color of the image, including contrast, brightness, etc.
Cutout [15]	Cut out part of the image using square masks
Box-level geometric transformation	Apply rotation, shearing, or translation to a random region within the bounding box
Image-level geometric transformation	Apply rotation, shearing, or translation to the whole image

Strong Augmentation Driven Consistency. The idea of consistency regularization is to generate a robust model by constraining the prediction to be consistent when adding noise to the data or model. In the proposed pipeline, we introduce a series of data augmentation as noise signal on unlabeled images to perform consistency regularization. There are two types of augmentation strategies in our pipeline, weak augmentation and strong augmentation. Weak augmentation consists of resizing and horizontal flipping. Strong augmentation extends from the RandAugment [16], which is adapted for object detection according to

[17]. The strong augmentation operations used in the proposed pipeline are listed in Table 2. Specifically, color transformation is firstly applied, then followed by the geometric transformations, including image-level and box-level operation, and cutout [15]. In our pipeline, to obtain accurate information from limited expert-labeled surgery videos, we adjust the augmentation strategy in [12] and apply weak augmentation to all labeled data and strong augmentation for all unlabeled data as shown in Fig. 2.

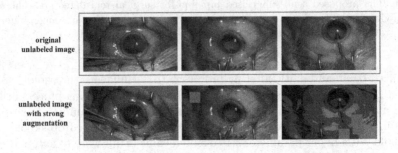

Fig. 2. Visualization of strong augmentation to the unlabeled video frames used in the proposed pipeline.

Loss Function. The loss function in the proposed pipeline consists of supervised loss $\mathcal{L}_s$ for labeled data and unsupervised loss $\mathcal{L}_u$ for unlabeled data. We implement tool detection using Faster R-CNN [9], so the two loss functions are formatted as Eq. 1 and Eq. 2:

$$\mathcal{L}_s = \frac{1}{N_{cls}} \sum_i \ell_{cls}(p_i, q_i) + \frac{\lambda}{N_{reg}} \sum_i q_i \ell_{reg}(s_i, t_i) \tag{1}$$

$$\mathcal{L}_u = \omega_i \left[\frac{1}{N_{cls}} \sum_i \ell_{cls}(p_i, q_i^*) + \frac{\lambda}{N_{reg}} \sum_i q_i \ell_{reg}(s_i, t_i^*) \right] \tag{2}$$

where i is the index of an anchor, p_i is the predicted probability of the anchor i being a surgical tool, and q_i is the matching binary label which equals 1 if i is indeed positive. s_i is a vector describing the location of the anchor i via four-dimensional coordinates, and t_i is the coordinates vector of the label corresponding to a positive anchor. The two terms of $\mathcal{L}_s$ are normalized by N_{cls} and N_{reg}, and then added together with a weight λ. q_i^* and t_i^* denotes the pseudo labels. ω_i is a binary parameter indicating whether the confidence score of the predicted bounding box is higher than the threshold τ.

When training the student model, the loss function is sum of the supervised loss and unsupervised loss weighted by the hyper-parameter λ_u as Eq. 3:

$$\mathcal{L}_{stu} = \mathcal{L}_s + \lambda_u \mathcal{L}_u \tag{3}$$

3   Experiments

We extensively validate the proposed semi-supervised pipeline on the CaDTD dataset. Three of the 25 labeled videos are used to evaluate the model performance. The backbone ResNet-50 in the Faster R-CNN model is implemented by Tensorpack [18] and initialized by weights pre-trained on ImageNet. During the student training process, we set $\tau = 0.9$ to rigorously filter the pseudo labels, and set $\lambda_u = 2$ as in [12].

3.1   Comparative Results

We evaluate the performance of our pipeline using mean average precision (mAP) of all tools, a commonly used metric of object detection. mAP_{50} indicates the mAP when the detection is considered correct if the intersection over union (IoU) between the predicted bounding box and its ground-truth reached 50%. Though most research on surgical tool detection take mAP_{50} as the evaluation metric, it is a saturated metric as mentioned in [12]. To be more convincing, $mAP_{50:95}$, which averages the values at a range of IoU between 50% to 95% is the major evaluation metric in this paper.

Table 3. The detection results for both tool setups on CaDTD (%).

	Tool Setup 1		Tool Setup 2	
	$mAP_{50:95}$	mAP_{50}	$mAP_{50:95}$	mAP_{50}
Supervised (Teacher)	67.6	88.3	69.4	91.0
Semi-supervised (Student)	71.9 (+4.3)	91.2 (+2.9)	72.1 (+2.7)	92.2 (+1.2)
p-value	0.0014	0.0510	0.0428	0.2994

Table 3 shows the comparison between the baseline supervised model and our semi-supervised model on both tool setups of CaDTD with the paired sample t-test. It can be seen that our semi-supervised model achieves an overall improvement on both tool setups, and the improvement of $mAP_{50:95}$ is significant ($p < 0.05$). The improvement of $mAP_{50:95}$ on tool setup 1 is 4.3%, which is greater than that of setup 2. Whereas, the general mAP of tool setup 2 is higher. Therefore, it can be inferred that when focusing on the whole body of tools, using unlabeled surgical videos to enlarge the training data via semi-supervised learning is more effective. In addition, Table 4 presents the detailed average precision of each surgical tool on setup 1. Except for class 1 (Phacoemulsifier Handpiece), our semi-supervised model shows significant improvement on all classes.

3.2   Ablation Study

We conduct ablation study on two strategies to show their effectiveness. We evaluate the surgical tool detectors with various strategies on the dataset of

Table 4. The average precision ($AP_{50:95}$) per class of tool setup 1 on CaDTD (%).

Class ID	1	2	3	4	5	6	7	8	9	10	11	12
Supervised (Teacher)	74.2	55.0	65.5	79.1	48.8	53.6	76.1	73.9	73.9	55.1	81.4	74.1
Semi-supervised (Student)	74.0	60.1	68.9	82.9	56.7	62.8	77.3	75.8	74.5	65.0	89.1	75.5

tool setup 1 in CaDTD, and the detection results are shown in Table 5. HC denotes highly confident pseudo labeling which filters pseudo labels with a high threshold ($\tau = 0.9$). SA denotes strong augmentation driven consistency. SSL, derived from semi-supervised learning, indicates a basic semi-supervised method with the teacher-student framework. To be more specific, in the SSL method, the threshold of confidence sore is set to a default value ($\tau = 0.5$), and both labeled and unlabeled videos are weakly augmented.

Table 5. The results of ablation experiments on tool setup 1 of CaDTD (%).

Model	$mAP_{50:95}$
Supervised	67.6
SSL	68.8 (+1.2)
SSL + HC	69.5 (+1.9)
SSL + SA	71.5 (+3.9)
SSL + HC + SA (Ours)	**71.9** (+4.3)

As shown in Table 5, the model performance improves when adding HC or SA alone, and reaches the best outcome when adding them both. This demonstrates that both key strategies in our method are effective. However, the SSL model is slightly greater (1.2%) than the supervised model, which further indicates that directly using the basic method of semi-supervised learning may not obtain much improvement since numerous unlabeled data with incorrect pseudo labels may bring noise to the network.

4   Conclusion

In this paper, we present a dataset CaDTD and propose a semi-supervised pipeline for surgical tool detection by introducing strategies of highly confident pseudo labeling and strong augmentation driven consistency. With the two strategies utilized, the proposed pipeline enables the tool detector to exploit unlabeled surgical videos without manual post-correction. We evaluate our method on CaDTD and show that our semi-supervised detector achieves a noteworthy overall improvement over the supervised detector. Future works include improving the detecting speed and further utilizing it in surgical workflow recognition.

Acknowledgments. This work was supported in part by the Guangdong Key Area Research and Development Program (2020B010165004), the Shenzhen Key Basic Science Program (JCYJ20180507182437217), the National Key Research and Development Program (2019YFC0118100 and 2017YFC0110903), the National Natural Science Foundation of China (12026602), and the Shenzhen Key Laboratory Program (ZDSYS201707271637577).

References

1. Cleary, K., Kinsella, A., Mun, S.K.: OR 2020 workshop report: operating room of the future. In: International Congress Series, vol. 1281, pp. 832–838. Elsevier (2005)
2. Padoy, N.: Machine and deep learning for workflow recognition during surgery. Minim. Invasive Ther. Allied Technol. **28**(2), 82–90 (2019)
3. Bouget, D., Allan, M., Stoyanov, D., et al.: Vision-based and marker-less surgical tool detection and tracking: a review of the literature. Med. Image Anal. **35**, 633–654 (2017)
4. Bhatia, B., Oates, T., Xiao, Y., et al.: Real-time identification of operating room-state from video. In: Proceedings of AAAI, vol. 2, pp. 1761–1766 (2007)
5. Sarikaya, D., Corso, J.J., Guru, K.A.: Detection and localization of robotic tools in robot-assisted surgery videos using deep neural networks for region proposal and detection. IEEE Trans. Med. Imaging **36**(7), 1542–1549 (2017)
6. Jin, A., Yeung, S., Jopling, J., et al.: Tool detection and operative skill assessment in surgical videos using region-based convolutional neural networks. In: Proceedings of WACV, pp. 691–699 (2018)
7. Kurmann, T., et al.: Simultaneous recognition and pose estimation of instruments in minimally invasive surgery. In: Descoteaux, M., Maier-Hein, L., Franz, A., Jannin, P., Collins, D.L., Duchesne, S. (eds.) MICCAI 2017. LNCS, vol. 10434, pp. 505–513. Springer, Cham (2017). https://doi.org/10.1007/978-3-319-66185-8_57
8. Zhang, B., Wang, S., Dong, L., et al.: Surgical tools detection based on modulated anchoring network in laparoscopic videos. IEEE Access **8**, 23748–23758 (2020)
9. Ren, S., He, K., Girshick, R., et al.: Faster R-CNN: towards real-time object detection with region proposal networks. IEEE Trans. Pattern Anal. Mach. Intell. **39**(6), 1137–1149 (2016)
10. van Engelen, J.E., Hoos, H.H.: A survey on semi-supervised learning. Mach. Learn. **109**(2), 373–440 (2019). https://doi.org/10.1007/s10994-019-05855-6
11. Yoon, J., Lee, J., Park, S.H., Hyung, W.J., Choi, M.-K.: Semi-supervised learning for instrument detection with a class imbalanced dataset. In: Cardoso, J., et al. (eds.) IMIMIC/MIL3ID/LABELS -2020. LNCS, vol. 12446, pp. 266–276. Springer, Cham (2020). https://doi.org/10.1007/978-3-030-61166-8_28
12. Sohn, K., Zhang, Z., Li, C. L., et al.: A simple semi-supervised learning framework for object detection. arXiv preprint. arXiv:2005.04757 (2020)
13. Al Hajj, H., Lamard, M., Conze, P.H., et al.: CATARACTS: challenge on automatic tool annotation for cataRACT surgery. Med. Image Anal. **52**, 24–41 (2019)
14. Grammatikopoulou, M., Flouty, E., Kadkhodamohammadi, A., et al.: CaDIS: cataract dataset for RGB-image segmentation. Med. Image Anal. **71**, 102053 (2021)
15. DeVries, T., Taylor, G. W.: Improved regularization of convolutional neural networks with cutout. arXiv preprint. arXiv:1708.04552 (2017)

16. Cubuk, E. D., Zoph, B., Shlens, J., et al.: Randaugment: practical automated data augmentation with a reduced search space. In: Proceedings of CVPR, pp. 702–703 (2020)
17. Zoph, B., Cubuk, E.D., Ghiasi, G., Lin, T.-Y., Shlens, J., Le, Q.V.: Learning data augmentation strategies for object detection. In: Vedaldi, A., Bischof, H., Brox, T., Frahm, J.-M. (eds.) ECCV 2020. LNCS, vol. 12372, pp. 566–583. Springer, Cham (2020). https://doi.org/10.1007/978-3-030-58583-9_34
18. Wu, Y., et al.: Tensorpack (2016). https://github.com/tensorpack

One-Shot Learning for Landmarks Detection

Zihao Wang[1]([✉]), Clair Vandersteen[2], Charles Raffaelli[2], Nicolas Guevara[2], François Patou[3], and Hervé Delingette[1]

[1] Université Côte d'Azur, Inria, Epione Team, Valbonne, France
zihao.wang@inria.fr
[2] Université Côte d'Azur, Nice University Hospital, Nice, France
[3] Oticon Medical, Vallauris, France

Abstract. Landmark detection in medical images is important for many clinical applications. Learning-based landmark detection is successful at solving some problems but it usually requires a large number of the annotated datasets for the training stage. In addition, traditional methods usually fail for the landmark detection of fine objects. In this paper, we tackle the issue of automatic landmark annotation in 3D volumetric images from a single example based on a one-shot learning method. It involves the iterative training of a shallow convolutional neural network combined with a 3D registration algorithm in order to perform automatic organ localization and landmark matching. We investigated both qualitatively and quantitatively the performance of the proposed approach on clinical temporal bone CT volumes. The results show that our one-shot learning scheme converges well and leads to a good accuracy of the landmark positions.

Keywords: One-shot learning · Landmarks detection · Deep learning

1 Introduction

Landmarks detection for target object localization plays a pivotal role in many imaging tasks. Automatic landmark detection can alleviate the challenges of image annotation by human experts and can also save time for many image processing tasks. The difficulty of landmark detection in clinical images may come from anatomical variability, or changes in body position which can lead to large differences of shape or appearance. The literature on automatic landmarks detection approaches can be roughly split into learning based versus non-learning based algorithms.

This work was partially funded by the French government, by the National Research Agency: ANR-15-IDEX-01, and by the grant AAP Sante 06 2017-260 DGADSH.

S. Engelhardt et al. (Eds.): DGM4MICCAI 2021/DALI 2021, LNCS 13003, pp. 163–172, 2021.
https://doi.org/10.1007/978-3-030-88210-5_15

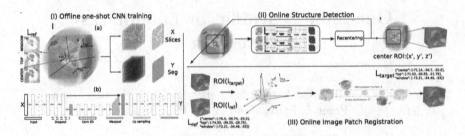

Fig. 1. Overview of the proposed framework.

Non-Learning Based Landmarks Detection. In [1] is proposed the augmentation of the scale-invariant feature transform (SIFT) to arbitrary n dimensions (n-SIFT) for 3D-MRI volumes. However, the computation cost for 3D SIFT features is heavy as their complexity is a cubic function of the image size. Wörz *et al.* [2] leverage parametric intensity models for image landmarks detection. Ricardo *et al.* [3] use log-Gabor filters to extract frequency features for 3D Phase Congruency (PC) applied to detect head and neck landmarks.

Learning Based Landmarks Detection. Probabilistic graphical models were used for bones landmark labelling in [4] and [5]. Potesil *et al.* [6] use joint spatial priors and parts based graphical models to improve the landmarks detection accuracy of organs. Shouhei *et al.* [7] proposed a Bayesian inference of landmarks through a parametric stochastic landmark detector of the candidates. Donner *et al.*[8] applied random forest and Markov Random Field (MRF) for vertebral body landmarks detection. Mothes *et al.* [9] proposed a one-shot SVM based landmarks tracking method for X-Ray image landmark detection. Suzani *et al.* [10] proposed to train a convolutional neural network (CNN) with an annotated dataset for automatic vertebrae detection and localization. Liang *et al.* [11] proposed a two-step based residual neural network for landmarks detection. Deep reinforcement learning for landmarks detection was investigated by Ghesu *et al.* [12] where landmarks localization is considered as a navigation problem.

The main drawback of the above deep learning based landmarks detection methods is that the creation of manually annotated dataset with 3D landmarks is time consuming and in practice very difficult to collect. To tackle this problem, Zhang *et al.* [13] proposed a deep learning based landmarks detection method that can be used a limited number of annotated medical images. Their framework consists of two CNNs: one for regressing the patches and the second to predict the landmark positions. Yet, this method like the rest of the learning-based methods are not suited when only one annotated image is available. Another source of difficulties is to detect landmarks that are concentrated on a small part of the image. A typical example is the detection of cochlear landmarks from CT images since the human cochlea is a tiny structure. In this paper, we tackle the problem of automatic determination of 3D landmarks in a volumetric image from a single example consisting of a reference image with its landmarks. We propose a one-shot learning approach that first localizes a Structure Of Interest (SOI) (e.g. the cochlea in a CT image of the inner ear) which lies next to the landmarks. A 2D CNN is trained offline by generating arbitrary oriented slices of a reference image with the binary mask of the SOI. Given a target image, the location of the SOI is iteratively estimated by applying the 2D CNN on 3 orthogonal sets of slices. After aligning the orientations of the two SOI on the target and reference images, a non-rigid registration algorithm is applied to propagate the landmarks to the target image. We apply this approach on 200 CT images of the temporal bone to locate 3 cochlear landmarks and show that the positioning error is within the intra-rater variability. To the best of our knowledge, this is the first one-shot learning method for landmark detection which makes it highly applicable for several clinical problems.

2   Method

2.1   Overview

The proposed approach is described in Fig. 1. The algorithm requires as input a reference image I_{ref} where a set of landmarks L_{ref} are positioned. In addition, we require that a binary mask of a visible anatomical or pathological structure $S_{ref} \subset I_{ref}$ including the landmarks $L_{ref} \in S_{ref}$ be provided. Given a target image I_{target}, landmarks L_{target} are estimated by applying an image registration algorithm between an image patch $P_{ref} \subset I_{ref}$ centered on the reference landmarks and an image patch $P_{target} \subset I_{target}$ extracted on the target image. The main challenge is to automatically extract the target image patch P_{target} such that it is roughly aligned in position and orientation with the reference image patch in order to ease the non-rigid image registration task. To extract the centered target image patch, we first train a 2D CNN to segment the mask S_{ref} on random slices of the reference image. This stage is performed offline and also requires an additional validation image I_{val} where the same visible structure S_{val} has been segmented. Given a target image, the localization stage extracts the target image patch P_{target} by iteratively applying the segmentation network to find the center of mass of the structure and by aligning its axis of inertia. The last stage applies a registration algorithm to estimate the position of landmarks L_{target}.

2.2   Offline One-Shot CNN Training

The objective is to train an algorithm that can roughly segment the structure of interest $S_{ref} \subset I_{ref}$. That structure must include the landmarks or must lie in the vicinity of the landmarks L_{ref}. It should also be present in all target images and must be easy to detect int the image with some visible borders. One issue of one-shot learning is the limited amount of training data that can easily lead to overfitting [14,15]. To this end, we chose to train a shallow 2D U-net f_ω segmentation network in order to segment the SOI that surrounds the landmarks. The training set consists of slices of the reference image I_{ref} along arbitrary rotations and translation offsets together with the associated binary masks created by slicing accordingly the reference segmentation S_{ref}. The 2D CNN is trained by minimizing the Binary Cross-Entropy (BCE) loss function. To limit the risk of overfitting, we use a validation set consisting of another volumetric image I_{val} and its segmentation S_{val}. The training is stopped when the segmentation performance of f_ω on the 3 orthogonal slices of I_{val} start to decrease. The details of the training procedure are provided in Algorithm 1. The CNN training can be performed offline and the 2D random image slices are

---

Algorithm 1: One-shot training of CNN

Inputs: image: $I_{\text{ref}}, I_{\text{val}}$, segmentation: $S_{\text{ref}}, S_{\text{ref}}$
Output: CNN parameters ω
Initialize: $f_\omega, \Delta T, \Delta R$;
while L_{val} *decreases* **do**

  $T \leftarrow (U(-1,1)\Delta_T)^3$; ;   // Uniform Random Translation
  $R \leftarrow (U(-1,1)\Delta_R)^3$;   // Uniform Random Rotation
  $I_{trans} \leftarrow \text{Resample}(I_{\text{ref}}, R, T)$;   // Transformed Image
  $S_{trans} \leftarrow \text{Resample}(S_{\text{ref}}, R, T)$;   // Transformed Segmentation
  for $i = 1;\ i < K;\ i++$ **do**
    $f_\omega \xleftarrow{\omega} I_{trans}[i] || S_{trans}[i]$;   // Train the CNN
  end
  $L_{val} \leftarrow loss(S_{\text{val}}, f_{cnn}(I_{\text{val}}))$;   // Validation loss

end

---

centered on the center of mass $\mathbf{C}_{\text{ref}}$ (for $T = 0$) of the segmented structure of interest S_{ref}. Furthermore, the 2D image size of the CNN input is chosen as to cope with the translation ΔT and rotation ΔR offsets such that random slices do not include any missing pixel values.

2.3   Online Structure Detection

Given an input image I_{target}, we seek to locate the structure of interest S_{target} with the proper translation and orientation offsets in order to ease the last image registration stage.

Translation offset estimation To determine the 3D translation offset between I_{target} and I_{ref}, we propose to align the centers of the mass corresponding to the structures of interest S_{target} and S_{ref}. We rely on the trained CNN $f_\omega()$ to determine S_{target} given I_{target}. However, with the limited training set of $f_\omega()$, we need to cope with its possible poor performance. To this end, we propose an iterative method described in Algorithm 2 and Fig. 2, where the estimation of the translation offset is progressively refined. We write as $f_\omega(I_{\text{target}}^x[k])[i,j]$ the output of the CNN applied on the slice k in the X direction of the volumetric image I_{target} which is a 2D probability map. We apply the CNN on the slices of

I_{target} extracted along the X,Y,Z directions. To improve the robustness of the center of mass estimation of I_{target}, we combine their output by multiplying the 3 probabilities outputs for each voxel. The joint output of the network at voxel $[i, j, k]$ is then written as :

$$p[i, j, k] = f_\omega(I^Z_{\text{target}}[k])[i, j] \cdot f_\omega(I^Y_{\text{target}}[j])[k, i] \cdot f_\omega(I^x_{\text{target}}[i])[j, k] \qquad (1)$$

The product of the 3 probability maps favors the pixels where the 3 outputs agree. This helps to filter out the false positive pixels produced by the network that are not correlated on the 3 slice orientations. The center of mass $\mathbf{C}_{\text{target}}$ is then simply estimated as the barycenter of the image voxels weighted by the joint probability $p[i, j, k]$:

$$\mathbf{C}_{\text{target}} = \frac{\left(\sum_{i,j,k} x[i,j,k] * p[i,j,k], \sum_{i,j,k} y[i,j,k] * p[i,j,k], \sum_{i,j,k} z[i,j,k] * p[i,j,k] \right)^T}{\sum_{i,j,k} p[i,j,k]} \qquad (2)$$

The target image is then cropped around the detected center $\mathbf{C}_{\text{target}}$ which is written as $\tilde{P}_{\text{target}}$. When the translation offset between the target and reference images is large, the CNN segmentation performances tend to degrade since it has been trained with slices roughly centered on the center of S_{ref}. This is why we propose to iteratively apply the same approach on I_{target} after being centered on $\mathbf{C}_{\text{target}}$. This way, we expect the centered image to be more and more accurately segmented by the neural network since it sees slices that resemble more and more to its training set. We stop the process when the changes in the detected center $\mathbf{C}_{\text{target}}$ become smaller than a threshold.

Algorithm 2: Iterative center of mass localization

Inputs: image: I_{target}, CNN: $f_\omega(\cdot)$
Output: Center of structure in target image $\mathbf{C}_{\text{target}}$
Initialize: ϵ;
$\mathbf{C}_{\text{target}} \leftarrow \mathbf{C}_{\text{ref}}$;
while $|\mathbf{C}_{\text{old}} - \mathbf{C}_{\text{target}}| < \epsilon$ **do**
$\quad \tilde{P}_{\text{target}} \leftarrow \text{Crop}(I_{\text{target}}, \mathbf{C}_{\text{target}})$;   // Patch centered on $\mathbf{C}_{\text{target}}$
$\quad$ **while** $o \in \{X, Y, Z\}$ **do**
$\quad\quad$ **for** $i = 1; i < K[o]; i + +$ **do**
$\quad\quad\quad out[o][i] \leftarrow f_{omega}(\tilde{P}^o_{\text{target}}[i])$;   // apply CNN on slices
$\quad\quad$ **end**
$\quad$ **end**
$\quad p \leftarrow out[X] \cdot out[Y] \cdot out[Z]$;   // Combine probability maps as Eq. 1
$\quad \mathbf{C}_{\text{old}} \leftarrow \mathbf{C}_{\text{target}}$;
$\quad \mathbf{C}_{\text{target}} \leftarrow$ Eq. 2;   // Update center of mass
end
$\tilde{P}_{\text{target}} \leftarrow \text{Crop}(I_{\text{target}}, \mathbf{C}_{\text{target}})$;   // Patch centered on $\mathbf{C}_{\text{target}}$

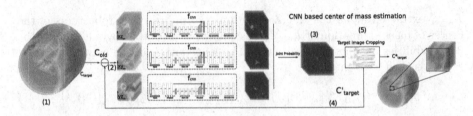

Fig. 2. Iterative determination of the center of mass of the structure of interest. Steps (1) - (2) show the 2D CNN segmentation of the structure of interest from the 3 set of orthogonal slices; (3) The probability maps of the 3 views are combined; (4) Update of the center of mass from the joint probability maps; (5) The target image is cropped around the center of mass.

Rotation offset estimation After having aligned the center of mass of the two structures of interest, the rotation offset is determined by aligning the moments of inertia of S_{ref} and S_{target}. More precisely, the matrix of inertia captures the ellipsoid appearance of each structure and it determines the structure orientation unambiguously if that structure does not have any axis of symmetry. Therefore the alignment of the matrices of inertia consists in applying a rotation to S_{target} such that the eigenvectors of the 2 matrices coincide [16,17] when they are sorted according to their eigenvalues. The moments of inertia of S_{target} are computed based on the combined probability $p[i,j,k]$ as computed in Eq. 1. Thus, after performing the eigenvalue decomposition of the 2 matrices, the rotation matrix centered on $\mathbf{C}_{\text{target}}$ is applied on the image patch $\tilde{P}_{\text{target}}$ to get the final target image patch P_{target}.

2.4   Online Image Patch Registration

After the two previous stages, the estimation of the landmarks L_{target} is achieved by performing a non-rigid registration of the reference image patch P_{ref} onto the target image patch P_{target}. The two image patches have the same size, are both centered on the structure of interest and their orientation roughly coincide. This is a good initialization for applying the standard diffeomorphic demons algorithm [18] as implemented in "itk::DiffeomorphicDemonsRegistrationFilter". This algorithm starts with a multi-resolution rigid registration followed by the non-rigid transformation parameterized by a stationary velocity field. It assumes that intensity distribution matches between the two images patches with a sum of square difference as similarity measure. The reference landmarks L_{ref} are then transported to the target image patch P_{target} through the estimated non-rigid transformation. Finally, the landmarks L_{target} on the original target image I_{target} are positioned by inverting the rigid transforms and cropping performed during the first two stages of the method.

3   Experiment

3.1   Dataset

The dataset consists of 200 volumetric CT images of the left temporal bones acquired by a GE LightSpeed CT scanner at the Nice University Center Hospital. The image dimensions are $(512, 512, 160)$ in 3D with corresponding spacing of $(0.25$ mm, 0.25 mm, 0.24 mm$)$. In this case, the structure of interest is the cochlea, a relatively small bone having a spiral shape similar to a snail shell and without any axis of symmetry. The cochlea is easily visible on CT images. Two volumetric images were randomly selected to serve as reference and validation images and their cochlea was then segmented by an expert in a semi-automatic fashion. Three landmarks corresponding the cochlea top, center and round window were manually set on the reference image as shown in Fig. 1.

3.2   Network Architecture and Training Details

We use a 2D U-net like network [19] for segmenting the cochlea in 2D images. The network structure is shown in Fig. 1 and is relatively shallow in order to minimize its complexity. The network input size is $[\cdot, 100, 100, 1]$ followed with 4 convolutional layers (shape: $[\cdot, 100, 100, 64]$). Feature maps are convoluted with a group of halved size layers but doubled in channels (shape: $[\cdot, 50, 50, 128]$). Up-sampling layer applied to recover the size of the feature maps to merged with the jump concatenates feature maps (shape:$[\cdot, 100, 100, 64 + 128]$). Finally, 5 convolutional layers (shape:$[\cdot, 100, 100, 64]$, $chn = 64$ for middle layers, $chn = 1$ for the last layer) are used for generating the final feature map. An Adam optimizer is used with a learning rate initialized to $lr = 0.1$ and decreasing with the number of epochs. The neural network was implemented with Tensorflow 2.0 framework and trained on one NVIDIA 1080 Ti GPU. The offline stage of the CNN takes less than 1h for training and the online stages takes around 30 s.

4   Results

The proposed approach was evaluated qualitatively and quantitatively. In Fig. 3(a), we show the position of the center of mass of the segmented cochlea C_{target} during three iterations of Algorithm 2. We see that the 3 points are getting closer to each other after each iteration thus demonstrating the convergence of the algorithm. In practice, we found that between 2 to 6 iterations are necessary to get a change of mass center position between two iterations less than 1 mm.

For a quantitative assessment of performance, an expert positioned twice the 3 landmarks on 20 additional volumes in order to estimate the positioning error and the intra-rater variability. In addition, we also try to employ a naive registration-based landmarks detection method without the iterative localization. The setup of the naive method shares the same registration conditions as the registration steps in the proposed framework.

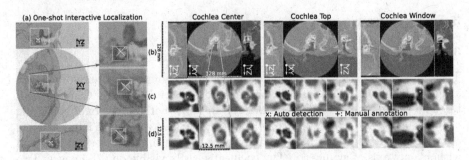

Fig. 3. (a) Positions of the center of mass of the cochlea during 3 iterations of the translation offset determination. The 3 cross marks in red, white, green correspond to the 1st, 2nd, 3rd iterations; Row (b) shows the result of the landmarks detection in the whole image I_{target}; Row (c) zooms on the detected landmarks before applying the last registration stage; Row (d) zooms on the generated landmarks ('x' marks) after the registration stage and the manually positioned landmarks ('+' marks) by an expert. (Color figure online)

In Fig. 3(d) we show the 3 landmarks generated by our algorithm with the same landmarks positioned by the expert. Clearly those points are very close to each other on the 3 views. In Table 1(top), we list the average position error of the 3 landmarks on the 20 images with respect to one set of landmarks manually positioned by an expert, and in the bottom rows, we show the corresponding results obtained by the naive registration based method.

In average, the position error of L_{target} is around 0.6mm which corresponds to a difference of position of 2 to 3 voxels. This result is satisfactory when considering the small size of the cochlea (width: 6.53 ± 0.35 mm, height: 3.26 ± 0.24 mm [20]) within the full CT volume (128 mm × 128 mm × 55 mm). *In contrast, the naive method is almost unusable for cochlea landmarks detection as the relative error (on average 9.48 mm) is too large in comparison with the size of the cochlea.* For a better assessment, we also provide the intra-expert landmark position error in Table 1(middle). It shows that the algorithm error is similar to the intra-expert variability, with a lower error for two (the center and window landmarks) out of the three landmarks. When computing the landmark position error with the second set of landmarks made by the expert, or with the average of the 2 annotations, we also found that the algorithm was performing similarly to the expert. Since the intra-rater variability is in most cases lower than inter-rater variability, we believe that the proposed method is an effective way to automate landmark positioning around the cochlea on CT images. Note that the mean landmark position errors reported by Zhang *et al.* [13] also correspond between 2.5 to 3 times the voxel size whereas Grewal *et al.* [21] after training on 168 scans reports errors between 2 to 9 times the voxel size (2–9 mm).

Table 1. Position errors of the 3 cochlear landmarks (centre, top and window) automatically generated landmarks (AUTO), a second set of manual (MANU) ones, and automatically generated landmarks of registration based naive method (REG).

Image ID	0	1	2	3	4	5	6	7	8	9	10	11	12	13	14	15	16	17	18	19	μ/σ
CEN AUTO	0.88	0.28	0.49	0.7	0.72	0.57	0.19	0.49	0.49	0.39	0.65	0.84	0.87	0.67	0.72	0.72	0.73	0.54	0.33	0.39	0.58±0.19 mm
TOP AUTO	0.7	1.33	0.56	0.73	0.72	0.37	0.31	0.78	0.35	0.2	1.63	1.15	1	0.26	1.04	0.67	0.8	1.23	0.55	0.39	0.73±0.39 mm
WIN AUTO	0.86	0.65	0.84	0.55	0.65	1.12	1.35	0.31	0.6	0.49	0.26	1.06	1.54	0.72	0.88	0.81	0.54	0.34	1.43	0.88	0.79±0.3 mm
CEN MANU	0.28	0.56	0.53	1.06	0.65	0.59	0.45	0.57	0.25	1.09	0.94	0.84	1.09	0.53	0.37	0.5	0.25	0.54	0.59	0.3	0.60±0.27 mm
TOP MANU	0.43	0.38	0.49	0.25	0.31	0.25	0.31	0.19	0.24	1.09	0	0.5	0.75	0.25	0.31	0.19	0.6	0.42	0.33	0.66	0.40±0.24 mm
WIN MANU	0.69	0.62	1.11	1.1	0.31	1.07	0.31	0.77	0.43	0.57	0.79	1.22	0.91	0.77	0.97	0.75	0.9	1.01	1.18	1.25	0.84±0.29 mm
CEN REG	4.42	10.95	15.78	16.49	12.83	13.04	14.28	15.09	9.66	16.21	6.82	12.91	11.06	6.96	22.69	6.16	2.22	11.68	17.79	9.74	11.84±4.97 mm
TOP REG	1.11	8.85	13.73	14.12	10.47	11.25	12.69	12.90	7.55	14.29	4.30	9.56	7.28	4.84	20.27	3.77	0.25	12.88	15.33	13.65	9.95±5.18 mm
WIN REG	2.21	2.82	8.62	9.73	4.15	5.67	5.37	7.44	2.51	7.89	1.09	6.42	3.46	1.45	15.33	1.77	4.91	16.20	9.40	16.77	6.66±4.8 mm

5 Conclusion

To the best of our knowledge, the proposed method is the first one-shot learning approach for 3D landmarks detection in volumetric images. We showed that the proposed approach was effective in localizing 3D landmarks in the cochlea from CT images of the inner ear. It relies on a segmentation stage and the registration of a single user-defined image patch which makes it easy explainable and interpretable. The approach is generic and could be applied to the detection of landmarks in CT imaging and other imaging modalities. In the future, we plan to use more complex image similarity measures in the final registration algorithm and to introduce more annotated data (few-shot learning) to address challenging landmark detection problems. Other network architectures proposed in the literature for one-shot deep learning such as [22–25] can be explored.

References

1. Cheung, W., et al.: N-sift: n-dimensional scale invariant feature transform. IEEE Trans. Image Process. **18**(9), 2012–2021 (2009)
2. Wörz, S., et al.: Localization of anatomical point landmarks in 3D medical images by fitting 3D parametric intensity models. Media **10**(1), 41–58 (2006)
3. Ferrari, R.J., Allaire, S., Hope, A., Kim, J., Jaffray, D., Pekar, V.: Detection of point landmarks in 3D medical images via phase congruency model. J. Braz. Comput. Soc. **17**(2), 117–132 (2011). https://doi.org/10.1007/s13173-011-0032-8
4. Schmidt, S., et al.: Spine detection and labeling using a parts-based graphical model. In: IPMI, pp. 122–133 (2007)
5. Corso, J., et al.: Lumbar disc localization and labeling with a probabilistic model on both pixel and object features. In: MICCAI, pp. 202–210 (2008)
6. Potesil, V., et al.: Personalization of pictorial structures for anatomical landmark localization. In: IPMI, pp. 333–345 (2011)
7. Shouhei, H., et al.: Automatic detection of over 100 anatomical landmarks in medical CT images. Media **35**, 192–214 (2017)
8. Donner, R., et al.: Global localization of 3D anatomical structures by prefiltered hough forests and discrete optimization. Media **17**, 1304–1314 (2013)
9. Mothes, O., et al.: One-shot learned priors in augmented active appearance models for anatomical landmark tracking. In: CVICG, pp. 85–104 (2019)

10. Suzani, A., et al.: Fast automatic vertebrae detection and localization in patholog-ical CT scans. In: MICCAI, vol. 9351 (2015)
11. Liang, X., et al.: A deep learning framework for prostate localization in cone beam CT-guided radiotherapy. Med. Phys. **47**(9), 4233–4240 (2020)
12. Ghesu, F., et al.: Multi-scale deep reinforcement learning for real-time 3D-landmark detection in CT scans. IEEE TPAMI **41**(1), 176–189 (2019)
13. Zhang, J., et al.: Detecting anatomical landmarks from limited medical imaging data using t2dl. IEEE TIP **26**(10), 4753–4764 (2017)
14. Wu, D., et al.: One shot learning gesture recognition from RGBD images. In: 2012 IEEE CVPR Workshops, pp. 7–12 (2012)
15. Oriol, V., et al.: Matching networks for one shot learning. In: NIPS, pp. 3630–3638 (2016)
16. Jaklic, A., et al.: Moments of superellipsoids and their application to range image registration. IEEE Trans. Cybern. **33**(4), 648–657 (2003)
17. Crisco, J.J., et al.: Efficient calculation of mass moments of inertia for segmented homogenous 3D objects. J. Biomech. **31**(1), 97–101 (1997)
18. Vercauteren, T., Pennec, X., Perchant, A., Ayache, N.: Non-parametric diffeomor-phic image registration with the demons algorithm. In: Ayache, N., Ourselin, S., Maeder, A. (eds.) MICCAI 2007. LNCS, vol. 4792, pp. 319–326. Springer, Heidel-berg (2007). https://doi.org/10.1007/978-3-540-75759-7_39
19. Ronneberger, O., Fischer, P., Brox, T.: U-Net: convolutional networks for biomed-ical image segmentation. In: Navab, N., Hornegger, J., Wells, W.M., Frangi, A.F. (eds.) MICCAI 2015. LNCS, vol. 9351, pp. 234–241. Springer, Cham (2015). https://doi.org/10.1007/978-3-319-24574-4_28
20. Devira, Z., et al.: Variations in cochlear size of cochlear implant candidates. Int. Arch. Otorhinolaryngol. **23**, 184–190 (2019)
21. Grewal, M., et al.: An end-to-end deep learning approach for landmark detection and matching in medical images. PBOI **11313**, 1131–1328 (2020)
22. Gregory, K., et al.: Siamese neural networks for one-shot image recognition. In: ICML Deep Learning Workshop (2015)
23. Amirreza, S., et al.: One-shot learning for semantic segmentation (2017)
24. Chen, Z., et al.: Image deformation meta-networks for one-shot learning. In: IEEE CVPR, June 2019
25. Shruti, J., et al.: Improving siamese networks for one shot learning using kernel based activation functions. ArXiv, abs/1910.09798, 2019

Compound Figure Separation of Biomedical Images with Side Loss

Tianyuan Yao[1], Chang Qu[1], Quan Liu[1], Ruining Deng[1], Yuanhan Tian[1], Jiachen Xu[1], Aadarsh Jha[1], Shunxing Bao[1], Mengyang Zhao[3], Agnes B. Fogo[2], Bennett A. Landman[1], Catie Chang[1], Haichun Yang[2], and Yuankai Huo[1(✉)]

[1] Vanderbilt University, Nashville, TN 37215, USA
yuankai.huo@vanderbilt.edu
[2] Vanderbilt University Medical Center, Nashville, TN 37215, USA
[3] Tufts University, Medford, MA 02155, USA

Abstract. Unsupervised learning algorithms (e.g., self-supervised learning, auto-encoder, contrastive learning) allow deep learning models to learn effective image representations from large-scale unlabeled data. In medical image analysis, even unannotated data can be difficult to obtain for individual labs. Fortunately, national-level efforts have been made to provide efficient access to obtain biomedical image data from previous scientific publications. For instance, NIH has launched the Open-i® search engine that provides a large-scale image database with free access. However, the images in scientific publications consist of a considerable amount of compound figures with subplots. To extract and curate individual subplots, many different compound figure separation approaches have been developed, especially with the recent advances in deep learning. However, previous approaches typically required resource extensive bounding box annotation to train detection models. In this paper, we propose a simple compound figure separation (SimCFS) framework that uses weak classification annotations from individual images. Our technical contribution is three-fold: (1) we introduce a new side loss that is designed for compound figure separation; (2) we introduce an intra-class image augmentation method to simulate hard cases; (3) the proposed framework enables an efficient deployment to new classes of images, without requiring resource extensive bounding box annotations. From the results, the SimCFS achieved a new state-of-the-art performance on the ImageCLEF 2016 Compound Figure Separation Database. The source code of SimCFS is made publicly available at https://github.com/hrlblab/ImageSeperation.

Keywords: Compound figures · Separation · Biomedical data · Deep learning

1 Introduction

Unsupervised learning algorithms [4,23] allow deep learning models to learn effective image representations from large-scale unlabeled data, such as self-supervised learning, auto-encoder, and contrastive learning [5]. However, even

© Springer Nature Switzerland AG 2021
S. Engelhardt et al. (Eds.): DGM4MICCAI 2021/DALI 2021, LNCS 13003, pp. 173–183, 2021.
https://doi.org/10.1007/978-3-030-88210-5_16

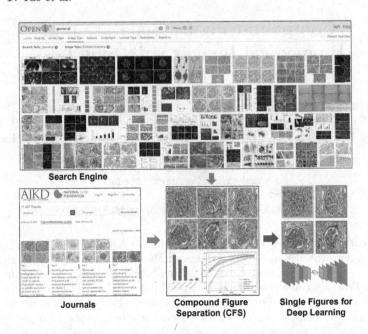

Fig. 1. Value of compound figure separation. This figure shows the hurdle (red arrow) of training self-supervised machine learning algorithms directly using large-scale biomedical image data from biomedical image databases (e.g., NIH OpenI) and academic journals (e.g., AJKD). When searching desired tissues (e.g., search "glomeruli"), a large amount of data are compound figures. Such data would advance medical image research via recent unsupervised learning algorithms, such as self-supervised learning, contrasting learning, and auto encoder networks [13] (Color figure online)

large-scale unannotated glomerular images can be difficult to obtain for individual labs [26]. Fortunately, many resources (e.g., NIH Open-i® [7] search engine, academic images released by journals) have provided the opportunity to obtain extra large-scale images. However, the images from such resources consist of a considerably large amount of compound figures with subplots (Fig. 1). To extract and curate individual subplots, compound figure separation algorithms can be applied [19].

Various compound figure separation approaches have been developed [1,6,12, 14,18,24,25], especially with recent advances in deep learning. However, previous approaches typically required resource extensive bounding box annotation to train detection models. In this paper, we propose a simple compound figure separation (SimCFS) framework that utilizes weak classification annotations from individual images for compound figure separation. Briefly, the contribution of this study are in three-fold:

- We propose a new side loss, an optimized detection loss for figure separation.
- We introduce an intra-class image augmentation method to simulate hard cases.

- The proposed framework enables an efficient deployment to new classes of images, without requiring resource extensive bounding box annotations.

We apply our technique to conduct compound figure separation for renal pathology. Glomerular phenotyping [17] is a fundamental task for efficient diagnosis and quantitative evaluations in renal pathology. Recently, deep learning techniques have played increasingly important roles in renal pathology to reduce the clinical working load of pathologists and enable large-scale population based research [3,8–10,16]. Due to the lack of publicly available annotated dataset for renal pathology, the related deep learning approaches are still limited on a small-scale [13]. Therefore, it is appealing to extract large-scale glomerular images from public databases (e.g., NIH Open-i® search engine) for downstream unsupervised or semi-supervised learning [13].

2   Related Work

In biomedical articles, about 40–60% of figures are multi-panel [15]. Several works have been proposed in the document analysis community, extracting figure and semantic information. For example, Huang et al. [12] presented their recognition results of textual and graphical information in literary figures. Davila et al. [6] presented a survey of approaches of several data mining pipelines for future research.

In order to collect scientific data massively and automatically, various approaches have been proposed by different researchers [19–21]. For example, Lee et al. (2015) [18] proposed an SVM-based binary classifier to distinguish complete charts from visual markers like labels, legend, and ticks. Apostolova et al. [1] proposed a figure separation method by capital index. These traditional computer vision approaches are commonly based on the figure's grid-based layout or visual information. Thus, the separation was usually accomplished by an x-y cut. However there are more complicated cases in compound figures like no white-space gaps or overlapped situations.

In the past few years, recent deep learning based algorithms using convolutional neural networks(CNNs) provided considerably better performance in extracting and processing textual and non-textual content from scholarly articles. Tsutsui and Crandall (2017) [25] proposed the first deep learning based approach to compound figure separation in which they applied a deep convolutional network to train the separator. They also implemented training on artificially-constructed datasets and reported superior performances on ImageCLEF data sets [11]. Shi et al. [24] developed a multi-branch output CNN to predict the irregular panel layouts and provided augmented data to drive learning; their network can predict compound figures of different sizes of structures with a better accuracy. Jiang et al. [14] combined the traditional vision method and high performance of deep learning networks by firstly detecting the sub-figure label and then optimizing the feature selection process in the sub-figure detection part. This improved the detection precision by 9%. In Tsutsui's study [25], they applied You Only Look Once (YOLO) Version 2 [22], a CNN based detection network. Deep learning based detection approaches utilized a single convolutional network to predict bounding boxes and class probabilities from full images simultaneously, which can achieve high speed detection and are in favor of sub-figure detection tasks.

3   Methods

The overall framework of the SimCFS approach is presented in Fig. 2. The training stage of SimCFS contains two major steps: (1) compound figure simulation, and (2) sub-figure detection. In the testing stage, only the detection network is needed.

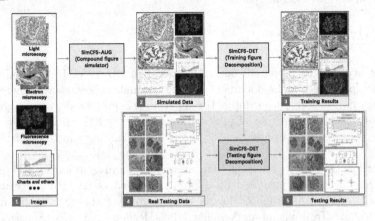

Fig. 2. The overall workflow of the proposed simple compound figure separation (SimCFS) workflow. In the training stage, SimCFS only requires single images from different categories. The pseudo compound figures are generated from the proposed augmentation simulator (SimCFS-AUG). Then, a detection network (SimCFS-DET) is trained to perform compound figure separation. In the testing stage (the gray panel), only the trained SimCFS-DET is required for separating the images.

3.1   Anchor Based Detection

YOLOv5, the latest version in the YOLO family [2], is employed as the backbone network for sub-figure detection. The rationale for choosing YOLOv5 is that the sub-figures in compound figures are typically located in horizontal or vertical orders. Herein, the grid-based design with anchor boxes is well adaptable to our application. A new side loss is introduced to the detection network that further optimizes the performance of compound figure separation.

3.2   Compound Figure Simulation

Our goal is to only utilize single images, which are non-compound images with weak classification labels in training a compound image separation method. In previous studies, the same task typically requires stronger bounding box annotations of subplots using real compound figures. In compound figure separation tasks, a unique advantage is that the sub-figures are not overlapped. Moreover, their spatial distributions are more ordered compared with natural images in object detection. Therefore, we propose to directly simulate compound figures from individual images as the training data for the downstream sub-figure detection.

Tsutsui et al. [25] proposed a compound figure synthesis approach (Fig. 3). The method first randomly samples a number of rows and random height for each row. Then a random number of single figures fills the empty template. However, the single figures are naively resized to fit the template, with large distortion (Fig. 3).

Inspired by prior arts [25], we propose a simple compound figure separation specific data augmentation strategy, called SimCFS-AUG, to perform compound figure simulation. Two groups of simulating compound figures are generated which are row-restricted and column-restricted. The length of each row or column is randomly generated within a certain range. Then, images from our database are randomly selected and concatenated together to fit in the preset space. As opposed to previous studies, the original ratio of individual images is kept in our SimCFS-AUG simulator, without distortion. Moreover, we introduce a new class within compound image separation augmentation to SimCFS-AUG so as to simulate the specific hard case in which all images belong to the same class.

3.3   Side Loss for Compound Figure Separation

For object detection on natural images, there is no specific preference between over detection and under detection as objects can be randomly located and even overlapped. In medical compound images, however, the objects are typically closely attached to each other, but not overlapping. In this case, over detection

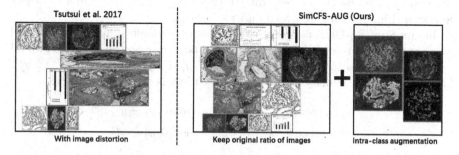

Fig. 3. Compound figure simulation. The left panel shows the current compound figure simulation strategy, which distorts the images with random ratios. The right panel presents the proposed SimCFS-AUG compound figure simulator. It keeps the original ratio of individual images. Moreover, intra-class augmentation is introduced to simulate the hard cases that the figures with similar appearances attach to each other.

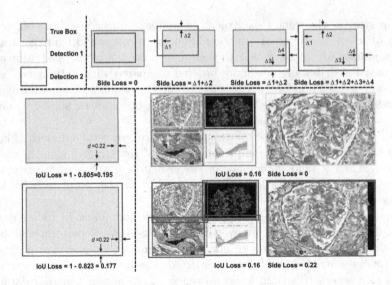

Fig. 4. Proposed Side loss for figure separation. The upper panel shows the principle of side loss, in which penalties only apply when vertices of detected bounding boxes are outside of true box regions. The lower left panel shows the bias of current IoU loss towards the over detection. When an under detection case (yellow box) and an over detection case (red box) have the same margins (d), from predicted to true boxes, the over detection has the smaller loss (larger IoU). The lower right panel shows that the under detection and over detection examples of the compound figure separation, with the same IoU loss. The Side loss is proposed to break the even IoU loss, given the results in the yellow boxes are less contaminated by nearby figures than the results in the red boxes (green arrows). (Color figure online)

would introduce undesired pixels from the nearby plots (Fig. 4), which are not ideal for downstream deep learning tasks. Unfortunately, the over detection is often encouraged by the current Intersection Over Union (IoU) loss in object detection (Fig. 4), compared with under detection.

In the SimCFS-DET network, we introduce a simple side loss, which will penalize over detection. We define a predicted bounding box as B^p and a ground truth box as B^g, with coordinates: $B^p = (x_1^p, y_1^p, x_2^p, y_2^p)$, $B^g = (x_1^g, y_1^g, x_2^g, y_2^g)$. The over detection penalty of vertices for each box is computed as:

$$
\begin{aligned}
x_1^{\mathcal{I}} &= \max(0, x_1^g - x_1^p), y_1^{\mathcal{I}} = \max(0, y_1^g - y_1^p) \\
x_2^{\mathcal{I}} &= \max(0, x_2^p - x_2^g), y_2^{\mathcal{I}} = \max(0, y_2^p - y_2^g)
\end{aligned}
\tag{1}
$$

Then, the side loss is defined as:

$$
\mathcal{L}_{side} = x_1^{\mathcal{I}} + y_1^{\mathcal{I}} + x_2^{\mathcal{I}} + y_2^{\mathcal{I}}
\tag{2}
$$

Side loss is combined with canonical loss functions in YOLOv5, including bounding box loss (L_{box}), object probability loss (L_{obj}), and classification loss (L_{cls}).

$\mathcal{L}_{total} = \lambda_1 L_{box} + \lambda_2 L_{obj} + \lambda_3 L_{cls} + \lambda_4 L_{side}$, where λ_1, λ_2, λ_3, λ_4 are constant weights to balance the four loss functions. Following the YOLOv5's implementation[1], the parameters were set as $\lambda_1 = box \times (3/nl)$, $\lambda_2 = obj \times (imgsize/640)^2 \times (3/nl)$, $\lambda_3 = (cls \times num\_cls/80) \times (3/nl)$, where num_cls was the number of classes, nl was the number of layers, and $imgsize$ was the image size. The λ_4 of the Side loss was empirically set to $\lambda_1/30$ across all experiments as the Side loss and Box loss are all based on the coordinates.

4   Data and Implementation Details

We collected two in-house datasets for evaluating the performance of different compound figure separation strategies. One compound figure dataset (called Glomeruli-2000) consisted of 917 training and 917 testing real figure plots from the American Journal of Kidney Diseases (AJKD), with the keywords "glomerular OR glomeruli OR glomerulus" as the keywords. Each figure was annotated manually with four classes, including glomeruli from (1) light microscopy, (2) fluorescence microscopy, and (3) electron microscopy, and (4) charts/plots.

To obtain single images to simulate compound figures, we downloaded 5,663 single individual images from other resources. Briefly, we obtained 1,037 images from Twitter, and obtained 4,626 images from the Google search engine, with five classes, including single images from (1) glomeruli with light microscopy, (2) glomeruli with fluorescence microscopy, (3) glomeruli with electron microscopy, (4) charts/plots, and (5) others. The individual images were combined using the SimCFS-AUG simulator to generate 9,947 pseudo training images. 2,000 of the pseudo images were simulated using intra-class augmentation, while 2,947 of them were simulated with only single sub-figures. The implementation of SimCFS-DET was based on YOLOv5 with PyTorch implementations. Google Colab was used to perform all experiments in this study.

In the experiment setting, the parameters are empirically chosen. We set the learning rate to 0.01, weight decay to 0.0005 and momentum to 0.937. The input image size was set to 640, box to 0.5, obj to 1, cls to 0.5, and number of layers to 3. For our in-house datasets, we trained 50 epochs using a batch size of 64. For the imageCLEF2016 dataset [11], we trained 50 epochs using a smaller batch size of 8.

---

[1] https://github.com/ultralytics/yolov5.

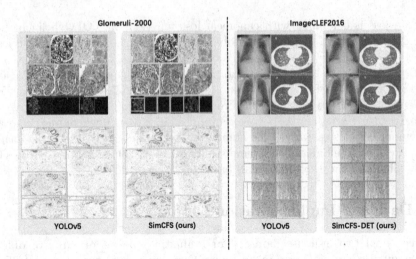

Fig. 5. Qualitative results. This figure shows the qualitative results of comparing proposed SimCFS approach with the YOLOv5 benchmark.

5 Results

5.1 Ablation Study

The Side loss is the major contribution to the YOLOv5 detection backbone. In this ablation study, we show the performance of using 917 real compound images with manual box annotations as training data (as "Real Training Images") in Table 1 and Fig. 5. This also shows the results of merely using simulated images as training data (as "Simulated Training Images"). The proposed side loss consistently improves the detection performance by a decent margin. The intra-class self-augmentation improves the performance when only using simulated training images.

5.2 Comparison with State-of-the-Art

We also compare CFS-DET with the state-of-the-art approaches including Tsuisui et al. [25] and Zou et al. [27] using the ImageCLEF2016 dataset [11]. Image-CLEF2016 is the commonly accepted benchmark for compound figure separation, including total 8,397 annotated multi-panel figures (6,783 figures for training and 1,614 figures for testing). Table 2 shows the results of the Image-CLEF2016 dataset. The proposed CFS-DET approach consistently outperforms other methods by considering evaluation metrics.

Table 1. The ablation study with different types of training data.

Training data	Method	Side loss	AUG	All	Light	Fluo.	Elec.	Chart
Real training images	YOLOv5 [2]			69.8	77.1	71.3	73.4	57.4
	SimCFS-DET (ours)	✓		79.2	86.1	**80.9**	84.2	**65.8**
Simulated training images	YOLOv5 [2]			66.4	79.3	62.1	76.1	48.0
	SimCFS (ours)	✓		69.4	77.6	67.1	84.1	48.8
	YOLOv5 [2]		✓	71.4	82.8	72.1	75.3	47.1
	SimCFS (ours)	✓	✓	**80.3**	**89.9**	78.7	**87.4**	58.8

*AUG is the intra-class self-augmentation. ALL is the Overall $mAP_{0.5:.95}$, which is reported for all classes, class Light, class Florescence and class Electron.

Table 2. The results on ImageCLEF2016 dataset.

Method	Backbone	$mAP_{0.5}$	$mAP_{0.5:.95}$
Tsutsui et al. [25]	YOLOv2	69.8	–
Tsutsui et al. [25]	Transfer	77.3	–
Zou et al. [27]	ResNet152	78.4	–
Zou et al. [27]	VGG19	81.1	–
YOLOv5 [2]	YOLOv5	85.3	69.5
SimCFS-DET (ours)	YOLOv5	**88.9**	**71.2**

6   Conclusion

In this paper, we introduce the SimCFS framework to extract images of interests from large-scale compounded figures with merely weak classification labels. The pseudo training data can be built using the proposed SimCFS-AUG simulator. The anchor-based SimCFS-DET detection achieves state-of-the-art performance by introducing a simple Side loss.

References

1. Apostolova, E., You, D., Xue, Z., Antani, S., Demner-Fushman, D., Thoma, G.R.: Image retrieval from scientific publications: text and image content processing to separate multipanel figures. J. Am. Soc. Inform. Sci. Technol. **64**(5), 893–908 (2013)
2. Bochkovskiy, A., Wang, C.Y., Liao, H.Y.M.: YOLOv4: optimal speed and accuracy of object detection. arXiv preprint arXiv:2004.10934 (2020)
3. Bueno, G., Fernandez-Carrobles, M.M., Gonzalez-Lopez, L., Deniz, O.: Glomerulosclerosis identification in whole slide images using semantic segmentation. Comput. Methods Programs Biomed. **184**, 105273 (2020)
4. Celebi, M.E., Aydin, K.: Unsupervised Learning Algorithms. Springer, Cham (2016). https://doi.org/10.1007/978-3-319-24211-8
5. Chen, T., Kornblith, S., Norouzi, M., Hinton, G.: A simple framework for contrastive learning of visual representations. In: International Conference on Machine Learning, pp. 1597–1607. PMLR (2020)

6. Davila, K., Setlur, S., Doermann, D., Bhargava, U.K., Govindaraju, V.: Chart mining: a survey of methods for automated chart analysis. IEEE Trans. Pattern Anal. Mach. Intell. (2020)
7. Demner-Fushman, D., Antani, S., Simpson, M., Thoma, G.R.: Design and development of a multimodal biomedical information retrieval system. J. Comput. Sci. Eng. **6**(2), 168–177 (2012)
8. Gadermayr, M., Dombrowski, A.K., Klinkhammer, B.M., Boor, P., Merhof, D.: CNN cascades for segmenting whole slide images of the kidney. arXiv preprint arXiv:1708.00251 (2017)
9. Ginley, B., et al.: Computational segmentation and classification of diabetic glomerulosclerosis. J. Am. Soc. Nephrol. **30**(10), 1953–1967 (2019)
10. Govind, D., Ginley, B., Lutnick, B., Tomaszewski, J.E., Sarder, P.: Glomerular detection and segmentation from multimodal microscopy images using a butterworth band-pass filter. In: Medical Imaging 2018: Digital Pathology, vol. 10581, p. 1058114. International Society for Optics and Photonics (2018)
11. García Seco de Herrera, A., Schaer, R., Bromuri, S., Müller, H.: Overview of the ImageCLEF 2016 medical task. In: Working Notes of CLEF 2016 (Cross Language Evaluation Forum), September 2016
12. Huang, W., Tan, C.L., Leow, W.K.: Associating text and graphics for scientific chart understanding. In: Eighth International Conference on Document Analysis and Recognition (ICDAR 2005), pp. 580–584. IEEE (2005)
13. Huo, Y., Deng, R., Liu, Q., Fogo, A.B., Yang, H.: AI applications in renal pathology. Kidney Int. **99**, 1309–1320 (2021)
14. Jiang, W., Schwenker, E., Spreadbury, T., Ferrier, N., Chan, M.K., Cossairt, O.: A two-stage framework for compound figure separation. arXiv preprint arXiv:2101.09903 (2021)
15. Kalpathy-Cramer, J., de Herrera, A.G.S., Demner-Fushman, D., Antani, S., Bedrick, S., Müller, H.: Evaluating performance of biomedical image retrieval systems–an overview of the medical image retrieval task at ImageCLEF 2004–2013. Comput. Med. Imaging Graph. **39**, 55–61 (2015)
16. Kannan, S., et al.: Segmentation of glomeruli within trichrome images using deep learning. Kidney Int. Rep. **4**(7), 955–962 (2019)
17. Koziell, A., et al.: Genotype/phenotype correlations of NPHS1 and NPHS2 mutations in nephrotic syndrome advocate a functional inter-relationship in glomerular filtration. Hum. Mol. Genet. **11**(4), 379–388 (2002)
18. Lee, P.-S., Howe, B.: Detecting and dismantling composite visualizations in the scientific literature. In: Fred, A., De Marsico, M., Figueiredo, M. (eds.) ICPRAM 2015. LNCS, vol. 9493, pp. 247–266. Springer, Cham (2015). https://doi.org/10.1007/978-3-319-27677-9_16
19. Lee, P.S., Howe, B.: Dismantling composite visualizations in the scientific literature. In: ICPRAM (2), pp. 79–91. Citeseer (2015)
20. Li, P., Jiang, X., Kambhamettu, C., Shatkay, H.: Compound image segmentation of published biomedical figures. Bioinformatics **34**(7), 1192–1199 (2017). https://doi.org/10.1093/bioinformatics/btx611
21. Li, P., Jiang, X., Kambhamettu, C., Shatkay, H.: Segmenting compound biomedical figures into their constituent panels. In: Jones, G.J.F., et al. (eds.) CLEF 2017. LNCS, vol. 10456, pp. 199–210. Springer, Cham (2017). https://doi.org/10.1007/978-3-319-65813-1_20
22. Redmon, J., Divvala, S., Girshick, R., Farhadi, A.: You only look once: unified, real-time object detection. In: Proceedings of the IEEE Conference on Computer Vision and Pattern Recognition, pp. 779–788 (2016)

23. Sathya, R., Abraham, A.: Comparison of supervised and unsupervised learning algorithms for pattern classification. Int. J. Adv. Res. Artif. Intell. **2**(2), 34–38 (2013)
24. Shi, X., Wu, Y., Cao, H., Burns, G., Natarajan, P.: Layout-aware subfigure decomposition for complex figures in the biomedical literature. In: 2019 IEEE International Conference on Acoustics, Speech and Signal Processing (ICASSP), ICASSP 2019, pp. 1343–1347. IEEE (2019)
25. Tsutsui, S., Crandall, D.J.: A data driven approach for compound figure separation using convolutional neural networks. In: 2017 14th IAPR International Conference on Document Analysis and Recognition (ICDAR), vol. 1, pp. 533–540. IEEE (2017)
26. Zhang, Y., Yang, L., Chen, J., Fredericksen, M., Hughes, D.P., Chen, D.Z.: Deep adversarial networks for biomedical image segmentation utilizing unannotated images. In: Descoteaux, M., Maier-Hein, L., Franz, A., Jannin, P., Collins, D.L., Duchesne, S. (eds.) MICCAI 2017. LNCS, vol. 10435, pp. 408–416. Springer, Cham (2017). https://doi.org/10.1007/978-3-319-66179-7_47
27. Zou, J., Thoma, G., Antani, S.: Unified deep neural network for segmentation and labeling of multipanel biomedical figures. J. Am. Soc. Inf. Sci. **71**(11), 1327–1340 (2020)

Data Augmentation with Variational Autoencoders and Manifold Sampling

Clément Chadebec$^{(\boxtimes)}$ and Stéphanie Allassonnière

Université de Paris, INRIA, Centre de recherche des Cordeliers, INSERM,
Sorbonne Université, Paris, France
{clement.chadebec,stephanie.allassonniere}@inria.fr

Abstract. We propose a new efficient way to sample from a Variational Autoencoder in the challenging low sample size setting (A code is available at https://github.com/clementchadebec/Data_Augmentation_with _VAE-DALI). This method reveals particularly well suited to perform data augmentation in such a low data regime and is validated across various standard and *real-life* data sets. In particular, this scheme allows to greatly improve classification results on the OASIS database where balanced accuracy jumps from 80.7% for a classifier trained with the raw data to 88.6% when trained only with the synthetic data generated by our method. Such results were also observed on 3 standard data sets and with other classifiers.

Keywords: Data augmentation · VAE · Latent space modelling

1 Introduction

Despite the apparent availability of always bigger data sets, the lack of data remains a key issue for many fields of application. One of them is medicine where practitioners have to deal with potentially very high dimensional data (*e.g.* functional Magnetic Resonance Imaging for neuroimaging) along with very low sample sizes (*e.g.* rare diseases or heterogeneous cancers) which make statistical analysis challenging and unreliable. In addition, the wide use of algorithms heavily relying on the deep learning framework [6] and requiring a large amount of data has made the need for data augmentation (DA) crucial to avoid poor performance or over-fitting [16]. As an example, a classic way to perform DA on images consists in applying simple transformations such as adding random noise, rotations etc. However, it may be easily understood that such augmentation techniques are strongly data dependent[1] and may still require the intervention of an expert assessing the relevance of the augmented samples. The recent development of generative models such as Generative Adversarial Networks (GAN) [7]

---

[1] Think of digits where rotating a *6* gives a *9* for example.

Electronic supplementary material The online version of this chapter (https://doi.org/10.1007/978-3-030-88210-5_17) contains supplementary material, which is available to authorized users.

© Springer Nature Switzerland AG 2021
S. Engelhardt et al. (Eds.): DGM4MICCAI 2021/DALI 2021, LNCS 13003, pp. 184–192, 2021.
https://doi.org/10.1007/978-3-030-88210-5_17

or Variational AutoEncoders (VAE) [9,14] paves the way for consideration of another way to augment the training data. While GANs have already seen some success [2,4,17] and even for medical data [11,15] VAEs have been of least interest. One limitation of the use of both generative models relies in their need of a large amount of data to be able to generate faithfully. In this paper, we argue that VAEs can actually be used to perform DA in challenging contexts provided that we amend the way we generate the data. Hence, we propose:

- A new non *prior-dependent* generation method using the learned geometry of the latent space and consisting in exploring it by sampling along geodesics.
- To use this method to perform DA in the small sample size setting on standard data sets and real data from OASIS database [13] where it allows to remarkably improve classification results.

2   Variational Autoencoder

Given a set of data $x \in \mathcal{X}$, a VAE aims at maximizing the likelihood of the associated parametric model $\{\mathbb{P}_\theta, \theta \in \Theta\}$. Assuming that there exist latent variables $z \in \mathcal{Z}$ living in a lower dimensional space $\mathcal{Z}$, the marginal distribution writes

$$p_\theta(x) = \int_{\mathcal{Z}} p_\theta(x|z)q(z)dz, \tag{1}$$

where q is a prior distribution over the latent variables and $p_\theta(x|z)$ is most of the time a simple distribution and is referred to as the *decoder*. A variational distribution q_φ (often taken as Gaussian) aiming at approximating the true posterior distribution and referred to as the *encoder* is then introduced. Using Importance Sampling allows to derive an unbiased estimate of $p_\theta(x)$ such that $\mathbb{E}_{z\sim q_\varphi}[\hat{p}_\theta] = p_\theta(x)$. Therefore, a lower bound on the logarithm of the objective function of Eq. (1) can be derived using Jensen's inequality:

$$\log p_\theta(x) \geq \mathbb{E}_{z\sim q_\varphi}\big[\log p_\theta(x, z) - \log q_\varphi(z|x)\big] = ELBO. \tag{2}$$

Using the reparametrization trick makes the ELBO tractable and so can be optimised with respect to both θ and φ, the *encoder* and *decoder* parameters. Once the model is trained, the decoder acts as a generative model and new data can be generated by simply drawing a sample using the prior q and feeding it to the decoder. Several axes of improvement of this model were recently explored. One of them consists in trying to bring geometry into the model by learning the latent structure of the data seen as a Riemannian manifold [3,5].

3   Some Elements on Riemannian Geometry

In the framework of differential geometry, one may define a Riemannian manifold $\mathcal{M}$ as a smooth manifold endowed with a Riemannian metric $\mathbf{G}$ which is a smooth inner product $\mathbf{G} : p \rightarrow \langle \cdot | \cdot \rangle_p$ on the tangent space $T_p\mathcal{M}$ defined at each point p of the manifold. The length of a curve γ between two points of the manifold $z_1, z_2 \in \mathcal{M}$ and parametrized by $t \in [0,1]$ such that $\gamma(0) = z_1$ and $\gamma(1) = z_2$ is given by $\mathcal{L}(\gamma) = \int\limits_{0}^{1} \|\dot{\gamma}(t)\|_{\gamma(t)} dt = \int\limits_{0}^{1} \sqrt{\langle\dot{\gamma}(t)|\dot{\gamma}(t)\rangle_{\gamma(t)}} dt$. Curves minimizing such a length are called geodesics. For any $p \in \mathcal{M}$, the exponential map at p, Exp_p, maps a vector v of the tangent space $T_p\mathcal{M}$ to a point of the manifold $\tilde{p} \in \mathcal{M}$ such that the geodesic starting at p with initial velocity v reaches $\tilde{p}$ at time 1. In particular, if the manifold is *geodesically complete*, then Exp_p is defined on the entire tangent space $T_p\mathcal{M}$.

4   The Proposed Method

We propose a new sampling method exploiting the structure of the latent space seen as a Riemannian manifold and independent from the choice of the prior distribution. The view we adopt is to consider the VAE as a tool to perform dimensionality reduction by extracting the latent structure of the data within a lower dimensional space. Having learned such a structure, we propose to exploit it to enhance the data generation process. This differs from the fully probabilistic view which uses the prior to generate. We believe that this is far from being optimal since the prior appears quite strongly data dependent. We will adopt the same setting as [5] and so use a RHVAE since the metric used by the authors is easily computable, constraints geodesic path to travel through most populated areas of the latent space and the learned Riemannian manifold is geodesically complete. Nonetheless, the proposed method can be used with different metrics as well as long as the exponential map remains computable. We now assume that we are given a latent space with a Riemannian structure where the metric has been estimated from the input data.

4.1   The Wrapped Normal Distribution

The notion of normal distribution may be extended to Riemannian manifolds in several ways. One of them is the *wrapped* normal distribution. The main idea is to define a classic normal distribution $\mathcal{N}(0, \Sigma)$ on the tangent space $T_p\mathcal{M}$ for any $p \in \mathcal{M}$ and pushing it forward to the manifold using the exponential map. This defines a probability distribution on the manifold $\mathcal{N}^W(p, \Sigma)$ called the *wrapped* normal distribution. Sampling from this distribution is straight forward and consists in drawing a velocity in the tangent space from $\mathcal{N}(0, \Sigma)$ and mapping it onto the manifold using the exponential map [12]. Hence, the *wrapped* normal allows for a latent space prospecting along geodesic paths. Nonetheless, this requires to compute Exp_p which can be performed with a numerical scheme (see. App. C). On the left of Fig. 1 are displayed some geodesic paths with respect to the metric and different starting points (red dots) and initial velocities (orange arrows). Samples from $\mathcal{N}^W(p, I_d)$ are also presented in the middle and the right along with the encoded input data. As expected this distribution takes into account the local geometry of the manifold thanks to the geodesic shooting steps. This is a very interesting property since it encourages the samples to remain close to the data as geodesics tend to travel through locations with the lowest volume element $\sqrt{\det \mathbf{G}(z)}$ and so avoid areas with very poor information.

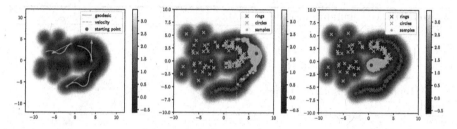

Fig. 1. *Left*: Geodesic *shooting* in a latent space learned by a RHVAE with different starting points (red dots) and initial velocities (orange arrows). *Middle and right*: Samples from the wrapped normal $\mathcal{N}^W(p, I_d)$. The log metric volume element $\log \sqrt{\det \mathbf{G}(z)}$ is presented in gray scale. (Color figure online)

4.2   Riemannian Random Walk

A natural way to explore the latent space of a VAE consists in using a random walk like algorithm which moves from one location to another with a certain probability. The idea here is to create a *geometry-aware* Markov Chain $(z^t)_{t \in \mathbb{N}}$ where z^{t+1} is sampled using the *wrapped* normal $z^{t+1} \sim \mathcal{N}^W(z^t, \Sigma)$. However, a drawback of such a method is that every sample of the chain is accepted regardless of its relevance. Nonetheless, by design, the learned metric is such that it has a high volume element far from the data [5]. This implies that it encodes in a way the amount of information contained at a specific location of the latent space. The higher the volume element, the less information we have. The same idea was used in [10] where the author proposed to see the inverse metric volume element as a maximum likelihood objective to perform metric learning. In our case the likelihood definition writes

$$\mathcal{L}(z) = \frac{\rho_S(z)\sqrt{\det \mathbf{G}^{-1}(z)}}{\int\limits_{\mathbb{R}^d} \rho_S(z)\sqrt{\det \mathbf{G}^{-1}(z)}dz}, \tag{3}$$

where $\rho_S(z) = 1$ if $z \in S$, 0 otherwise, and S is taken as a compact set so that the integral is well defined. Hence, we propose to use this measure to assess the samples quality as an *acceptance-rejection* rate α in the chain where $\alpha(\widetilde{z}, z) = \min\left(1, \frac{\sqrt{\det \mathbf{G}^{-1}(\widetilde{z})}}{\sqrt{\det \mathbf{G}^{-1}(z)}}\right)$, z is the current state of the chain and $\widetilde{z}$ is the proposal obtained by sampling from the *wrapped* Gaussian $\mathcal{N}^W(z, \Sigma)$. The idea is to compare the relevance of the proposed sample to the current one. The ratio is such that any new sample improving the likelihood metric $\mathcal{L}$ is automatically accepted while a sample degrading the measure is more likely to be rejected in the spirit of Hasting-Metropolis sampler. A pseudo-code is provided in Algorithm 1.

Algorithm 1. Riemannian random walk

Input: z_0, Σ
  for $t = 1 \to T$ **do**
    Draw $v_t \sim \mathcal{N}(0, \Sigma)$
    $\widetilde{z}_t \leftarrow \mathrm{Exp}_{z_{t-1}}(v_t)$
    Accept the proposal $\widetilde{z}_t$ with probability α
  end for

4.3 Discussion

It may be easily understood that the choice of the covariance matrix Σ in Algorithm 1 has quite an influence on the resulting sampling. On the one hand, a Σ with strong eigenvalues will imply drawing velocities of potentially high magnitude allowing for a better prospecting but proposals are more likely to be rejected. On the other hand, small eigenvalues involve a high acceptance rate but it will take longer to prospect the manifold. An adaptive method where Σ depends on **G** may be considered and will be part of future work.

Remark 1. *If Σ has small enough eigenvalues then Algorithm 1 samples from Eq. (3).*

For the following DA experiments we will assume that Σ has small eigenvalues and so will sample directly using this distribution. See App. A for sampling results using the aforementioned method.

5 Data Augmentation Experiments for Classification

In this section, we explore the ability of the method to enrich data sets to improve classification results.

5.1 Augmentation Setting

We first test the augmentation method on three reduced data sets extracted from *well-known* databases MNIST and EMNIST. For MNIST, we select 500 samples applying either a balanced split or a random split ensuring that some classes are far more represented. For EMNIST, we select 500 samples from 10 classes such that they are composed of both lowercase and uppercase characters so that we end up with a small database with strong variability within classes. These data sets are then split such that 80% is allocated for training (referred to as the *raw data*) and 20% for validation. For a fair comparison, we use the original test set (*e.g.* ~1000 samples per class for MNIST) to test the classifiers. This ensures statistically meaningful results while assessing the generalisation power on unseen data. We also validate the proposed DA method on the OASIS database which represents a nice example of day-to-day challenges practitioners have to face and is a benchmark database. We use 2D gray scale MR Images (208×176) with a

Table 1. Summary of OASIS database demographics, mini-mental state examination (MMSE) and global clinical dementia rating (CDR) scores.

Data set	Label	Obs.	Age	Sex M/F	MMSE	CDR
OASIS	CN	316	45.1 ± 23.9	119/197	29.1 ± 1.1	0: 316
	AD	100	76.8 ± 7.1	41/59	24.3 ± 4.1	0.5: 70, 1: 28, 2: 2
Train	CN	220	45.6 ± 23.6	86/134	29.1 ± 1.2	0: 220
	AD	70	77.4 ± 6.8	29/41	23.7 ± 4.3	0.5: 47, 1: 21, 2: 2
Val	CN	30	48.9 ± 24.1	11/19	29.2 ± 0.8	0: 30
	AD	12	75.4 ± 7.2	4/8	25.8 ± 4.2	0.5: 7, 1: 5, 2: 0
Test	CN	66	41.7 ± 24.3	22/44	29.0 ± 1.0	0: 66
	AD	18	75.1 ± 7.5	8/10	25.8 ± 2.7	0.5: 16, 1: 2, 2: 0

mask notifying brain tissues and are referred to as the *masked T88 images* in [13]. We refer the reader to their paper for further image preprocessing details. We consider the binary classification problem consisting in trying to detect MRI of patients having been diagnosed with Alzheimer Disease (AD). We split the 416 images into a training set (70%) (*raw data*), a validation set (10%) and a test set (20%). A summary of demographics, mini-mental state examination (MMSE) and global clinical dementia rating (CDR) is made available in Table 1. On the one hand, for each data set, the train set (*raw data*) is augmented by a factor 5, 10 and 15 using classic DA methods (random noise, cropping etc.). On the other hand, VAE models are trained individually on each class of the *raw data*. The generative models are then used to produce 200, 500, 1k or 2k synthetic samples per class with either the classic generation scheme (*i.e.* the prior) or the proposed method. We then train classifiers with 5 independent runs on 1) the *raw data*; 2) the augmented data using basic transformations; 3) the augmented data using the VAE models; 4) only the synthetic data generated by the VAEs. A DenseNet model[2] [8] is used for the toy data while we also train hand made MLP and CNN models on OASIS (See App. E). The main metrics obtained on the test set are reported in Tables 2 and 3.

5.2   Results

Toy Data. As expected generating new samples using the proposed method improves their relevance. The method indeed allows for a quite impressive gain in the model accuracy when synthetic samples are added to the real ones (leftmost column of Table 2). This is even more striking when looking at the rightmost column where only synthetic samples are used to train the classifier. For instance, when only 200 synthetic samples per class for MNIST are generated with a VAE and used to train the classifier, the classic method fails to produce meaningful samples since a loss of 20 pts in accuracy is observed when compared to the

---

[2] We use the code in [1] (See App. E).

raw data. Interestingly, our method seems to avoid such an effect. Even more impressive is the fact that we are able to produce synthetic data sets on which the classifier outperforms greatly the results observed on the *raw data* (3 to 6 pts gain in accuracy) while keeping a relatively low standard deviation (see gray cells). Secondly, this example also shows why geometric DA is still questionable and remains data dependent. For instance, augmenting the *raw data* by a factor 10 (including flips and rotations) does not seem to have a notable effect on the MNIST data sets but still improves results on EMNIST. On the contrary, our method seems quite **robust to data set changes**.

OASIS. Balanced accuracy obtained on OASIS with 3 classifiers is made available in Table 3. In this experiment, using the new generation scheme again improves overall the metric for each classifier when compared to the *raw data* and other augmentation methods. Moreover, the strong relevance of the created samples is again supported by the fact that the classifiers are again able to strongly outperform the results on the *raw data* even when trained only with synthetic ones. Finally, the method appears **robust to classifiers** and can be used with high-dimensional complex data such as MRI.

Table 2. DA on *toy* data sets. Mean accuracy and standard deviation across 5 independent runs are reported. In gray are the cells where the accuracy is higher on synthetic data than on the *raw data*.

DATA SETS	MNIST	MNIST**	EMNIST**	MNIST	MNIST**	EMNIST**
RAW DATA	89.9 (0.6)	81.6 (0.7)	82.6 (1.4)	-	-	-
	RAW + SYNTHETIC			SYNTHETIC ONLY		
AUG. (X5)	92.8 (0.4)	86.5 (0.9)	85.6 (1.3)	-	-	-
AUG. (X10)	88.3 (2.2)	82.0 (2.4)	85.8 (0.3)	-	-	-
AUG. (X15)	92.8 (0.7)	85.9 (3.4)	86.6 (0.8)	-	-	-
VAE-200*	88.5 (0.9)	84.1 (2.0)	81.7 (3.0)	69.9 (1.5)	64.6 (1.8)	65.7 (2.6)
VAE-500*	90.4 (1.4)	87.3 (1.2)	83.4 (1.6)	72.3 (4.2)	69.4 (4.1)	67.3 (2.4)
VAE-1K*	91.2 (1.0)	86.0 (2.5)	84.4 (1.6)	83.4 (2.4)	74.7 (3.2)	75.3 (1.4)
VAE-2K*	92.2 (1.6)	88.0 (2.2)	86.0 (0.2)	86.6 (2.2)	79.6 (3.8)	78.9 (3.0)
RHVAE-200*	89.9 (0.5)	82.3 (0.9)	83.0 (1.3)	76.0 (1.8)	61.5 (2.9)	59.8 (2.6)
RHVAE-500*	90.9 (1.1)	84.0 (3.2)	84.4 (1.2)	80.0 (2.2)	66.8 (3.3)	67.0 (4.0)
RHVAE-1K*	91.7 (0.8)	84.7 (1.8)	84.7 (2.4)	82.0 (2.9)	69.3 (1.8)	73.7 (4.1)
RHVAE-2K*	92.7 (1.4)	86.8 (1.0)	84.9 (2.1)	85.2 (3.9)	77.3 (3.2)	68.6 (2.3)
OURS-200*	91.0 (1.1)	84.1 (2.0)	85.1 (1.1)	87.2 (1.1)	79.5 (1.6)	77.1 (1.6)
OURS-500*	92.3 (1.1)	87.7 (0.9)	85.1 (1.1)	89.1 (1.3)	80.4 (2.1)	80.2 (2.0)
OURS-1K*	93.3 (0.8)	**89.7 (0.8)**	87.0 (1.0)	90.2 (1.4)	86.2 (1.8)	82.6 (1.3)
OURS-2K*	**94.3 (0.8)**	89.1 (1.9)	**87.6 (0.8)**	**92.6 (1.1)**	**87.6 (1.3)**	**86.0 (1.0)**

* NUMBER OF GENERATED SAMPLES ** UNBALNCED DATA SETS

Table 3. DA on OASIS data base. Mean balanced accuracy on independent 5 runs with several classifiers.

Networks	MLP		CNN		Densenet	
Raw data	80.7 (4.1)	-	72.5 (3.5)	-	77.4 (3.3)	-
	Raw + Synthetic	Synthetic Only	Raw + Synthetic	Synthetic Only	Raw + Synthetic	Synthetic Only
Aug. (X5)	84.3 (1.3)	-	80.0 (3.5)	-	73.9 (5.1)	-
Aug. (X10)	76.0 (2.8)	-	82.8 (3.7)	-	78.3 (4.1)	-
Aug. (X15)	78.7 (5.3)	-	80.3 (3.7)	-	76.6 (1.1)	-
VAE-200*	80.7 (1.5)	77.8 (1.3)	79.4 (3.6)	65.0 (12.3)	76.5 (3.2)	74.0 (3.0)
VAE-500*	79.7 (1.4)	77.4 (1.5)	72.6 (7.0)	70.2 (5.0)	74.9 (4.3)	72.8 (1.8)
VAE-1000*	81.3 (0.0)	76.5 (0.6)	74.4 (9.4)	73.0 (3.3)	73.5 (1.3)	74.9 (2.6)
VAE-2000*	80.7 (0.3)	78.1 (1.6)	71.1 (4.9)	76.9 (2.6)	74.0 (4.9)	73.3 (3.4)
Ours-200*	84.3 (0.0)	86.7 (0.4)	76.4 (5.0)	75.4 (6.6)	78.2 (3.0)	74.3 (4.8)
Ours-500*	**87.2 (1.2)**	**88.6 (1.1)**	81.8 (4.6)	81.8 (3.7)	80.2 (2.8)	84.2 (2.8)
Ours-1000*	84.2 (0.3)	84.4 (1.8)	83.5 (3.2)	79.8 (2.8)	82.2 (4.7)	76.7 (3.8)
Ours-2000*	85.3 (1.9)	84.2 (3.3)	**84.5 (1.9)**	**83.9 (1.9)**	**82.9 (1.8)**	73.6 (5.8)

* Number of generated samples

6 Conclusion

In this paper, we proposed a new way to generate new data from a Variational Autoencoder which has learned the latent geometry of the input data. This method was then used to perform DA to improve classification tasks in the low sample size setting on both toy and real data and with different kind of classifiers. In each case, the method allows for a impressive gain in the classification metrics (*e.g.* balanced accuracy jumps from 80.7 to 88.6 on OASIS). Moreover, the relevance of the generated data was supported by the fact that classifiers were able to perform better when trained with only synthetic data than on the *raw data* in all cases. Future work would consist in using the method on even more challenging data such as 3D volumes and using smaller data sets.

Acknowledgment. The research leading to these results has received funding from the French government under management of Agence Nationale de la Recherche as part of the "Investissements d'avenir" program, reference ANR-19-P3IA-0001 (PRAIRIE 3IA Institute) and reference ANR-10-IAIHU-06 (Agence Nationale de la Recherche-10-IA Institut Hospitalo-Universitaire-6). Data were provided in part by OASIS: Cross-Sectional: Principal Investigators: D. Marcus, R, Buckner, J, Csernansky J. Morris; P50 AG05681, P01 AG03991, P01 AG026276, R01 AG021910, P20 MH071616, U24 RR021382.

References

1. Amos, B.: bamos/densenet.pytorch (2020). https://github.com/bamos/densenet. pytorch. Original-date: 2017-02-09T15:33:23Z
2. Antoniou, A., Storkey, A., Edwards, H.: Data augmentation generative adversarial networks. arXiv:1711.04340 [cs, stat] (2018-03-21)

3. Arvanitidis, G., Hansen, L.K., Hauberg, S.: Latent space oddity: on the curvature of deep generative models. In: 6th International Conference on Learning Representations, ICLR 2018 (2018)

4. Calimeri, F., Marzullo, A., Stamile, C., Terracina, G.: Biomedical data augmentation using generative adversarial neural networks. In: Lintas, A., Rovetta, S., Verschure, P.F.M.J., Villa, A.E.P. (eds.) ICANN 2017. LNCS, vol. 10614, pp. 626–634. Springer, Cham (2017). https://doi.org/10.1007/978-3-319-68612-7_71

5. Chadebec, C., Mantoux, C., Allassonnière, S.: Geometry-aware Hamiltonian variational auto-encoder. arXiv:2010.11518 (2020)

6. Goodfellow, I., Bengio, Y., Courville, A., Bengio, Y.: Deep Learning, no. 2, vol. 1. MIT Press, Cambridge (2016)

7. Goodfellow, I., et al.: Generative adversarial nets. In: Advances in Neural Information Processing Systems, pp. 2672–2680 (2014)

8. Huang, G., Liu, Z., Van Der Maaten, L., Weinberger, K.Q.: Densely connected convolutional networks. In: 2017 IEEE Conference on Computer Vision and Pattern Recognition (CVPR), pp. 2261–2269. IEEE (2017)

9. Kingma, D.P., Welling, M.: Auto-encoding variational Bayes. arXiv:1312.6114 [cs, stat] (2014)

10. Lebanon, G.: Metric learning for text documents. IEEE Trans. Pattern Anal. Mach. Intell. **28**(4), 497–508 (2006)

11. Liu, Y., Zhou, Y., Liu, X., Dong, F., Wang, C., Wang, Z.: Wasserstein GAN-based small-sample augmentation for new-generation artificial intelligence: a case study of cancer-staging data in biology. Engineering **5**(1), 156–163 (2019)

12. Mallasto, A., Feragen, A.: Wrapped gaussian process regression on Riemannian manifolds. In: 2018 IEEE/CVF Conference on Computer Vision and Pattern Recognition, pp. 5580–5588. IEEE (2018)

13. Marcus, D.S., Wang, T.H., Parker, J., Csernansky, J.G., Morris, J.C., Buckner, R.L.: Open access series of imaging studies (OASIS): cross-sectional MRI data in young, middle aged, nondemented, and demented older adults. J. Cogn. Neurosci. **19**(9), 1498–1507 (2007)

14. Rezende, D.J., Mohamed, S., Wierstra, D.: Stochastic backpropagation and approximate inference in deep generative models. In: International Conference on Machine Learning, pp. 1278–1286. PMLR (2014)

15. Sandfort, V., Yan, K., Pickhardt, P.J., Summers, R.M.: Data augmentation using generative adversarial networks (CycleGAN) to improve generalizability in CT segmentation tasks. Sci. Rep. **9**(1), 16884 (2019)

16. Shorten, C., Khoshgoftaar, T.M.: A survey on image data augmentation for deep learning. J. Big Data **6**(1), 1–48 (2019). https://doi.org/10.1186/s40537-019-0197-0

17. Zhu, X., Liu, Y., Qin, Z., Li, J.: Data augmentation in emotion classification using generative adversarial networks. arXiv:1711.00648 [cs] (2017)

Medical Image Segmentation with Imperfect 3D Bounding Boxes

Ekaterina Redekop and Alexey Chernyavskiy$^{(\boxtimes)}$

Philips AI Research, Moscow, Russia
alexey.chernyavskiy@philips.com

Abstract. The development of high quality medical image segmentation algorithms depends on the availability of large datasets with pixel-level labels. The challenges of collecting such datasets, especially in case of 3D volumes, motivate to develop approaches that can learn from other types of labels that are cheap to obtain, e.g. bounding boxes. We focus on 3D medical images with their corresponding 3D bounding boxes which are considered as series of per-slice non-tight 2D bounding boxes. While current weakly-supervised approaches that use 2D bounding boxes as weak labels can be applied to medical image segmentation, we show that their success is limited in cases when the assumption about the tightness of the bounding boxes breaks. We propose a new bounding box correction framework which is trained on a small set of pixel-level annotations to improve the tightness of a larger set of non-tight bounding box annotations. The effectiveness of our solution is demonstrated by evaluating a known weakly-supervised segmentation approach with and without the proposed bounding box correction algorithm. When the tightness is improved by our solution, the results of the weakly-supervised segmentation become much closer to those of the fully-supervised one.

Keywords: Weakly-supervised image segmentation · Bounding box · Noisy labels · Computed tomography

1 Introduction

Automatic solutions for medical image segmentation are designed to increase the work efficiency of medical practitioners, as manual segmentation is an error-prone and time-consuming process. Deep convolutional neural networks (CNN) are known to achieve state-of-the-art performance for this task. However, their success highly depends on the availability of large collections of pixel-level annotations performed by experts. Drawing masks for a 2D image typically requires $\sim 8x$ more time than delineating a bounding box, and $\sim 78x$ more time than assigning an image-level label [2]. For 3D medical images, the need to have high-quality pixel-level labels makes the manual annotation even more tedious. This motivates to develop methods that leverage large amounts of data labelled by weak annotations that are cheaper to obtain. There exist various forms of weak

© Springer Nature Switzerland AG 2021
S. Engelhardt et al. (Eds.): DGM4MICCAI 2021/DALI 2021, LNCS 13003, pp. 193–200, 2021.
https://doi.org/10.1007/978-3-030-88210-5_18

labels, including image-level tags, scribbles, bounding boxes. We focus on the latter ones as they are simple, cheap in terms of annotation cost and, moreover, they provide the localization information about an object. Weak labels can be used individually in the context of weak supervision, or in combination with a small set of accurate pixel-level annotations for semi-supervised learning.

In the 2D case, bounding boxes can be defined by the coordinates of two opposing corners. In 3D, bounding boxes can be defined either as a series of coordinates of two corners for each slice along a chosen axis, or by three corner coordinates for the entire 3D volume. As we will show, the first alternative is suitable for applying existing weakly- or semi-supervised 2D image segmentation methods. The downside is that the delineation of a bounding box in each 2D layer of the entire volume is time-consuming. The second alternative allows to obtain bounding boxes easily, by quick inspection of a region of interest in three dimensions, but the quality of segmentation approaches can drop increasingly when they are trained using bounding boxes that are far from being tight.

Our contribution is three-fold. First, we show the limitation of current weakly-supervised approaches that use 2D bounding boxes as weak labels, when applied to medical image segmentation in cases when the bounding boxes are not tight. Second, we propose a bounding box correction framework which shrinks the bounding boxes closer to the actual size of the object cross-section in each slice of the 3D volume. Finally, we demonstrate that the proposed solution allows increasing the accuracy of 3D computed tomography (CT) segmentation algorithms trained using pseudo-annotations generated from weak labels.

2   Related Work

Weakly-supervised learning methods can significantly reduce the cost of annotation that is needed to collect a training set. The methods differ by the type of weak annotation they rely on, such as image-level labels [12], points [2], partial labels [13] or global image statistics [1]. In this work, we build upon the recent papers that have focused on training neural networks using pseudo-annotations generated from bounding boxes. In [15] and [14], Xu et al. formulated the weakly-labeled segmentation as a sparse boundary point detection task solved by training a CNN that predicts the offsets from the given bounding box to the true object boundary. In [5], Kervadec et al. proposed an image segmentation approach based on global constraints derived from bounding box annotations, including the deep tightness prior and background emptiness constraint. The use of these priors allowed the authors of [5] to significantly outperform DeepCut [9] which also relied on bounding boxes for supervision. In [6], bounding boxes are treated as noisy labels, and per-class attention maps are produced to guide the cross-entropy loss to focus on foreground pixels.

Semi-supervised learning is the ability of neural networks to derive information from limited sets of labeled data. The authors of [7] proposed an algorithm for semi-supervised semantic image segmentation based on adversarial training with a feature matching loss to learn from unlabeled images. The approach of

Ouali et al. [8] to the same task is based on cross-consistency, where the general idea of consistency loss is to encourage smooth predictions of the same data under different perturbations. Ibrahim et al. [3] proposed to train a primary segmentation model on a small fully-labeled dataset with the aid of an ancillary model that generates segmentation labels for a larger weakly-labeled dataset. In this work, we also use the advantage of a small set of accurately labeled data to train a bounding box correction framework.

3   Methodology

3.1   Bounding Box Correction

Consider a three-dimensional object within a volume. It is straightforward to produce a 3D bounding box of the object by finding its extreme points in the three coordinate axes. While this 3D bounding box will be tight in the 3D sense, its rectangular cross-sections will not, in general, remain tight with respect to the planar cross-sections of the volume. Figure 1 illustrates such a case for the task of liver segmentation in a CT volume. In the Experiments section we show that the success of existing 2D weakly-supervised segmentation methods relies on the bounding boxes being tight and therefore the tightness of the individual 2D bounding boxes should be corrected before training and applying a segmentation CNN.

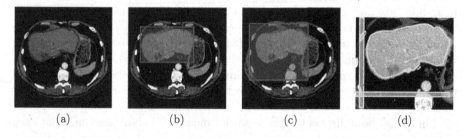

(a)   (b)   (c)   (d)

Fig. 1. (a) Ground truth mask, (b) tight bounding box for ground truth mask, (c) non-tight bounding box for a 2D slice of the 3D volume, (d) breaking of the bounding box tightness assumption (see Sect. 3.2).

We propose a method to improve the tightness of bounding boxes by using a patch-based classification neural network. The network is trained on a limited subset of ground truth data which is accurately annotated on a pixel level. The proposed solution consists of four steps shown in Fig. 2. First, a non-tight bounding box is cropped from each slice of the 3D image. Second, each crop is divided into patches of size $p \times p$ pixels, with an overlap equal to $p/2$. Each patch is assigned a binary label y: $y = 1$ if the foreground object occupies more than 50% of the patch area; otherwise $y = 0$. Third, during training and inference, the classification neural network assigns a label for every patch inside the cropped bounding box area. Finally, the patches that the classification network has labeled as foreground determine the extent of the corrected bounding box. During inference, we apply this neural network to data annotated by non-tight bounding boxes, and, following the classification step, obtain more accurate and more tight bounding boxes.

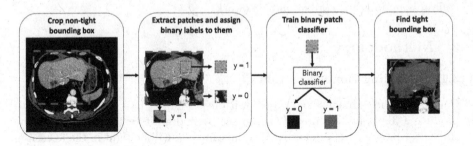

Fig. 2. Bounding box correction framework.

3.2 Bounding Boxes for Weakly Supervised Segmentation

We test the bounding box correction method in combination with the novel weakly-supervised framework proposed by Kervadec et al. [5]. The authors perform medical image segmentation by deriving several global constrains from bounding box annotations. In order to regularize the output of the network, they leverage the bounding box tightness prior which was reformulated as a set of global constrains. At the same time, in order to enforce the network to predict no foreground outside the bounding box, the authors add a global background emptiness constraint. The training of a neural network is performed using a sequence of unconstrained losses based on an extension of the log-barrier method.

The global bounding box tightness prior mentioned above assumes that each of side of the box is sufficiently close to the target region. This means that for any region shape, each vertical or horizontal line inside the bounding box will cross at least one pixel belonging to the target region. This condition does not hold when the provided annotation comes as a 3D bounding box which is represented as a series of per-slice non-tight 2D bounding boxes. In this case, there will exist vertical or horizontal lines shown as stripes in Fig. 1(d), that will lie outside of the actual object boundary. In the Experiments section we demonstrate the poor performance of the weakly-supervised approach from [5] when the user-provided bounding box is much wider than the true object of interest.

3.3 Implementation Details

For patch classification that is used for correcting the bounding boxes, we train a VGG-16 [10] CNN using cross-entropy loss. After the bounding boxes are corrected, we use a residual version of a standard UNet [5] neural network which we trained in a 2.5D manner by taking a stack of three neighboring slices as input and outputting a segmentation for the single central slice of the stack. This approach allows to take advantage of richer spatial information compared to 2D, while requiring less computations compared to 3D CNNs. Following [5], we trained the segmentation model using the tightness prior in combination with constraint on the global size and masked cross-entropy. We performed three-fold cross-validation to study the variability of image segmentation.

Both the classification and segmentation neural networks are trained using Adam optimizer with learning rate equal to 10^{-4}. The mini-batch size and the number of epochs are set to 32 and 50 respectively. We set the bounding box tightness prior parameters following [5].

4 Experiments and Discussion

We validate the proposed bounding box correction method, followed by the weakly-supervised segmentation framework, on the liver segmentation dataset provided by the organizers of the Medical Segmentation Decathlon [11]. The data consist of 131 3D contrast-enhanced CT images and was divided into training and validation sets in the proportion 100:31. We normalize the CT data as suggested in [4]. First, the intensity values of pixels that fall under the segmentation masks are collected for the whole training set. Second, the intensity values for the entire dataset are clipped to the [0.5, 99.5] percentiles of the collected values. Third, z-score normalization is applied based on the statistics of the collected values.

To train the bounding box correction framework, we further divided the training set into a small subset of accurate pixel-level and a larger subset of weak bounding box annotations, with the size of the small subset equal to 5%, 10% and 20% of the whole training set. We also studied the effect of patch size on the tightness of corrected bounding boxes and on the segmentation accuracy, which was measured as the Dice similarity coefficient between the CNN outputs and ground truth masks.

4.1 Weakly-Supervised Segmentation of 3D CT Volume Using Bounding Box Correction

In Table 1 we compare the performance of fully- and weakly-supervised training strategies for the liver CT dataset, where 3D voxel-level segmentation masks are available for each 3D CT scan. In this case one can easily obtain a 3D bounding box for the entire object of interest, or a series of tight 2D bounding boxes corresponding to each individual cross-section of the object. The first alternative implies the absence of the bounding box tightness property on most of the 2D slices of the volume.

The weakly-supervised approach [5] can be efficiently applied to slice-wise 2D image segmentation of a 3D object, as long as the ground truth labels are given as tight bounding boxes for each image slice (Table 1). If, instead of a series of 2D bounding boxes, the ground truth segmentation labels are given as

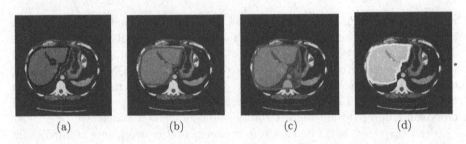

      (a)   (b)   (c)   (d)

Fig. 3. (a) Ground truth mask. Segmentation results of a 2.5D UNet trained on: (b) 2D tight bounding boxes, (c) 3D non-tight bounding boxes, (d) 3D corrected bounding boxes.

a 3D bounding box computed over the entire object of interest embedded within a 3D image, then, depending on the shape of the object, the edges of many rectangular 2D cross-sections of a 3D bounding box will be quite distant from the boundaries of the object. In this case, the performance of the semi-supervised approach drops considerably. In order to boost the performance of segmentation networks trained on this kind of weak and noisy labels, we use the advantage of the proposed bounding box correction framework described in Sect. 3.1 and pictured in Fig. 2. In Table 2 we show the bounding box tightness computed as the intersection over union (IoU) between the tight bounding box generated from a 2D ground truth mask, and the bounding box coming from a 3D box before and after applying our correction procedure on the liver CT dataset. As shown in Table 1 ('3D corrected'), the improvement of bounding box tightness using our approach results in higher segmentation accuracy of models trained with weak supervision. The experiments also show that a smaller patch size ($p = 16$) used for correcting the bounding boxes results in higher segmentation accuracy. The amount of accurately labeled data that is used to train the bounding box correction network also plays the role in the final segmentation accuracy. By providing 20 examples one can achieve the quality that is comparable to the performance of segmentation models trained using tight 2D bounding boxes (Figs. 3 and 4).

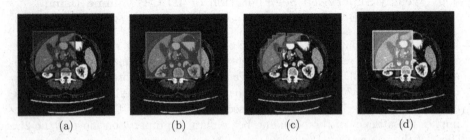

      (a)   (b)   (c)   (d)

Fig. 4. (a) Tight bounding box for ground truth mask, (b) non-tight bounding box for a 2D slice of the 3D volume, (c) output of patch classification CNN, (d) corrected bounding box for a 2D slice of the 3D volume.

Table 1. Segmentation of 3D liver CT images: Dice scores obtained by training segmentation models using weak supervision provided as tight, non-tight and corrected bounding boxes. The Dice score for full supervision (voxel-level masks) is 0.92 ± 0.03.

2D tight	3D non-tight		3D corrected number of images for box correction		
			5	10	20
0.90 ± 0.01	0.28 ± 0.06	$p = 16$	0.83 ± 0.02	0.86 ± 0.01	**0.90 ± 0.04**
		$p = 32$	0.82 ± 0.04	0.84 ± 0.03	0.89 ± 0.02

Table 2. Mean IoU values before and after 3D bounding box correction computed with respect to tight slice-wise ground truth bounding boxes.

3D non-tight bounding box		3D corrected number of images for box correction		
		5	10	20
0.15 ± 0.01	$p = 16$	0.87 ± 0.02	0.89 ± 0.04	**0.93 ± 0.04**
	$p = 32$	0.89 ± 0.01	0.89 ± 0.02	0.92 ± 0.01

5 Conclusions and Discussions

We have addressed the main limitation of a known approach to weakly-supervised 2D and 3D medical segmentation that assumes that the labels, coming in the form of two-dimensional bounding boxes, are tight. We have shown that in a practical case when a single 3D bounding box is provided for the whole object, the tightness of 2D slice-wise bounding boxes deteriorates, which results in poor segmentation accuracy of neural networks trained with this type of supervision. We have proposed a bounding box correction framework that improves the tightness by using a patch-based classification network trained on a small subset of pixel-level annotated data. By producing higher quality annotations out of weak labels, our approach allows to increase the accuracy of 3D medical weakly-supervised segmentation.

Since the performance of the proposed approach may depend on the soft tissue contrast, its applicability for segmentation of other organs is yet to be investigated. The patch size and the share of fully-supervised samples used for bounding box correction may play an important role.

References

1. Bateson, M., Kervadec, H., Dolz, J., Lombaert, H., Ayed, I.B.: Constrained domain adaptation for segmentation. In: Shen, D. (ed.) MICCAI 2019. LNCS, vol. 11765, pp. 326–334. Springer, Cham (2019). https://doi.org/10.1007/978-3-030-32245-8_37

2. Bearman, A., Russakovsky, O., Ferrari, V., Fei-Fei, L.: What's the point: semantic segmentation with point supervision. In: Leibe, B., Matas, J., Sebe, N., Welling, M. (eds.) ECCV 2016. LNCS, vol. 9911, pp. 549–565. Springer, Cham (2016). https://doi.org/10.1007/978-3-319-46478-7_34

3. Ibrahim, M.S., Vahdat, A., Ranjbar, M., Macready, W.G.: Semi-supervised semantic image segmentation with self-correcting networks. In: Proceedings of the IEEE/CVF Conference on Computer Vision and Pattern Recognition, pp. 12715–12725 (2020)
4. Isensee, F., et al.: nnU-Net: self-adapting framework for U-Net-based medical image segmentation. arXiv preprint arXiv:1809.10486 (2018)
5. Kervadec, H., Dolz, J., Wang, S., Granger, E., Ayed, I.B.: Bounding boxes for weakly supervised segmentation: global constraints get close to full supervision. In: Medical Imaging with Deep Learning, pp. 365–381. PMLR (2020)
6. Kulharia, V., Chandra, S., Agrawal, A., Torr, P., Tyagi, A.: Box2Seg: attention weighted loss and discriminative feature learning for weakly supervised segmentation. In: Vedaldi, A., Bischof, H., Brox, T., Frahm, J.-M. (eds.) ECCV 2020. LNCS, vol. 12372, pp. 290–308. Springer, Cham (2020). https://doi.org/10.1007/978-3-030-58583-9_18
7. Mittal, S., Tatarchenko, M., Brox, T.: Semi-supervised semantic segmentation with high-and low-level consistency. IEEE Trans. Pattern Anal. Mach. Intell. **43**, 1369–1379 (2019)
8. Ouali, Y., Hudelot, C., Tami, M.: Semi-supervised semantic segmentation with cross-consistency training. In: Proceedings of the IEEE/CVF Conference on Computer Vision and Pattern Recognition, pp. 12674–12684 (2020)
9. Rajchl, M., et al.: DeepCut: object segmentation from bounding box annotations using convolutional neural networks. IEEE Trans. Med. Imaging **36**(2), 674–683 (2016)
10. Simonyan, K., Zisserman, A.: Very deep convolutional networks for large-scale image recognition. In: Bengio, Y., LeCun, Y. (eds.) 3rd International Conference on Learning Representations, ICLR 2015, San Diego, CA, USA, May 7–9, 2015. Conference Track Proceedings (2015)
11. Simpson, A.L., et al.: A large annotated medical image dataset for the development and evaluation of segmentation algorithms. arXiv preprint arXiv:1902.09063 (2019)
12. Wei, Y., Xiao, H., Shi, H., Jie, Z., Feng, J., Huang, T.S.: Revisiting dilated convolution: a simple approach for weakly-and semi-supervised semantic segmentation. In: Proceedings of the IEEE Conference on Computer Vision and Pattern Recognition, pp. 7268–7277 (2018)
13. Xu, J., Schwing, A.G., Urtasun, R.: Learning to segment under various forms of weak supervision. In: Proceedings of the IEEE Conference on Computer Vision and Pattern Recognition, pp. 3781–3790 (2015)
14. Xu, X., Meng, F., Li, H., Wu, Q., Ngan, K.N., Chen, S.: A new bounding box based pseudo annotation generation method for semantic segmentation. In: 2020 IEEE International Conference on Visual Communications and Image Processing (VCIP), pp. 100–103. IEEE (2020)
15. Xu, X., Meng, F., Li, H., Wu, Q., Yang, Y., Chen, S.: Bounding box based annotation generation for semantic segmentation by boundary detection. In: 2019 International Symposium on Intelligent Signal Processing and Communication Systems (ISPACS), pp. 1–2. IEEE (2019)

Automated Iterative Label Transfer Improves Segmentation of Noisy Cells in Adaptive Optics Retinal Images

Jianfei Liu, Nancy Aguilera, Tao Liu, and Johnny Tam[✉]

National Eye Institute, National Institutes of Health, Bethesda, MD, USA
johnny@nih.gov

Abstract. High quality data labeling is essential for improving the accuracy of deep learning applications in medical imaging. However, noisy images are not only under-represented in training datasets, but also, labeling of noisy data is low quality. Unfortunately, noisy images with poor quality labels are exacerbated by traditional data augmentation strategies. Real world images contain noise and can lead to unexpected drops in algorithm performance. In this paper, we present a nontraditional, purposeful data augmentation method to specifically transfer high quality automated labels into noisy image regions for incorporation into the training dataset. The overall approach is based on the use of paired images of the same cells in which variable image noise results in cell segmentation failures. Iteratively updating the cell segmentation model with accurate labels of noisy image areas resulted in an improvement in Dice coefficient from 77% to 86%. This was achieved by adding only 3.4% more cells to the training dataset, showing that local label transfer through graph matching is an effective augmentation strategy to improve segmentation.

Keywords: Data labels · Data augmentation · Graph matching · Cell segmentation · U-Net

1 Introduction

Conventional data augmentation has been shown to be effective for enlarging training data to improve the performance of deep neural networks [15], and is especially important for medical imaging applications. Most simply, data augmentation starts by translating, rotating, and flipping of images, or by changing intensity values. Beyond this, image fusion has also been demonstrated as a method to augment training data to improve the performance of brain glioma segmentation [2]. However, applying arbitrary transformations or changes can lead to the introduction of unrealistic images [5].

Generative adversarial networks (GANs) are able to generate realistic images that more closely resemble real world images [4], and can therefore be used as an effective strategy to augment medical imaging datasets in which there is limited

© Springer Nature Switzerland AG 2021
S. Engelhardt et al. (Eds.): DGM4MICCAI 2021/DALI 2021, LNCS 13003, pp. 201–208, 2021.
https://doi.org/10.1007/978-3-030-88210-5_19

training data available, such as for tumor classification where tumor images for training are relatively rare [3]. For image segmentation applications, in which segmentation masks are needed alongside training data, translation-to-translation conditional GANs have been proposed [7]. Conditional GANs have also been applied to augment medical data [6], such as retinal cell segmentation [10]. However, noisy images that are a key reason for drops in algorithm performance are seldom generated through GANs methods.

Unfortunately, it is challenging to accurately label noisy images, which motivates us to develop a targeted strategy to incorporate high quality labeling, specifically from noisy images, into training datasets to improve the performance of deep learning based algorithms. In optical imaging, image noise is often nonuniform, with neighboring cells containing different amounts of noise (Fig. 1). In this example, a pair of images showing the same cells imaged on two different days illustrates the effect of image noise on cell segmentation obtained using U-Net [11]. We consider the use of actual noisy images of cells, with the goal to obtain objective labels of these noisy images (Fig. 1 arrows) so that they can be introduced into the training data to improve the overall cell segmentation model. By iteratively refining the cell segmentation model with the addition of high quality labels of noisy cells, the overall performance of the cell segmentation model on native images of noisy cells can be improved.

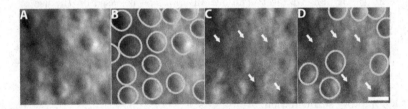

Fig. 1. Adaptive optics (AO) retinal images [14] of cone photoreceptor cells. (A-D) show the same cells imaged across two separate days. In the image from the first day (A), all of the cells are correctly segmented (B). Some cells are noisier on the second visit (C), leading to errors in segmentation (D) (white arrows). Scale bar, 10 μm.

2   Methodology

2.1   Cell Segmentation Initialization

Using images of cells that are not noisy, the cell segmentation model, implemented using U-Net, is initialized. The U-Net takes AO images and predicts both cell centroid and region maps. For this initial cell segmentation model, the combination of dice coefficient and binary cross entropy is used as the loss function to train the model. Following a previously-developed cell segmentation model [11], cell centroids are extracted by thresholding the predicted centroid maps. Next, cell regions are extracted using Otsu's method [13] on the predicted region maps, which is in turn combined with the cell centroids to separate each individual cell regions through watershed segmentation [12].

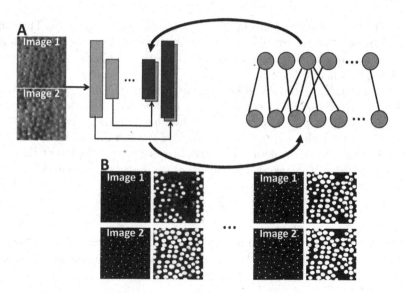

Fig. 2. Overview of iterative label transfer method to improve cell segmentation. (A) A pair of AO images of the same exact cells, acquired over two separate visits, is segmented using a U-Net model which predicts cell centroid and region maps for the two AO images. After using a graph matching framework to determine cell-to-cell correspondences, segmentation labels of centroid and region maps for noisy cells within the image pair are bidirectionally transferred to add additional labels to the images. The updated segmentation maps are used to re-train and update the segmentation model. This process is then repeated. (B) Iteratively, as the training dataset is augmented with additional examples of noisy cells, the overall cell segmentation accuracy improves.

2.2 Cell-to-Cell Correspondence Using Graph Matching

The initial cell segmentation model is applied to a pair of AO images of the same retinal region in which the same exact cells are imaged, but where cells are noisier in one image compared to the other (Fig. 1). Before labels can be transferred from one cell to its noisier pair, the cell-to-cell correspondence has to be solved. Mathematically, we aim to determine a set of unique one-to-one correspondences between two sets of cell centroids (C_1 and C_2), which we solve using bipartite graph matching. Relative image deformation between two AO images is estimated as an affine transform using a spatial transformer network [8], and the bipartite graph G is constructed by transforming C_1 and connecting possible corresponding points in C_2 that are within 30 pixels of the transformed C_1 (this distance corresponds to approximately the diameter of one cell).

We begin with the set of all possible cell connections at G which can be expressed as $C_1 \times C_2$, from which we define C as the reduced subset of cell connections restricted to a local neighborhood around each cell (discarding far away cells). Solving for the unique cell-to-cell correspondences can be formulated as finding a binary-valued vector $\mathbf{m} = \{0, 1\}^C$ that represents matched cell connections between C_1 and C_2, subject to the following criterion: if a cell connection

α is an actual cell correspondence, $m_\alpha = 1$; otherwise, $m_\alpha = 0$. Three constraints are used to ensure that $\mathbf{m}$ represents true cell-to-cell correspondences: including visual similarity, topological configuration similarity, and one-to-one cell correspondence.

Visual similarity is measured by comparing visual feature representations at cell regions. Given two cell centroids, $p_1 \in C_1$ and $p_2 \in C_2$, the visual feature vectors $\mathbf{f}_1$ and $\mathbf{f}_2$ at p_1 and p_2 are established by vectorizing two $16 \times 16 \times 16$ image regions from the 16-layer feature space of the last convolution layer of the U-Net (the later layers in neural networks are assumed to represent the object features). Based on this, we define the visual similarity constraint as

$$E^v(\mathbf{m}) = \sum_{\alpha \in C} \|\mathbf{f}_1 - \mathbf{f}_2\|_1 \cdot m_\alpha \tag{1}$$

Topological configuration similarity evaluates if two matched cells have similar topologies with their neighbored cells. We define S to contain all adjacent cell pairs to a given correspondence, given by

$$S = \{\langle (p_1, p_2), (q_1, q_2) \rangle \in C \times C | p_i \in N(q_i) \wedge q_i \in N(p_i), i = 1, 2\} \tag{2}$$

where N is the 6-nearest neighborhood in C_1 and C_2 based on cell hexagonal packing [9]. The topological constraint includes both distant and angular components.

$$E^t(\mathbf{m}) = \sum_{(\alpha,\beta) \in S} \left(\exp\left(\delta_{\alpha,\beta}^2\right) - 1\right) + \left(\exp\left(\gamma_{\alpha,\beta}^2\right) - 1\right) \cdot m_\alpha \cdot m_\beta$$

$$\delta_{\alpha,\beta} = \|\|p_1 - q_1\| - \|p_2 - q_2\|\|/(\|p_1 - q_1\| + \|p_2 - q_2\|)$$

$$\gamma_{\alpha,\beta} = \arccos((p_1 - q_1)/\|p_1 - q_1\|, (p_2 - q_2)/\|p_2 - q_2\|) \tag{3}$$

The last constraint term is to ensure unique one-to-one correspondence and is given by

$$E^p(\mathbf{m}) = 1 - \sum_{\alpha \in A} m_\alpha / \min\{|P_1|, |P_2|\} \tag{4}$$

The overall constraint function is thus given by the sum of E^v, E^t, and E^p. Dual decomposition approach [9] is applied to minimize this function to obtain the cell-to-cell correspondence.

2.3   Data Augmentation Through Iterative Label Transfer

Once the cell-to-cell correspondence is found, it can be used to guide the transfer of labels across images (Fig. 2A). Starting with a pair of input AO images I_1 and I_2, we illustrate the process to update the segmentation mask of I_1 by transferring segmentation labels from I_2. The transfer is bi-directional as the mirror process is used to update the segmentation mask of I_2 from I_1. The average cell size $\bar{s}$ and standard deviation σ are computed for the segmentation mask of I_1. The cells are kept if their size s is within two σ of the expected size

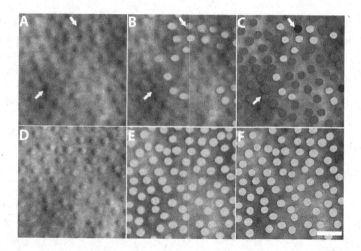

Fig. 3. Automated segmentation label transfer facilitated by cell-to-cell correspondence. (A, D): AO image pair from two visits in which the top visit has noisier cells. (B, E) Predicted cell region masks from the initial cell segmentation model. (C, F) Updated cell region masks guided by cell-to-cell correspondence. In (C), cell regions in green are those that are kept from the original predicted region maps; red denotes segmentation labels that were not previously present and were transferred from a matched cell; blue denotes a segmentation labels that was previously under-segmented from the initial cell segmentation model, which was subsequently modified by a matched cell. Scale bar, 25 μm. (Color figure online)

$\bar{s} - 2\sigma \leq s \leq \bar{s} + 2\sigma$. Otherwise, the cell is treated as falsely-segmented, and the segmentation labels from the mask of I_2 are transferred (e.g. blue regions, Fig. 3C). We also search for isolated cells in I_2 that are missing cell correspondences in I_1 and transfer in the missing segmentation masks from I_2, which occurs when segmentation labels are missing due to noisy cells (red regions, Fig. 3C). The updated segmentation masks are sent back to retrain the U-Net segmentation model. Iteratively, the U-Net model is improved to better segment noisy cells (Fig. 2B).

2.4 Data Collection and Validation Methods

Sixteen subjects (age: 28.4 ± 8.1 years, mean $\pm$ std) were recruited to perform AO imaging across two non-contiguous days from which 1,138 AO images were extracted (333×333 pixels). From this dataset, 386 AO images that did not have substantial image degradation were selected and manually marked to compose a labeled training data, which is used to initialize the U-Net based cell segmentation model. From the remaining 752 images, 604 unlabeled images (corresponding to 302 two-visit image pairs) which had noisy regions were imported into the graph matching framework to augment the training data for improving the U-Net model. The remaining 148 AO images were used as the test dataset to validate cell segmentation improvement.

Six metrics were used to evaluate segmentation accuracy: area overlap (AP), Dice coefficient (DC), area difference (AD), average symmetric contour ditance (ASD), symmetric room mean square contour distance (RSD), and maximum symmetric absolute contour distance (MSD) [11]. The segmentation model trained on 386 non-noisy AO images was considered to be the state-of-the-art model for comparison against the proposed iterative label transfer based models.

3   Experimental Results

3.1   Iterative Cell Segmentation in Noisy Images

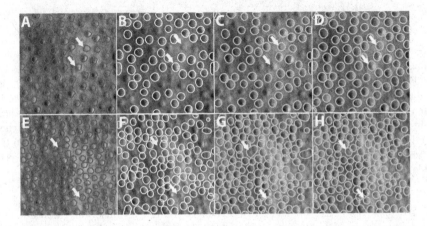

Fig. 4. Examples of cone segmentation improved through label transfer on two subjects corresponding to each row. Overall, segmentation improved from iteration 1 to 5. In the first subject, some cells in noisy image regions are missed (B), which are recovered after one iteration except for one cell which was initially missing after iteration 1 but recovered after iteration 5 (D). The second subject has dense cell packing, which causes missed cells in the noisy regions. Segmentation accuracy is improved after the iterations. Scale bar, 25 μm.

Overall, the iterative label transfer strategy successfully segmented cells that were missed by the U-Net baseline model. These cells were located in noisy image regions. Figure 4 illustrates two examples from two subjects (corresponding to each row). In the first subject, cell over-segmentation and extra cell segmentation were observed (white arrows, Fig. 4B). After one iteration, the improved cell segmentation model could identify some cells but with the cost of missing some cells that were identified by the baseline (Fig. 4C). These were all successfully segmented after five iterations (Fig. 4D). The image from the second subject, with denser cell packing, showed some missed cell segmentations (Fig. 4F). After one iteration, the segmentation model could identify all these cells but there were examples of over-segmentation (Fig. 4G). The over-segmentation was fixed

Table 1. Evaluation of image generation comparing real and generated images.

Model	AP (%)	DC (%)	AD (%)	ASD (Pix)	RSD (Pix)	MSD (Pix)
U-Net	62.6 ± 10.9	74.0 ± 10.4	48.7 ± 67.7	3.7 ± 1.9	4.1 ± 1.9	7.3 ± 3.3
Iter. 1	65.8 ± 10.0	77.2 ± 10.4	48.4 ± 55.7	3.1 ± 1.9	3.4 ± 1.6	6.5 ± 2.9
Iter. 2	69.5 ± 9.5	80.7 ± 8.5	48.5 ± 45.4	2.4 ± 1.3	2.8 ± 1.4	5.6 ± 2.4
Iter. 3	74.3 ± 10.0	82.0 ± 8.5	40.5 ± 40.7	2.2 ± 1.2	2.6 ± 1.3	5.2 ± 2.3
Iter. 4	78.5 ± 10.4	85.3 ± 8.5	33.3 ± 37.7	2.2 ± 1.1	2.6 ± 1.2	4.1 ± 2.2
Iter. 5	79.2 ± 10.8	85.9 ± 8.5	30.9 ± 36.0	2.2 ± 1.1	2.6 ± 1.2	4.0 ± 2.2

after five iterations (Fig. 4H). These two examples show that the iterative cell segmentation model is particularly useful for improving the segmentation results of noisy image areas which are often discarded in analyses (Table 1).

3.2 Purposeful Data Augmentation Improves Training Results

Quantification of the cell segmentation accuracy in the test dataset showed an improvement over the baseline U-Net model which continued to improve with additional iterations. The segmentation accuracy of the model with five iterations was significantly better than the baseline model ($p < 0.05$, two-tailed paired t-test). Importantly, the number of cells in noisy image regions that was added over the iterations was small. Out of a total of 47,195 cells in the training dataset in iteration 1, only 1,612 new cells were added by iteration 5 (3.4% increase). Despite this relatively small increase, there was a substantial improvement in accuracy from iteration 1 to 5, demonstrating that adding even a very small amount of accurately labeled noisy image regions can be very effective for improving cell segmentation training.

4   Conclusion and Future Work

Using graph matching to transfer segmentation labels to noisy image regions improved the ability of the cell segmentation model to handle noisy image regions without adversely affecting the non-noisy image regions. In contrast to existing augmentation methods that focus on realistic image generation, our method enhances training data with noisy images which often result in segmentation failures. Our strategy complements existing data augmentation strategies and can be deployed where longitudinal data is collected, often available in most medical imaging modalities [1]. The incorporation of non-perfect images broadens the scope of the available training data to better encapsulate the full range of image quality that is more typically encountered in medical imaging. Once trained, the overall improvement in segmentation on single-visit noisy images can be substantially improved, making the overall medical image analysis pipeline more robust and applicable to real-world situations.

References

1. Brown, R.A., Fetco, D., et al.: Deep learning segmentation of orbital fat to calibrate conventional MRI for longitudinal studies. Neurocomputing **208**, 116442 (2020)
2. Eaton-Rosen, Z., Bragman, F., Ourselin, S., Cardoso, M.J.: Improving data augmentation for medical image segmentation. In: 1st Conference on Medical Imaging with Deep Learning (2018)
3. Frid-Adar, M., Diamant, J., Klang, E., et al.: GAN-based synthetic medical image augmentation for increased CNN performance in liver lesion classification. Neurocomputing **321**, 321–331 (2018)
4. Goodfellow, I., Pouget-Abadie, J., Mirae, M., et al.: Generative adversarial networks. In: NIPS, pp. 2672–2680 (2014)
5. Hussain, Z., Gimenez, F., Yi, D., Rubin, D.: Differential data augmentation techniques for medical imaging classification tasks. In: AMIA Annual Symposium Proceedings, pp. 979–984 (2017)
6. Iqbal, T., Ali, H.: Generative adversarial network for medical images (MI-GAN). J. Med. Syst. **42**(11), 1–11 (2018)
7. Isola, P., Zhu, J., Zhou, T., Efros, A.: Image-to-image translation with conditional adversarial networks. In: IEEE CVPR (2017)
8. Jaderberg, M., Simonyan, K., Zisserman, A., Kavukcuoglu, K.: Spatial transformer networks. In: NIPS, pp. 2672–2680 (2014)
9. Liu, J., Jung, H.W., Tam, J.: Accurate correspondence of cone photoreceptor neurons in the human eye using graph matching applied to longitudinal adaptive optics images. In: Descoteaux, M., Maier-Hein, L., Franz, A., Jannin, P., Collins, D.L., Duchesne, S. (eds.) MICCAI 2017. LNCS, vol. 10434, pp. 153–161. Springer, Cham (2017). https://doi.org/10.1007/978-3-319-66185-8_18
10. Liu, J., Shen, C., Liu, T., Aguilera, N., Tam, J.: Active appearance model induced generative adversarial network for controlled data augmentation. In: Shen, D., et al. (eds.) MICCAI 2019. LNCS, vol. 11764, pp. 201–208. Springer, Cham (2019). https://doi.org/10.1007/978-3-030-32239-7_23
11. Liu, J., Shen, C., Liu, T., Aguilera, N., Tam, J.: Deriving visual cues from deep learning to achieve subpixel cell segmentation in adaptive optics retinal images. In: Fu, H., Garvin, M.K., MacGillivray, T., Xu, Y., Zheng, Y. (eds.) OMIA 2019. LNCS, vol. 11855, pp. 86–94. Springer, Cham (2019). https://doi.org/10.1007/978-3-030-32956-3_11
12. Najman, L., Schmitt, M.: Geodesic saliency of watershed contours and hierarchical segmentation. IEEE Trans. Pattern Anal. Mach. Intell. **18**(12), 1163–1173 (1996)
13. Otsu, N.: A threshold selection method from gray-level histograms. IEEE Trans. Syst. Man Cybern. **9**(1), 62–66 (1979)
14. Roorda, A., Duncan, J.: Adaptive optics ophthalmoscopy. Ann. Rev. Vis. Sci. **11045**, 146–154 (2015)
15. Shorten, C., Khoshgoftaar, T.M.: A survey on image data augmentation for deep learning. J. Big Data **6**, 60 (2019)

How Few Annotations are Needed for Segmentation Using a Multi-planar U-Net?

William Michael Laprade[1,2]([✉]) [ID], Mathias Perslev[1] [ID], and Jon Sporring[1,2] [ID]

[1] Department of Computer Science, University of Copenhagen (DIKU),
Copenhagen, Denmark
{wl,map,sporring}@di.ku.dk
[2] Center for Quantification of Image Data from MAX IV (QIM),
Technical University of Denmark (DTU), Lyngby, Denmark

Abstract. U-Net architectures are an extremely powerful tool for segmenting 3D volumes, and the recently proposed multi-planar U-Net has reduced the computational requirement for using the U-Net architecture on three-dimensional isotropic data to a subset of two-dimensional planes. While multi-planar sampling considerably reduces the amount of training data needed, providing the required manually annotated data can still be a daunting task. In this article, we investigate the multi-planar U-Net's ability to learn three-dimensional structures in isotropic sampled images from sparsely annotated training samples. We extend the multi-planar U-Net with random annotations, and we present our empirical findings on two public domains, fully annotated by an expert. Surprisingly we find that the multi-planar U-Net on average outperforms the 3D U-Net in most cases in terms of dice, sensitivity, and specificity and that similar performance from the multi-planar unit can be obtained from half the number of annotations by doubling the number of automatically generated training planes. Thus, sometimes less is more!

Keywords: 3D imaging · Segmentation · Deep learning · U-Net · Sparse annotations

1 Introduction

Deep learning methods for the segmentation of 3D image data typically require large quantities of manually labeled data for training. Often, similar structures in the images are labeled repeatedly, even when the model could learn from fewer samples. In this paper, we investigate how well the Multi-planar U-Net [5] extension of the 2D U-net [6] can learn to segment 3D images from only sparsely annotated label maps in comparison with the 3D U-Net [8].

Typical approaches to 3D image segmentation include fitting a 3D segmentation model to the data directly, or a 2D model to 2D image-slices along a single axis. The former approach is computationally demanding and requires

© Springer Nature Switzerland AG 2021
S. Engelhardt et al. (Eds.): DGM4MICCAI 2021/DALI 2021, LNCS 13003, pp. 209–216, 2021.
https://doi.org/10.1007/978-3-030-88210-5_20

more training data, while the latter is inherently limited in its ability to account for 3D structures.

In contrast, the multi-planar U-Net [5] makes use of the computational and statistical efficiency of the 2D U-Net while including resampled, off-plane training images. This is a welcome improvement to standard data augmentation since it results in improved estimates of the true underlying distribution of image patches as opposed to merely guessing, as often done when augmenting image data. For some datasets, such as brain tissue imaged using electron microscopy, there are no intrinsic orientations of the imaged objects within the volume at medium scale, and hence, it should be expected that the multi-planar U-Net performs well, since resampling planes at any angle will result in statistically similar images.

In this paper, we investigate the hypothesis that multi-planar sampling [5] allows the model to maximize the use of the available information in sparsely labeled datasets thus facilitating a stronger learning signal. Our paper is organized as follows: First, we motivate and formalize our method and the sampling technique. Then we present an empirical investigation on the relation between the number of annotated pixels and the Dice score on two public domain, expert-annotated 3-dimensional datasets (electron microscopy and cardiac MRI), and we compare the multi-planar U-Net and 3D U-Net's performance on similarly sparse datasets. Finally, we give our conclusions.

2   Methods

In this paper, we study the effect of sparsity on the multi-planar U-Net in comparison with the 3D U-Net. Therefore, we randomly select training-planes in which we randomly select annotation lines that are annotated by an expert. We analyze the performance of the 2D U-Net and the 3D U-Net for varying numbers of training-planes and annotation lines.

In practice, we simulate sparse, random annotations by sampling from fully annotated 3D images. Specifically, we randomly select a set of annotation-planes and a set of training-planes. We produce fully annotated 2D images by intersecting the annotation-planes with the fully annotated 3D-image, and we produce sparse training-images by intersecting the training-planes with the annotation-planes. The process is exemplified in Fig. 1.

We used trilinear interpolation and nearest-neighbor interpolation for sampling the image and the label sets, respectively. An example of a training-plane overlaid with 4 random annotation lines is shown in Fig. 2.

The multi-planar U-Net [5] is a 2D U-Net for 3D images, which as input takes resampled 2D images from multiple angles across a single 3D image and as output aggregates the resulting 2D images into a single 3D image. At the center of the multi-planar U-Net is the popular U-Net [6] architecture modified to use nearest neighbor up-sampling blocks [4] and batch normalization [1] layers. It has 4 down-sampling and 4 up-sampling blocks and a total of 31,044,289 trainable parameters. No information is supplied to the model regarding the position and

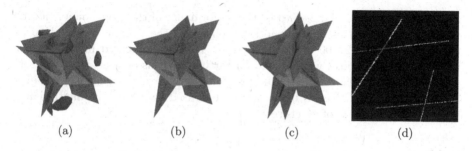

(a)   (b)   (c)   (d)

Fig. 1. (a) Mitochondria (green) and 4 annotation-planes, (b) sparsely-labeled volume used for training-plane generation, (c) an example training-plane (red) that slices through the sparsely-labeled volume, (d) the example training-plane, where only non-black pixels contribute to the loss-function. (Color figure online)

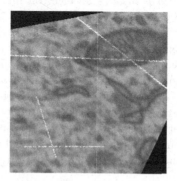

Fig. 2. An example of a training-plane taken from the electron microscopy mitochondria dataset.

orientation of training and prediction-planes. We aggregate the prediction-planes as the thresholded average across all such planes. For simplicity in our experiments, we choose $K = 3$ random prediction-angles and produce K sequences of parallel prediction planes such that all voxels are predicted by exactly K differently oriented prediction-planes. In practice, we rotate the volume to each orientation and present all 2D slices along the x-axis to the 2D U-Net, thus generating a rotated prediction volume P_k. Each prediction volume is then rotated back and the data is aggregated into single volume $P = \text{mean}(\{P_k\})$ which is considered the final segmentation.

We compare the multi-planar U-Net to a 3D U-Net implemented as in [8] with an input shape of $64 \times 64 \times 64$.

For both networks, we performed data augmentations on the fly and include a combination of 2D and 3D equivalent flips, rotations, scaling, brightness, contrast, gamma adjustments, and random noise. We minimize a masked cross-entropy loss function in which only annotated pixels in the input images contribute to the loss and backpropagation of gradients. We used the Adam optimizer [2] with a learning rate of $\eta = 10^{-5}$ and default $\beta_1 = 0.9$, $\beta_2 = 0.999$ and

$\epsilon = 10^{-7}$ parameters. We monitored the performance of the model on a held-out validation with half as many annotation-planes as the training set. Optimization continued for 22400 gradient updates. The best observed model (as per validation performance) was selected for further analysis on a held-out test set.

For the multi-planar U-Net, the loss function is evaluated on batches of 4 slices. For the 3D U-Net, a batch size of 1 is used to match the multi-planar version in the number of pixels seen.

3 Datasets

We consider 2 datasets: the publicly available mitochondria [3] and cardiac [7] datasets. Both are fully annotated by experts and have been used as benchmark datasets. The *mitochondria dataset* is an electron microscope image of a 5 × 5 × 5 μm section of the CA1 hippocampus region of a rodent brain with two annotated 165 × 768 × 1024 sub-volumes (a training and testing volume) with a voxel resolution of approximately 5 × 5 × 5 nm. The training volume is split into 4 sub-volumes of dimensions 165 × 448 × 448. The evaluation set contains four 165 × 165 × 165 sub-volumes taken from the testing volume.

The *cardiac dataset* consists of 20 mono-modal MRIs of dimension 320 × 320 × z, where z varies between 90 and 130 depending on the scan. Out of the 20 available MRIs, 4 were set aside and used as the held-out test set. The remaining 16 volumes were divided into 12 training volumes and 4 validation volumes. New train-validation splits were made in each experimental repetition.

4 Experiments and Results

We tested the performance of the multi-planar U-Net and the 3D U-Net as a function of two optimization hyperparameters: The number of annotation- and sampling-planes, which we denote n_a and n_t.

We fit both models using all combinations of $n_a \in \{4, 8, 12, 16\}$ and $n_t \in \{0, 128, 256, 384, 512\}$ and repeated each experiment 12 times with new annotation- and training-planes in each repetition. When the number of sampled training-planes $n_t = 0$, the model is fit to only the n_a fully annotated planes without resampling. In all other cases, each model observes *only* the sampled sparse training-planes. Some of the resulting statistics for $n_t > 0$ was fitted to

$$f(n_a, n_t) = \alpha_a \log(n_a) + \alpha_t \log(n_t) + \beta \tag{1}$$

using Matlab's Statistics and Machine Learning Toolbox function `fitlm` and we performed ANOVA tests using the `anova` function from the same toolbox. We report the F-score and the p-values of the fitted model against the models $\alpha_a = 0$, $\alpha_t = 0$, and $\alpha_a = \alpha_t = 0$. In the following, we will discuss the segmentation problem of the *mitochondria* and the *cardiac dataset*.

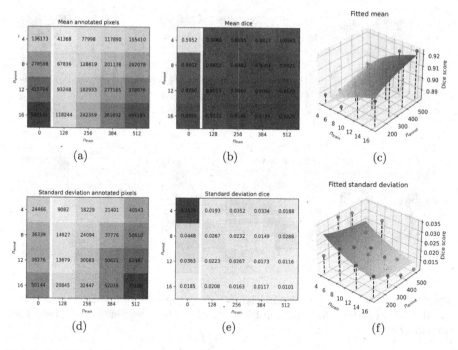

Fig. 3. (a) Mean number of annotated pixels in the set used to train each model, (b) Mean Dice scores on the held-out test set, (c) Best fit plane through mean Dice scores using log of n_a and n_t, (d) Standard deviation of the number of pixels used to train each model, (e) Standard deviation of Dice scores on the held-out test set, (f) Best fit plane through the standard deviations of Dice scores using the log of n_a and n_t.

Mitochondrial Segmentation: For the multi-planar U-Net, we evaluated the performance of each experimental configuration on the held-out test set. We report the mean and standard deviation Dice scores and the mean and standard deviation of the number of annotated pixels across the 12 repetitions. As expected, the performance for the mitochondria segmentation task improves as both the number of initial annotation planes and the number of generated training planes increases, see Fig. 3.

The models trained with $n_a = 4$ and $n_t = 0$ have the highest variability (standard deviation of 0.2826 and median absolute deviation of 0.1167) and lowest scores (mean of 0.5952 and median of 0.6929). This suggests a difficulty during optimization and is the result of some of the experimental runs overfitting to the 4 samples in the training set, thus performing poorly on the held-out test set.

Surprisingly, we see that we can achieve nearly the same performance with $n_a = 4$ and $n_t = 512$ as we can with $n_a = 16$ and $n_t = 0$ (0.8949 and 0.8955 respectively). Consequently, with this dataset we can achieve roughly the performance of 16 manually annotated planes with only 25% of the manual segmentation effort. Interestingly, models trained with $n_a = 4$ and $n_t = 512$ have, on

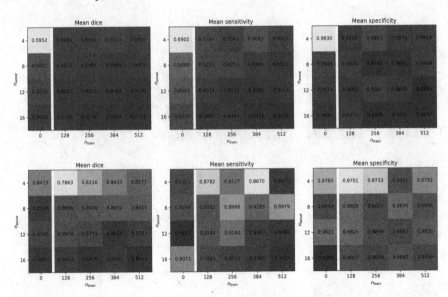

Fig. 4. Except for $n_t = 0$, the multi-planar U-Net (top row) consistently outperforms the 3D U-Net (bottom row) on mean Dice, sensitivity, and specificity on the mitochondria dataset.

average, more than 70% fewer annotated pixels in the dataset than those trained with $n_a = 16$ and $n_t = 0$ yet the model still performs just as well. Impressively, the highest Dice score achieved, 0.9228, (when $n_a = 16$ and $n_t = 512$) is comparable to the score of 0.9288 achieved with full supervision in [3].

We fitted (1) to the Dice scores resulting in $(\alpha_a, \alpha_t, \beta) = (0.019, 0.010, 0.80)$, see Fig. 3c. The (F-score, p-value) for $\alpha_a = 0$, $\alpha_t = 0$, and $\alpha_a = \alpha_t = 0$ was $(36, 9.1e - 9)$, $(9.9, 0.0019)$, and $(23, 1.1e - 9)$ respectively. Thus, for the mitochondria dataset, we observe that the Dice score has a statistically significant increase linearly in $\log(n_a)$ and $\log(n_t)$, and further by Fig. 3d, that the standard deviation also decrease linearly in $\log(n_a)$ and $\log(n_t)$.

For comparison, we also trained the 3D U-Net models with equivalent sparsity and 12 times for each combination. The mean Dice, sensitivity, and specificity scores for the multi-planar and the 3D U-Net are shown in Fig. 4. Fitting (1) to 3D U-Net's Dice scores gave $(\alpha_a, \alpha_t, \beta) = (0.042, 0.016, 0.68)$, and the (F-score, p-value) for $\alpha_a = 0$, $\alpha_t = 0$, and $\alpha_a = \alpha_t = 0$ was $(44.7, 2.6e - 10)$, $(6.2, 0.14)$, and $(25, 1.7e - 10)$ respectively. Thus, we observe that the Dice score for the 3D U-Net also has a statistically significant linear increase in $\log(n_a)$, and that this is slightly worse than the multi-planar U-Net. We further note that the mean Dice, sensitivity, and specificity fluctuates vary more in comparison to the multi-planar U-Net suggesting added optimization challenges in spite the reduced number of parameters for the 3D U-Net.

The training time for the 3D model takes approximately 3 h and 20 min, while the multi-planar version takes 1 h and 5 min. The longer training time observed in the 3D model is partially due to the data augmentations performed on-the-fly during training being more computationally intensive in 3D than 2D. The prediction times for the multi-planar and the 3D U-Net were 45 and 30 s per volume respectively, where the prediction time of the multi-planar U-Net depends on number of prediction planes to be aggregated.

Cardiac Segmentation: We performed a similar set of experiments on the cardiac dataset.

For the multi-planar U-Net, the mean Dice scores for $n_t > 0$ fell in the range $[0.74, 0.83]$. For $n_t = 0$, we observed mean dice scores in the range $[0.57, 0.75]$. Fitting (1) to the Dices scores for $n_t > 0$ gave $(\alpha_a, \alpha_t, \beta) = (0.045, -0.00075, 0.69)$. The (F-score, p-value) for $\alpha_a = 0$, $\alpha_t = 0$, and $\alpha_a = \alpha_t = 0$ was $(26, 8.2e - 7)$, $(0.0071, 0.93)$, and $(13, 5.1e - 6)$ respectively. We observe that $|\alpha_t| \approx 0$, and in comparison with $\alpha_t = 0$ the F-score is low and the p-value is high, hence, little new is learned when increasing the number of training planes above 128. The opposite is the case for α_a, and we further observe that the exponential increase in the number of annotation planes gives about twice the increase in Dice score as observed in the mitochondria dataset.

For the 3D U-Net, the mean Dice scores for $n_t > 0$ fell in the range $[0.36, 0.57]$. For $n_t = 0$, we observed mean dice scores in the range $[0.47, 0.5]$. Fitting (1) to the Dices scores for $n_t > 0$ gave we got $(\alpha_a, \alpha_t, \beta) = (0.011, 0.022, 0.35)$. The (F-score, p-value) for $\alpha_a = 0$, $\alpha_t = 0$, and $\alpha_a = \alpha_t = 0$ was $(0.20, 0.65)$, $(0.79, 0.38)$, and $(0.5, 0.61)$ respectively. We observe that the mean Dice scores are very low and that all the F-scores are low and all the corresponding p-value are high, hence, the fitted model did not learn the essential features of the cardiac dataset and the fit is not statistically significantly different from the constant model.

Comparing the multi-planar and the 3D U-Net, the training time was the same as for the mitochondria dataset for both models. The prediction times for the multi-planar and the 3D U-Net were 83 and 45 s per volume respectively. We observe that the multi-planar U-Net has learned essential features of the cardiac dataset, but the 3D U-Net has not.

5   Discussion

Segmenting data in 3D is usually the first step in analyzing 3D medical data. As such, we must do this in a way that is both accurate and time-efficient both in training/evaluation time as well as annotation time.

With the multi-planar U-Net [5] we can achieve a good segmentation of a 3D volume via a 2D U-Net from only sparsely annotated samples. The multi-planar U-Net has two important advantages over traditional 2D U-Nets applied to 3D data: 1) We can learn rotational invariance by resampling, where the added knowledge is from real data rather than from guessing by augmentation. 2) Sparse sampling reduces the annotations needed, while still statistically representing a large area in an image.

Comparing the multi-planar and 3D U-Net models it seems that the multi-planar method performs slightly better with faster training time but slower prediction time. The improved performance may be caused by the multi-planar U-Net having more parameters, but it is in spite of the multi-planar U-Net only having access to 2D planar views of the 3D data. Future investigation into improvements to the standard 3D U-Net may improve its performance.

For the multi-planar U-Net, we conclude that the combination of sparse annotation and a high number of random training-planes significantly lessens the annotation burden. E.g., comparing all entries in Fig. 3a and 3b for $n_a > 4$, we observe that the mean dice score is higher for all $n_t > 0$ as compared to $n_t = 0$ even though that the corresponding mean number of annotated pixels is lower.

In summary, we provide evidence that the multi-planar U-Net outperforms the standard 3D U-Net and that with a small initial set of samples, we can increase the segmentation performance by generating more unique datasets with fewer annotated pixels per sample, but with more variation in viewing angles.

References

1. Ioffe, S., Szegedy, C.: Batch normalization: accelerating deep network training by reducing internal covariate shift. In: International Conference on Machine Learning (ICML), pp. 448–456. PMLR (2015)
2. Kingma, D.P., Ba, J.: Adam: a method for stochastic optimization. In: International Conference on Learning Representations (ICLR) (2015)
3. Lucchi, A., Li, Y., Fua, P.: Learning for structured prediction using approximate subgradient descent with working sets. In: Proceedings of the IEEE Computer Society Conference on Computer Vision and Pattern Recognition (2013). https://doi.org/10.1109/CVPR.2013.259
4. Odena, A., Dumoulin, V., Olah, C.: Deconvolution and checkerboard artifacts. Distill (2016). https://doi.org/10.23915/distill.00003. http://distill.pub/2016/deconv-checkerboard
5. Perslev, M., Dam, E.B., Pai, A., Igel, C.: One network to segment them all: a general, lightweight system for accurate 3D medical image segmentation. In: Shen, D., et al. (eds.) MICCAI 2019. LNCS, vol. 11765, pp. 30–38. Springer, Cham (2019). https://doi.org/10.1007/978-3-030-32245-8_4
6. Ronneberger, O., Fischer, P., Brox, T.: U-net: convolutional networks for biomedical image segmentation. In: Navab, N., Hornegger, J., Wells, W.M., Frangi, A.F. (eds.) MICCAI 2015. LNCS, vol. 9351, pp. 234–241. Springer, Cham (2015). https://doi.org/10.1007/978-3-319-24574-4_28
7. Simpson, A.L., et al.: A large annotated medical image dataset for the development and evaluation of segmentation algorithms, February 2019
8. Çiçek, Ö., Abdulkadir, A., Lienkamp, S.S., Brox, T., Ronneberger, O.: 3D U-net: learning dense volumetric segmentation from sparse annotation. In: Ourselin, S., Joskowicz, L., Sabuncu, M.R., Unal, G., Wells, W. (eds.) MICCAI 2016. LNCS, vol. 9901, pp. 424–432. Springer, Cham (2016). https://doi.org/10.1007/978-3-319-46723-8_49

FS-Net: A New Paradigm of Data Expansion for Medical Image Segmentation

Xutao Guo[1], Yanwu Yang[1,2], and Ting Ma[1,2,3,4]($\boxtimes$)

[1] Department of Electronic and Information Engineering,
Harbin Institute of Technology (Shenzhen), Shenzhen, China
tma@hit.edu.cn
[2] Peng Cheng Laboratory, Shenzhen, Guangdong, China
[3] National Clinical Research Center for Geriatric Disorders,
Xuanwu Hospital Capital Medical University, Beijing, China
[4] Advanced Innovation Center for Human Brain Protection,
Capital Medical University, Beijing, China

Abstract. Pre-training can alleviate the requirement of labeling data for a new task. However, Pre-training as a sequential learning typically suffers in fact from forgetting the older tasks. Especially in complex medical image segmentation tasks, this problem is more prominent. To solve above problem, we propose a network structure based on feature space transformation (FS-Net) for data expansion of medical image segmentation. FS-Net share parameters during training to help exploiting regularities present across tasks and improving the performance by constraining the learned representation. In the experiment, we use M&Ms as the extended dataset of HVSMR, these two tasks have the same segmentation target (heart). The segmentation accuracy of FS-Net is up to 7.12% higher than the baseline network, which is significantly better than Pre-training. In addition, we use Brats2019 as expansion dataset on WMH, and the segmentation accuracy is improved by 0.77% compared with the baseline network. And Brats2019 (glioma) and WMH (white matter hyperintensities) have different segmentation targets.

Keywords: Data expansion · Pre-training · Medical image segmentation · Deep learning

1 Introduction

Segmentation and quantitative evaluation of region of interest in medical images are of great importance in formulating therapeutic strategies, monitoring the disease's progress and predicting the prognosis of patients [1]. Data-driven methods such as deep convolution neural network (DCNN) have recently achieved state-of-the-art performance in medical image segmentation tasks [2–7]. As we all know, one of the basic facts contributing to this success is a large amount of

S. Engelhardt et al. (Eds.): DGM4MICCAI 2021/DALI 2021, LNCS 13003, pp. 217–225, 2021.
https://doi.org/10.1007/978-3-030-88210-5_21

labeled training data [8,9]. However, it is incredibly difficult to construct large medical image datasets due to the rarity of diseases, patient privacy, medical experts' requirements for labels, and the cost and labor required for medical image processing.

Pre-training [10,11] has received much attention throughout the history of deep learning, which can alleviate the requirement of labeling data for a new task to a certain extent. Since many vision tasks are related, a pre-trained model on one dataset can help another. This is very helpful for downstream tasks to obtain "general features" in advance on large-scale image dataset [12]. Suppose taking natural image domain for example, it is now common practice to pre-train the backbones of object detection and segmentation models on ImageNet [13] classification. However, Pre-training as a sequential learning typically suffers in fact from forgetting the older tasks, a phenomenon aptly referred to as "catastrophic forgetting" in [14]. On the other hand, compared with natural images, medical images have huge differences among different datasets due to different organs, different types of diseases, different data acquisition methods and different data sizes, and their task complexity and difficulty are significantly improved. This makes it more difficult for pre-training to work.

To solve above problem, we propose a new paradigm (FS-Net) of data expansion for medical image segmentation. FS-Net share parameters during training to help exploiting regularities present across tasks and improving the performance by constraining the learned representation [15,16]. In order to solve "catastrophic forgetting" of Pre-training, FS-Net optimize one single network based on datasets of many different segmentation tasks at the same time, so that the network can extract more common features and greatly enhance the network capacity. FS-Net is to supersede the common paradigm of addressing different image understanding problems independently, using ad-hoc solutions and learning different and largely incompatible models for each of them. Just like the human brain is capable of addressing a very large number of different image analysis tasks, so it should be possible to develop models that address well and efficiently a variety of different computer vision problems, with better efficiency and generalization than individual networks. In order to train the dataset of different medical image segmentation tasks in the same network, we adopt two strategies: (1) Images channel coding and re-encoding the ground truth; (2) The feature transformation module is proposed to map different domains to the same separable feature space. The above two strategies can transform the data of different segmentation tasks into the form that a single network can train at the same time. Finally, we introduce the weighted loss function to optimize the whole network. The weighted loss mainly realizes the optimization of the target datset in the training process, and other dataset only play the role of constraint regularization [15].

In the experiment, we use M&Ms [17] as the extended dataset of HVSMR [18], these two tasks have the same segmentation target (heart). The segmentation accuracy of FS-Net is up to 7.12% higher than the baseline network, which is significantly better than Pre-training. In addition, we use Brats2019 [19–21]

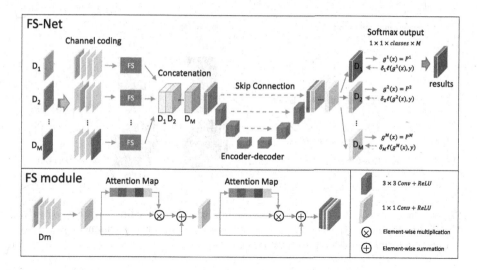

Fig. 1. An overview of our proposed FS-Net. After channel coding and feature transformation, the outputs of different FS branches are concatenated and then input to the segmentation network to realize simultaneous training of different datasets (concatenation is performed on batch size). The network output is split according to the previous concatenation, and the weighted loss function is used to calculate the loss value. Our implementation is available at https://github.com/********

as expansion dataset on WMH, and the segmentation accuracy is improved by 0.77% compared with the baseline network. Among them, Brats2019 and WMH have different segmentation targets (glioma, white matter hyperintensities). Especially the use of completely different datasets for a segmentation task to assist in improving the performance of the target task has very practical value.

2 Proposed FS-Net

The whole network consists of three parts. Firstly, the original data from different segmentation tasks are channel coded to further improve the discrimination between different tasks. At the same time, we make unified one-hot coding for the labels of different tasks to enable one single decoding branch to decode different classes of different tasks at the same time. Then, the encoded data is transformed by FS module to solve the problem that it is difficult for one single network to train data of different segmentation tasks at the same time. Finally, the transformed data of different tasks are concated to train the network, and an weighted loss function is introduced to optimize the whole network. The weighted loss function makes the network update the network mainly according to the main task, and other tasks only has similar regularization effect.

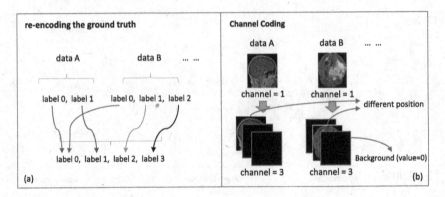

Fig. 2. An overview of the images channel coding and re-encoding the ground truth. The figure only gives the general form of implementation. Such as channel coding, We can also uniformly encoded data into two channels.

2.1 Images Channel Coding and Re-Encoding the Ground Truth

FS-Net share parameters during training to help exploiting regularities present across tasks and improving the performance by constraining the learned representation. In FS-Net, we use one single U-Net to train multiple datasets at the same time, as shown in Fig. 1. Please note that we have shared the encoding branch and decoding branch parameters at the same time. However, different datasets labels have overlapping problems, for example, all different datasets will contains label 1. In order to enable a single decoding branch to decode different classes of different datasets at the same time, we first make unified one-hot coding for different datasets, as shown in Fig. 2(a). In this way, different classes of different datasets have unique labels during simultaneous training.

On the other hand, different classes of different datasets may have similar texture features on the original images. At this time, if different data is directly input into FS-Net, it will be difficult for the network to distinguish the two classes. To solve this problem, we have performed channel coding on the original image to further improve the discrimination between different data of different tasks, as shown in Fig. 2(b). Although the channel coding changes the number of original images, the new channels are only simple background with the value of 0, so it does not destroy the texture information of the original images.

2.2 FS Module

Because there are certain differences in data distribution in different tasks or in different scenarios of the same task, domain adaptation is required in order to be able to train multiple datasets in a single network at the same time. To solve this problem, we design a feature space transformation module (FS), which can transform different data into the same feature space with the same dimension. Then input the converted data into the same segmentation network for training at the same time. The effect of this feature transformation is learned adaptively by the network under the supervision of the loss function. FS module is implemented by 1×1 convolution and channel attention mechanism. Channel attention mechanism can effectively reduce the information redundancy caused by previous image channel coding. The structure of channel attention module is illustrated in Fig. 1, which adopts the realization of [22]. Firstly, we calculate the channel attention map $\mathbf{X} \in R^{C \times C}$ from the inputting features $\mathbf{A} \in R^{C \times H \times W}$. We reshape $\mathbf{A}$ to $R^{C \times N}$, and then perform a matrix multi-plication between $\mathbf{A}$ and the transpose of $\mathbf{A}$. Finally, we apply a softmax layer to obtain the channel attention map $\mathbf{X}$. We perform a matrix multiplication between the transpose of $\mathbf{X}$ and $\mathbf{A}$ and reshape their result to $R^{C \times H \times W}$. Then we perform an element-wise sum operation with $\mathbf{A}$ to obtain the final output $\mathbf{E} \in R^{C \times H \times W}$. The final feature of each channel is a weighted sum of the features of all channels and original features, which models the long-range semantic dependencies between feature maps. It helps to boost feature discriminability.

2.3 Weighted Loss

We introduce the weighted loss function to optimize the whole network. The network output is split according to the previous concatenation, and the weighted loss function is used to calculate the loss value, as shown in Fig. 1. Let x and y represents training images and its label, respectively. The loss function is as following:

$$\Re(g(x), y) = \sum_{m=1}^{M} \delta_m \ell(g^m(x), y) \tag{1}$$

Where $g^m(x)$ is the softmax output of the m'th dataset, δ_m is the optimal weight. The weighted loss function sets a larger weight for the target task, which mainly realizes the optimization of the target data in the training process, and other data only play the role of constraint regularization. In this paper, we set the weight of the main task to 0.95 and the weight of the auxiliary task to 0.05.

3 Experiments

3.1 Datasets

HVSMR: The dataset is provided by the MICCAI Workshop on Whole-Heart and Great Vessel Segmentation from 3D Cardiovascular MRI in Congenital Heart Disease 2016 (HVSMR) [18]. HVSMR2016 provides three types of data. Here we only use full-volume images data for verification. Labels contain three different classes: Background, blood pool (BP) and ventricular myocardium (VM). All images are randomly split into training (6 images), test (4 images).

M&Ms: The dataset provides by Multi-Centre, Multi-Vendor & Multi-Disease Cardiac Image Segmentation Challenge (M&Ms) [17]. The training sets contain

75 training data with labels. The labels include contours for the left (LV, label 1) and right ventricle (RV, label 3) blood pools, as well as for the left ventricular myocardium (MYO, label 2).

WMH: This is our private data, which contain 53 annotated images. For each subject, a 3D T1 weighted image and a 3D T2-FLAIR image were provided. All images have been segmented by experienced clinicians, including Background, WMH (WM). All images randomly assign into a training set (33 images), a validation set (10 images), and a test set (10 images).

BraTS2019: The dataset is provided by the MICCAI Brain Tumor Segmentation Challenge 2019 (BraTS2019) [19–21]. Here we only use HGG for verification. Each case provides a T1, a T1Gd, a T2, and a FLAIR volume and the corresponding annotation results. Four different tissues were combined into three sets: (1) the whole tumor (label 1, 2, 4); (2) the tumor core (label 1, 4); (3) tumor enhanced area (label 4).

Table 1. The network architectures and training parameters used in our study.

Task	U-Net			Patch	Batch	lr	2D/3D	Loss
	Conv_s	Conv_e	depth					
HVSMR	16	128	4	$128 \times 128 \times 6$	6	0.0005	3D	Dice
M&Ms	16	128	4	$128 \times 128 \times 6$	6	0.0005	3D	Dice
WMH	16	128	4	$128 \times 128 \times 32$	4	0.0001	3D	Dice+ce
BraTS2019	16	128	4	$128 \times 128 \times 32$	4	0.0001	3D	Dice+ce

"Conv_s" and "Conv_e" represent the width of the first and last convolution layer of the coding layer, respectively. "ce" represent cross entropy loss function

Table 2. Results of FS-Net, U-Net and Pre-training in two data expansion scenarios.

Architecture	HVSMR		WMH
	BP	VM	WM
U-Net	74,90	85.59	78.25
Pre-training	76.13	86.57	78.31
FS-Net	**82.02**	**89.03**	**79.02**

3.2 Baselines and Implementation

The details of the network architectures and training parameters are shown in Table 1. Without loss of generality, we use U-Net [8] as the segmentation network in FS-Net, which is popular for bio-medical segmentation. Besides, we did not add too many extra tricks and post-processing techniques in order to exclude other interference factors and show the effect of FS-Net simply and directly. U-Net has 4 encoder and decoder steps. Each encoder step is made by two convolutions layer with stride 1. The decoder is made by transpose convolution with stride 2, followed by a convolution with stride 1. Each convolution has kernel size 3, He weights initialization [23] and 1 padding. The number of filters are doubled at each encoder step.

Our experiments are implemented with Pytorch. Adam used as the optimizer with different learning rate. To ensure the experiment's objectivity, we strictly guarantee that the training parameters are consistent. All the experiments are performed using four NVIDIA TESLA V-100 (Pascal) GPUs with 32 GB memory each. Dice-coefficient is used to evaluate segmentation performance.

3.3   Results

Here, we divide data expansion into two scenarios: First, different datasets with the same segmentation target, such as HVSMR (heart) and M&Ms (heart); Second, different datasets have different segmentation targets, such as Brats2019 (glioma) and WMH (white matter hyperintensities); There are certain differences in the data distribution of different tasks or different scenarios of the same task. Table 2 shows we use M&Ms as the extended dataset of HVSMR, the segmentation accuracy of FS-Net is up to 7.12% higher than the baseline network. In addition, we use Brats2019 as expansion dataset on WMH, and the segmentation accuracy is improved by 0.77% compared with the baseline network. Especially the use of completely different datasets for a segmentation task to assist in improving the performance of the target task has very practical value. M&Ms and Brats2019 are used as expansion data without calculating their segmentation accuracy.

Table 3. Ablation study for image channel coding, FS-module, and weighted loss function

HVSMR	BP	VM
FS-Net	**82.02**	**89.03**
Joint training	76.22	87.75
FS-Net w/0 channel coding	81.77	88.38
FS-Net w/0 FS-module	81.22	88.08
FS-Net w/0 weighted loss	77.65	88.17

3.4   Ablation Study

In this section, we perform ablation experiments to investigate the effect of different components in FS-Net. These experiments provide more insights into FS-Net. All following experiments were performed on HVSMR dataset. Table 3 shows image channel coding, FS-module, and weighted loss function can improve the accuracy of the network. Among them, the effect of weighted loss function is the most obvious. We also compared the joint training, and its performance is also lower than FS-Net.

4 Conclusions

In this paper, we proposed a new paradigm (FS-Net) of data expansion for medical image segmentation. FS-Net effectively solves the problems of Pre-training as a sequential learning typically suffers in fact from forgetting the older tasks. FS-Net shares encoding and decoding parameters at the same time during training to help exploiting regularities present across tasks and improving the performance by constraining the learned representation. We show that FS-Net has a significant improvement over pre-traing through two sets of experiments in different scenarios. Especially the use of completely different data for a segmentation task to assist in improving the performance of the target task has very practical value.

Acknowledgment. This study is supported by grants from the National Key Research and Development Program of China (2018YFC1312000) and Basic Research Foundation of Shenzhen Science and Technology Stable Support Program (GXWD20201230155 427003-20200822115709001).

References

1. Wolz, R., et al.: Automated abdominal multi-organ segmentation with subject-specific atlas generation. IEEE Trans. Med. Imaging **32**(9), 1723–1730 (2013)
2. Sarker, M.M.K., et al.: SLSDeep: skin lesion segmentation based on dilated residual and pyramid pooling networks. In: Frangi, A.F., Schnabel, J.A., Davatzikos, C., Alberola-López, C., Fichtinger, G. (eds.) MICCAI 2018. LNCS, vol. 11071, pp. 21–29. Springer, Cham (2018). https://doi.org/10.1007/978-3-030-00934-2_3
3. Ibtehaz, N., Sohel Rahman, M.: MultiResUNet: rethinking the U-Net architecture for multimodal biomedical image segmentation. arXiv:1902.04049 (2019). http://arxiv.org/abs/1902.04049
4. Jiang, H., et al.: Improved cGAN based linear lesion segmentation in high myopia ICGA images. Biomed. Opt. Express **10**(5), 2355 (2019)
5. Trullo, R., Petitjean, C., Ruan, S., Dubray, B., Nie, D., Shen, D.: SEG-mentation of organs at risk in thoracic CT images using a SharpMask architecture and conditional random fields. In: Proceedings of IEEE 14th International Symposium on Biomedical Imaging (ISBI), pp. 1003–1006, April 2017
6. Chiu, S.J., Allingham, M.J., Mettu, P.S., Cousins, S.W., Izatt, J.A., Farsiu, S.: Kernel regression based segmentation of optical coherence tomography images with diabetic macular edema. Biomed. Opt. Express **6**(4), 1172 (2015)
7. Venhuizen, F.G., et al.: Deep learning approach for the detection and quantification of intraretinal cystoid fluid in multivendor optical coherence tomography. Biomed. Opt. Express **9**(4), 1545 (2018)
8. Ronneberger, O., Fischer, P., Brox, T.: U-net: convolutional networks for biomedical image segmentation. In: Navab, N., Hornegger, J., Wells, W.M., Frangi, A.F. (eds.) MICCAI 2015. LNCS, vol. 9351, pp. 234–241. Springer, Cham (2015). https://doi.org/10.1007/978-3-319-24574-4_28
9. Halevy, A., Norvig, P., Pereira, F.: The unreasonable effectiveness of data. IEEE Intell. Syst. **24**(2), 8–12 (2009)

10. Girshick, R., Donahue, J., Darrell, T., Malik, J.: Rich feature hierarchies for accurate object detection and semantic segmentation. In: CVPR (2014)
11. Long, J., Shelhamer, E., Darrell, T.: Fully convolutional networks for semantic segmentation. In: CVPR (2015)
12. Zoph, B., Ghiasi, G., Lin, T.Y., et al.: Rethinking pre-training and self-training (2020)
13. Russakovsky, O., et al.: ImageNet large scale visual recognition challenge. Int. J. Comput. Vis. **115**(3), 211–252 (2015)
14. French, R.M.: Catastrophic forgetting in connectionist networks. Trends Cogn. Sci. **3**(4), 128–135 (1999)
15. Bilen, H., Vedaldi, A.: Universal representations: the missing link between faces, text, planktons, and cat breeds. arXiv preprint arXiv:1701.07275 (2017)
16. Caruana, R.: Multitask learning. Mach. Learn. **28**, 41–75 (1997)
17. Campello, M., Lekadir, K.: Multi-centre multi-vendor & multi-disease cardiac image segmentation challenge (M&Ms). In: Medical Image Computing and Computer Assisted Intervention (2020)
18. Pace, D.F., Dalca, A.V., Geva, T., Powell, A.J., Moghari, M.H., Golland, P.: Interactive whole-heart segmentation in congenital heart disease. In: Navab, N., Hornegger, J., Wells, W.M., Frangi, A.F. (eds.) MICCAI 2015. LNCS, vol. 9351, pp. 80–88. Springer, Cham (2015). https://doi.org/10.1007/978-3-319-24574-4_10
19. Menze, B.H., et al.: The multimodal brain tumor image segmentation benchmark (BRATS). IEEE Trans. Med. Imaging **34**(10), 199–2024 (2014)
20. Bakas, S., et al.: Advancing the cancer genome atlas glioma MRI collections with expert segmentation labels and radiomic features. Sci. Data **4**, 170117 (2017)
21. Bakas, S., et al.: Identifying the best machine learning algorithms for brain tumor segmentation, progression assessment, and overall survival prediction in the BRATS challenge. arXiv preprint arXiv:1811.02629 (2018)
22. Fu, J., et al.: Dual attention network for scene segmentation. In: 2019 IEEE/CVF Conference on Computer Vision and Pattern Recognition (CVPR). IEEE (2020)
23. He, K., et al.: Delving deep into rectifiers: surpassing human-level performance on ImageNet classification. In: Proceedings of the IEEE International Conference on Computer Vision (2015)

An Efficient Data Strategy for the Detection of Brain Aneurysms from MRA with Deep Learning

Youssef Assis[1]([✉]), Liang Liao[2], Fabien Pierre[1], René Anxionnat[2,3], and Erwan Kerrien[1]

[1] University of Lorraine, CNRS, LORIA, 54000 Nancy, France
youssef.assis@loria.fr
[2] Department of Diagnostic and Therapeutic Neuroradiology, University of Lorraine, Nancy, France
[3] IADI, INSERM U1254, University of Lorraine, Nancy, France

Abstract. The detection of intracranial aneurysms from Magnetic Resonance Angiography images is a problem of rapidly growing clinical importance. In the last 3 years, the raise of deep convolutional neural networks has instigated a streak of methods that have shown promising performance. The major issue to address is the very severe class imbalance. Previous authors have focused their efforts on the network architecture and loss function. This paper tackles the data. A rough but fast annotation is considered: each aneurysm is approximated by a sphere defined by two points. Second, a small patch approach is taken so as to increase the number of samples. Third, samples are generated by a combination of data selection (negative patches are centered half on blood vessels and half on parenchyma) and data synthesis (patches containing an aneurysm are duplicated and deformed by a 3D spline transform). This strategy is applied to train a 3D U-net model, with a binary cross entropy loss, on a data set of 111 patients (155 aneurysms, mean size 3.86 mm $\pm$ 2.39 mm, min 1.23 mm, max 19.63 mm). A 5-fold cross-validation evaluation provides state of the art results (sensitivity 0.72, false positive count 0.14, as per ADAM challenge criteria). The study also reports a comparison with the focal loss, and Cohen's Kappa coefficient is shown to be a better metric than Dice for this highly unbalanced detection problem.

Keywords: Brain aneurysm detection · Data sampling · CNN

1 Introduction

Intracranial aneurysms are local dilations of the cerebral blood vessels. Their rupturing accounts for 85% of subarachnoid hemorrhages (SAH), and is related to high mortality and morbidity rates [13]. The generalization of radiologic examinations in the diagnostic process has exposed the detection of unruptured aneurysms as a problem of increasing clinical importance. The need to browse through 3D Computed Tomography Angiography (CTA) or Magnetic

S. Engelhardt et al. (Eds.): DGM4MICCAI 2021/DALI 2021, LNCS 13003, pp. 226–234, 2021.
https://doi.org/10.1007/978-3-030-88210-5_22

Resonance Angiography (MRA) data in an ever increasingly time-constrained clinical setting however leads to inevitable errors. The innocuity of 3D Time-of-Flight (TOF) MRA makes it particularly suited for screening, even though the detection of small aneurysms (<5 mm) may be challenging [8]. A reliable automated method would be a valuable asset to assist radiologists in their clinical routine.

The first computer-assisted detection (CAD) system reported in the literature [1] was based on traditional image processing. Recently, convolutional neural networks (CNNs) have proven their superior performance in visual tasks, including medical image analysis. Detecting brain aneurysms is very challenging because aneurysms are scarce (a few tens to hundreds of positive voxels among millions in the MRA data), and their number is indefinite a priori. Therefore, it has been but very recently that deep learning approaches have been investigated in that context. First, several 2D approaches have been proposed in the literature [12,17], but all most recent approaches are fully 3D. The performance of the dual-path multiscale DeepMedic model was deemed promising [14] but as a complement to an expert reader [6]. Last year, the ADAM challenge enabled an objective comparison of a variety of other 3D approaches. The leading 3 methods for the detection task were based on 3D U-net [3] to evade the problem of the indefinite number of aneurysms through the generation of a heat map. The scarcity, and thereby high class imbalance, was tackled either through the loss function and/or the model. The Dice similarity coefficient, and the Binary Cross Entropy (BCE) and TopK losses were combined to form an ensemble loss in [11]. A different ensemble loss approach was taken in [18]. Fours models based on the No New-Net were trained and the final segmentation was decided in a majority voting amongst these models. The leader method [2] focused on the model architecture with a Retina U-net model that aggregates an encoder network and a feature pyramid network to guide the high-resolution detection with strong semantic features at low resolution. The actual impact of all these variants was questioned by the recent emergence of a new leader with a vanilla 3D U-net model trained with a combined Dice and BCE loss, and a prediction based on an ensemble of 5 models [19]. We believe a major difference dwells in the samples generation. If the first 3 methods used large patches ($\{192, 224, 256\} \times 256 \times 56$ voxels), this last one used 128^3 patches with a rich data augmentation process. Another limitation that we see is the small size of the dataset and the drudgery of the voxel-wise annotation.

The current study focuses on the data strategy to generate input samples that are better designed to tackle the class imbalance and reduced data set issues. We used a vanilla 3D U-net in all our experiments but small patches (48^3) were used. Aneurysm annotation is approximate but fast. This approximation is claimed to be precise enough for the detection task. A combination of guided sample selection and sample synthesis is also proposed, and BCE and focal losses are compared. Finally, we advocate for the use of Cohen's Kappa as a better metric than Dice in this class imbalance situation.

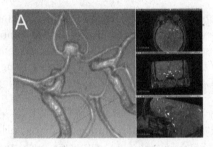

Fig. 1. Aneurysm annotation as an approximate sphere with 2 points in Slicer

2   Materials and Methods

2.1   Dataset and Data Annotation

A total of 111 TOF-MRA examinations (56 females, 55 males) were collected at our medical institution between April 2015 and January 2020. The criterion for inclusion was the presence of an aneurysm. All aneurysms are saccular. Criteria for exclusion were any pre-treated aneurysm and large aneurysms (>20 mm). The images were acquired on a 3T scanner (GE Healthcare) with the following parameters: $TR = 28$ ms, $TE = 3.4$ ms, slice thickness $= 0.8$ mm, $FOV = 24$, flip angle $= 17°$, 4 slabs (54 slices/slab), acquisition time $= 6$ min 28 s, resulting in $512 \times 512 \times 254$ volumes with a $0.47 \times 0.47 \times 0.4$ mm^3 voxel size. Each DICOM data was anonymized and converted to NIfTI format on the clinical site before processing. Each examination contained from one (81/111) to five aneurysms (1 case) for a total of 155 aneurysms with a mean diameter of 3.86 mm $\pm$ 2.39 mm (min: 1.23 mm, max: 19.63 mm). These were mostly small aneurysms since 60 were below 3 mm and 66 between 3–5 mm, which makes it a challenging dataset.

Previous works rely on databases where aneurysms have been segmented voxel-wise. This annotation is both tedious and tainted with intra- and inter-rater variability. Since we only aim at detecting aneurysms, we deployed a less accurate but much faster annotation: each aneurysm was annotated, by a radiologist with 10 years of experience, by placing two points, one at the center of the neck and the other at the dome, so as to define a sphere that approximated the aneurysm sack. 3D Slicer software was used [7] to place points in volume rendering view (see Fig. 1).

2.2   Model Implementation

Our software code was written in Python (3.8.5) using Keras (2.4.3). It was based on D.G. Ellis's open-source 3D U-net implementation [5] with 4 layers. Our first investigations with the available regular patch sampling did not show convergence of the model. As a consequence, the patch generator described below was plugged as input to the model. The following hyperparameters were used: 100 epochs, constant learning rate $= 10^{-4}$, BCE loss, Adam optimizer, batch

Table 1. Comparison of model variations: *Model0* is our proposed model. Dice and κ (Kappa) coefficients were evaluated on the validation set at the end of training. Other ADAM and patch-wise metrics were measured on the test set (11 patients).

Model	Validation set		ADAM metrics		Patch-wise metrics			
	Dice	κ	Sensitivity	FPs/case	TP	FP	FN	TN
Model0	0.339	0.665	0.970	0.454	14	5	2	2194
Model1	0.089	0.527	0.803	0.190	12	2	4	2197
Model2	0.038	−1.21e-8	0	0	0	0	16	2199
Model3	0.434	0.772	0.879	1.545	14	11	2	2188
Model4	0.245	0.589	0.833	1.0	13	8	3	2191

size $= 10$, with batch normalization. Each input volume was normalized between 0 and 1.

A full volume could be predicted by patch reconstruction: the initial volume was resampled to an isotropic 0.4 mm voxel size; predictions were computed for patches that cover the entire volume; and the resulting volume was resampled to the original resolution. To avoid border effects due to convolutions on small patches, an overlap of 8 voxels was considered between neighboring patches and only the central $32 \times 32 \times 32$ part of the patches were juxtaposed to cover the final volume.

2.3 Patch Generation and Data Augmentation

The discriminative power of a classifier depends on its capacity to statistically model both the background (negative samples) and the foreground (positive samples). In large patches approaches, an aneurysm is present in most patches, which requires healthy patients in the database. In our small patch approach, negative (aneurysm-free) patches are very common outside the aneurysm surroundings, but multiple instances need to be extracted from each patient data to build reliable statistics on the background. On the opposite side, only one single positive patch exists for each aneurysm. Adapted data sampling is an efficient strategy to handle this problem [9]. Our first sampling strategy then consists in duplicating 50 times each positive patch, centered on each aneurysm. A variety of shapes are synthesized by applying a random distortion to each duplicate: each control point on a $3 \times 3 \times 3$ lattice enclosing the patch, except the central point, is moved randomly by 4 mm in all 3 space directions and patch voxel locations in the original volume are computed using cubic spline interpolation. But class imbalance also emanates from the vascular information that only represents 3 to 5% of the background signal. In order to guide the model to discriminate between healthy and pathological vessels, our second sampling strategy consists in taking half of the negative samples centered on a blood vessel. The 100 brightest voxels were selected as patch centers. The other 100 centers were randomly selected within voxel values between the 20th and the 80th percentiles. Patch overlap was avoided by enforcing a minimum 20 mm distance between any two patch centers.

As a result we used 200 negative patches, and 50 positive patch duplicates. We used $48 \times 48 \times 48$ patches with an isotropic voxel size of 0.4 mm, closest to the nominal resolution, so that patches were cubes with a side length of 19 mm. Data augmentation was applied in the process with random rotations by 0 to 180° and shifts by 10 mm in all 3 space directions.

2.4 Metrics and Performance Evaluation

The training was monitored with the Dice coefficient. However, since the aneurysms are scarce and small, this metric lacks sensitivity to detection errors. We also computed Cohen's Kappa coefficient (κ) [4], that is more robust to class imbalance. These metrics are computed voxel-wise on the collection of input patches.

The performance of a model was evaluated on a test set, using mean sensitivity and FP count/case scores as defined for Task 1 in the ADAM challenge [16]. The connected components (CC) in both ground truth and predicted full volumes were labeled. A True Positive (TP) is a CC in the ground truth that contains the center of gravity of a predicted CC. A False Negative (FN) is a CC in the ground truth with no such predicted CC. A False Positive (FP) is a predicted CC whose center is not contained in any ground truth CC.

However, the above metrics do not enable True Negative count, and thereby prevent from computing specificity. Thereafter, we also computed patch-wise statistics (no positive duplicate): a patch is considered positive if it contains a positive voxel, else it is negative. This enables to compute a full confusion matrix.

3 Experiments and Results

3.1 Ablation Study

A first set of experiments aimed at evaluating the relevance of various parts of the model and the patch sampling strategy described in Sects. 2.2 and 2.3, which will be denoted as *Model0*. 4 variants were tested.

BCE is very sensitive to class imbalance [15]. In order to see the effectiveness of our data strategy to counter class imbalance, *Model1* was trained the same as *model0* but using the focal loss [10], that was designed to focus the training on the minority class. *Model2* only generates 5 duplicates (instead of 50) for each positive patch. *Model3* uses 50 duplicates but without random distortion applied. *Model4* only considers 100 background patches (instead of 200, 50 on vessels, 50 outside).

The data set was split into 3 sets used for: training (78 cases, 70%), validation (22 cases, 20%) and test (11 cases, 10%). All models were trained with the training set, and were monitored with the validation set. Table 1 reports the results

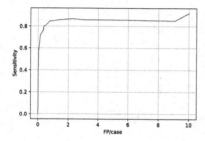

Method	Sensi-tivity	FP count /case
abc [19]	0.68	0.40
mibaumgartner [2]	0.67	0.13
joker [18]	0.63	0.16
junma [11]	0.61	0.18
Our model	0.72	0.14

Fig. 2. (left) FROC curve for our model: AUC = 85.24%. (right) Comparison with 4 leading methods in ADAM challenge (in decreasing order).

of this study. Dice and κ are observed on the validation set at the end of training. Other performance metrics (see Sect. 2.4) are computed on the test data set.

No real improvement could be observed with the focal loss (*Model1*), and BCE is even more sensitive, which demonstrates the efficacy of our sampling strategy. The Dice score remained very low because the predicted CC were very small. However, κ could better capture the relatively good performance of this model. *Model2* did not converge: class imbalance is indeed an issue. *Model3* provided good results but with too many FPs, due the lack of diversity in the aneurysm shapes shown to the model during training. The excess in positive voxels leads to larger TPs, which explains the better Dice and κ scores. Finally, the class balance is improved in *Model4*, but it underperforms *Model0* because the sample size for the negative patches is too small to reliably model background statistics.

3.2 5-Fold Validation

The global performance of our proposed model (*Model0*) was assessed using 5-fold validation. 5 models were trained, each time with 4 subsets for training and leaving one subset for test. Predictions were generated for each patient in each test set, providing a prediction for each patient. The mean aneurysm diameters in the 5 splits were: 3.82 mm, 3.74 mm, 3.96 mm, 3.84 mm and 3.93 mm.

Figure 2 displays the Free-response Receiver Operating Characteristics (FROC) curve: It reports the sensitivity and FP count scores, as per ADAM, computed on all 111 patients in our dataset for various detection thresholds. By adjusting the detection threshold, and by comparison with abc method, our method achieves a sensitivity of 0.80 @ 0.40 FP/case, and, with mibaumgartner method, a sensitivity of 0.70 @ 0.13 FP/case. The Area Under Curve (AUC) is 85.24%. The optimal detection threshold was determined as the closest point to the upper left corner. Our model reaches a sensitivity of 0.72@0.14 FP/case.

4   Discussion

In this study, in order to determine the impact of our data strategy we voluntarily used a vanilla 3D U-net model with a BCE score and simple optimization process

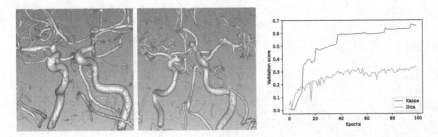

Fig. 3. (left) Branching of small arteries (here, ophtalmic artery, see arrow) may be mistaken for an aneurysm (predicted CC are in red, annotation points are present). (middle) Example of an overlooked aneurysm (arrow). (right) Typical surge of κ is a good predictor of the final convergence (around epoch 10).

(e.g. fixed learning rate and number of epochs). The focus was put on the data to assess the impact of various aspects of their preparation on learning.

First, a rough, but fast annotation was employed, which enables to rapidly label a large number of MRA volumes. Besides, a small patch approach was chosen. Small, non-intersecting patches are assumed independent, which allows for an efficient exploitation of even a small set of original MRA images (111). Second, we proposed an adapted data sampling process in two steps. On one side, guided sampling: The negative (aneurysm-free) patches are extracted by half centered on blood vessels and the other half elsewhere. We have shown that 200 patches were better able than 100 to capture the background statistics (*Model0* vs *Model4*). On the other side, data synthesis: The positive patches are duplicated 50 times, which enables to counter the high class imbalance (*Model0* vs *Model2*), and various shapes are synthesized by applying random non-rigid distortions, which describes the foreground statistics more accurately and enables a reduction of FPs (*Model0* vs *Model3*).

The proposed model has a sensitivity of 0.72, with a FP count/case of 0.14. FROC analysis showed that it is competitive with the best current leading methods in the ADAM challenge. Our method will have to be adapted to the ADAM challenge conditions for a definite comparison to be made. Future efforts will aim at further reducing FPs. Tests with the Focal Loss (*Model1*) generated smaller CC, which reduced the FP score but at the expense of sensitivity. Besides cases that are easy for a radiologist to discard, the most challenging FPs are located where a small artery, close to the resolution limit, branches onto a large artery. These are mistaken for small aneurysms (see Fig. 3, left). Indeed the performances of our model are lower on small aneurysms. Of all 155 aneurysms, 34 were not detected (FN). But 18 of these FNs had a diameter below 2 mm and 10 more were below 3 mm. The sensitivity of our model is 0.53 for aneurysms smaller than 2 mm, and reaches 0.89 for larger aneurysms. But this difficulty in detecting small aneurysms is inherent to MRA [8]. Note that during the visual review of the results by a radiologist with 30 years of experience, 8 FPs proved to be actual aneurysms that had been overlooked during the initial annotation (see Fig. 3, middle).

In our experiments, we observed a typical sudden surge in κ score that was correlated to a satisfactory convergence of the training phase (see Fig. 3, right). We interpreted it as a better sensitivity of κ over Dice to even small intersections between prediction and ground truth volumes.

5 Conclusion

In this paper, we presented an efficient data sampling strategy to detect intracranial aneurysms from MRA images, that is able to reach a state of the art sensitivity of 0.72 at 0.14 FP/case. Joining forces will hopefully decrease the number of FPs to design a more specific classifier. A future extension of this work is combining this data strategy with more sophisticated architectures and loss functions whose efficacy has been demonstrated, in particular by the ADAM challenge. Furthermore, a current work in progress investigates the κ score as a loss function, to leverage its capacity to assess the quality of a classifier despite the highly class imbalance problem.

Acknowledgments. We want to thank the Grand Est region and the regional and university hospital center (CHRU) of Nancy in France for funding this work. Experiments presented in this paper were carried out using the Grid'5000 experimental testbed, being developed under the INRIA ALADDIN development action with support from CNRS, RENATER and several Universities as well as other funding bodies (see https://www.grid5000.fr).

References

1. Arimura, H., Li, Q., Korogi, Y., et al.: Computerized detection of intracranial aneurysms for three-dimensional MR angiography: feature extraction of small protrusions based on a shape-based difference image technique. Med. Phys. **33**(2), 394–401 (2006)
2. Baumgartner, M., Jaeger, P., Isensee, F., et al.: Retina U-Net for aneurysm detection in MR images. In: Automatic Detection and SegMentation Challenge (ADAM) (2020). https://adam.isi.uu.nl/results/results-miccai-2020/participating-teams-miccai-2020/ibbm/
3. Çiçek, Ö., Abdulkadir, A., Lienkamp, S.S., Brox, T., Ronneberger, O.: 3D U-Net: learning dense volumetric segmentation from sparse annotation. In: Ourselin, S., Joskowicz, L., Sabuncu, M.R., Unal, G., Wells, W. (eds.) MICCAI 2016. LNCS, vol. 9901, pp. 424–432. Springer, Cham (2016). https://doi.org/10.1007/978-3-319-46723-8_49
4. Cohen, J.: A coefficient of agreement for nominal scales. Educ. Psychol. Measur. **20**(1), 37–46 (1960)
5. Ellis, D.: 3D U-Net convolution neural network with Keras (2017). https://github.com/ellisdg/3DUnetCNN (legacy branch, commit dc2d0604499298266e7aaf1db68603288bd34577)
6. Faron, A., Sichtermann, T., Teichert, N., et al.: Performance of a deep-learning neural network to detect intracranial aneurysms from 3D TOF-MRA compared to human readers. Clin. Neuroradiol. **30**(3), 591–598 (2020)

7. Fedorov, A., Beichel, R., Kalpathy-Cramer, J., et al.: 3D Slicer as an image computing platform for the quantitative imaging network. Magn. Reson. Imaging **30**(9), 1323–1341 (2012). https://slicer.org. pMID: 22770690

8. Jang, M., Kim, J., Park, J., et al.: Features of "false positive" unruptured intracranial aneurysms on screening magnetic resonance angiography. PloS One **15**(9), e0238597 (2020)

9. Johnson, J.M., Khoshgoftaar, T.M.: Survey on deep learning with class imbalance. J. Big Data **6**(1), 1–54 (2019). https://doi.org/10.1186/s40537-019-0192-5

10. Lin, T.Y., Goyal, P., Girshick, R., et al.: Focal loss for dense object detection. In: Proceedings of the IEEE International Conference on Computer Vision (ICCV), pp. 2980–2988 (2017)

11. Ma, J., An, X.: Loss ensembles for intracranial aneurysm segmentation: an embarrassingly simple method. In: Automatic Detection and SegMentation Challenge (ADAM) (2020). https://adam.isi.uu.nl/results/results-miccai-2020/participating-teams-miccai-2020/junma-2/

12. Nakao, T., Hanaoka, S., Nomura, Y., et al.: Deep neural network-based computer-assisted detection of cerebral aneurysms in MR angiography. J. Magn. Reson. Imaging **47**(4), 948–953 (2018)

13. Shi, Z., Hu, B., Schoepf, U., et al.: Artificial intelligence in the management of intracranial aneurysms: current status and future perspectives. Am. J. Neuroradiol. **41**(3), 373–379 (2020)

14. Sichtermann, T., Faron, A., Sijben, R., et al.: Deep learning-based detection of intracranial aneurysms in 3D TOF-MRA. Am. J. Neuroradiol. **40**(1), 25–32 (2019)

15. Asgari Taghanaki, S., Abhishek, K., Cohen, J.P., Cohen-Adad, J., Hamarneh, G.: Deep semantic segmentation of natural and medical images: a review. Artif. Intell. Rev. **54**(1), 137–178 (2020). https://doi.org/10.1007/s10462-020-09854-1

16. Taha, A., Hanbury, A.: Metrics for evaluating 3D medical image segmentation: analysis, selection, and tool. BMC Med. Imaging **15**(1), 1–28 (2015)

17. Ueda, D., Yamamoto, A., Nishimori, M., et al.: Deep learning for MR angiography: automated detection of cerebral aneurysms. Radiology **290**(1), 187–194 (2018)

18. Yang, Y., Lin, Y., Li, Y., et al.: Automatic aneurysm segmentation via 3D U-Net ensemble. In: Automatic Detection and SegMentation Challenge (ADAM) (2020). https://adam.isi.uu.nl/results/results-miccai-2020/participating-teams-miccai-2020/joker/

19. Yu, H., Fan, Y., Shi, H.: Team ABC. In: Automatic Detection and SegMentation Challenge (ADAM) (2020). https://adam.isi.uu.nl/results/results-live-leaderboard/abc/

Evaluation of Active Learning Techniques on Medical Image Classification with Unbalanced Data Distributions

Quok Zong Chong[1,2]($\boxtimes$), William J. Knottenbelt[1], and Kanwal K. Bhatia[2]

[1] Department of Computing, Imperial College London, London, UK
qzc17@imperial.ac.uk
[2] Metalynx Ltd., London, UK

Abstract. In supervised image classification, convolutional deep neural networks have become the dominant methodology showing excellent performance in a number of tasks. These models typically require a very large number of labelled data samples to achieve required performance and generalisability. While data acquisition is relatively easy, data labelling, particularly in the case of medical imaging where expertise is required, is expensive. This has led to the investigation of active learning methods to improve the effectiveness of choosing which data should be prioritised for labelling. While new algorithms and methodologies continue to be introduced for active learning, each reporting improved performance, one key aspect that can be overlooked is the underlying data distribution of the dataset. Many active learning papers are benchmarked using curated datasets with balanced class distributions. This is not representative of many real-world scenarios where the data acquired can be heavily skewed towards a certain class. In this paper, we evaluate the performance of several established active learning techniques on an unbalanced dataset of 15153 chest X-Ray images, forming a more realistic scenario. This paper shows that the unbalanced dataset has a significant impact on the performance of certain algorithms, and should be considered when choosing which active learning strategy to implement.

1 Introduction

Deep learning has become the primary methodology for tackling image classification tasks, ever since AlexNet took the top result in the ImageNet 2012 challenge. Since then, subsequent leader-boards on the ImageNet challenges have been filled with different and novel deep learning implementations [5], which cements deep convolutional models as the go-to method to deal with image classification tasks.

Deep learning has shown a number of successes in medical image classification tasks in recent years. However, these methods often require large amounts of labelled data to achieve high performance. For example, in the detection of Diabetic Retinopathy, over 128,000 retinal fundus photographs were used for training [3]. Similarly, 80,000 Optical Coherence Tomography image slices

© Springer Nature Switzerland AG 2021
S. Engelhardt et al. (Eds.): DGM4MICCAI 2021/DALI 2021, LNCS 13003, pp. 235–242, 2021.
https://doi.org/10.1007/978-3-030-88210-5_23

were needed to classify age-related macular degeneration in [8]. Other examples include skin cancer detection [2] (130,000 images) and lung cancer stratification [10]. This last dataset additionally typifies a common property of medical imaging where class numbers are unbalanced due to varying prevalence of disease: the training data used consisted of 14,761 benign nodules (5,972 patients) but only 932 malignant nodules (575 patients). In such cases, important decisions have to be taken about how to curate the most effective training dataset. Given the high cost of expert labelling of medical images, being able to efficiently choose which data to label can improve efficiency. Moreover, when data collection results in unbalanced distributions, prioritising which data is used for training is key to ensuring good performance across all important classes.

The field of active learning [13] aims to increase the efficiency and effectiveness of the labelling process by intelligently selecting which data to label. Over the active learning cycle, this allows the model to choose the data it deems most effective for it, instead of picking unlabelled samples at random. Redundant images not only waste annotator time as they add data instances that do not benefit in the model's performance, but also needlessly increases the time it takes to train the model. Furthermore, allowing data imbalances in training can have negative impacts on end results [7]. Active learning algorithms free us from having to manually analyse and select the most effective data instances to label and avoid redundancy.

Despite this, active learning is not a panacea. As a relatively new field tied to a lot of fast-moving research, papers have reported wildly different baseline results ([17] at 86% accuracy whereas [14] at 74%). Many active learning papers are tested on balanced datasets, which is a prior assumption on the unlabelled data distribution that should not be assumed.

1.1 Active Learning in Medical Imaging

Active learning has gained traction in the medical domain particularly due to the high cost of labelling by trained medical professionals. There have been several medical specific active learning papers, summarised in [1] and references therein. A recent paper [16], integrates active learning into a hybrid lung segmentation and classification model. Here, the authors sample from an available dataset which has already been curated to have a balance of 300 patients in each of the three classes being predicted. The evaluation of different active learning strategies on balanced and bespoke datasets in prior art is one of the fundamental challenges to evaluating the benefits of such methods in practical applications.

The aim of this paper is not to produce a bespoke active learning methodology for a particular dataset, but to evaluate different fundamental active learning strategies in the context of both balanced and unbalanced datasets. Evaluating these strategies on a unbalanced dataset, which has not been deliberately curated, more closely reflects the reality of acquisition of medical image data for deep learning classification. As [1] concludes, "developing baseline human-in-the-loop methods to compare to will be vital to assess the contributions of individual works in each area and to better understand the influences of competing improvements in these areas". This is the goal of our current paper.

1.2 Active Learning Methodology

Active learning can be split into two stages, the scoring stage and the sampling stage. The scoring function is the process in which the unlabelled data is scored based on its informativeness score and ranked accordingly. In the sampling stage, an algorithm is employed to sample from the unlabelled data utilising the informativeness scores.

For informativeness scoring, the uncertainty scoring method introduced in the classical setting [13] was also explored in the deep learning context [15] by utilising several metrics to predict model uncertainty, such as utilising the confidence of the top predicted class, utilising the top two predicted classes (margin sampling) or by only sampling the positive classes (positive sampling). Other than utilising model uncertainty, another method of scoring the informativeness of data is to maximising expected model change (EMC) [6].

2 Methods

In this section, we outline several techniques and established active learning algorithms from recent AL papers [14,17] that will be tested on the unbalanced dataset, as well as the parameters and scale of the experiments.

2.1 Datasets

The most common datasets used to benchmark deep learning with active learning and image classification papers are CIFAR and ImageNet, which are all datasets where class distributions are balanced. Even medical domain specific papers like [16], which employs active learning on COVID-19 Chest CT images, choose a subset of the full dataset which is balanced. In the real world, no assumption should be placed on the unlabelled prior distribution, and this paper evaluates several active learning techniques on this basis. In fact, medical image datasets tend to be unbalanced, as pathology occurs far more rarely than normal controls.

This paper utilises the COVID-19 dataset from [11], which features an unbalanced dataset of 15153 Chest X-Ray images, with 10192 samples normal control images (NC), 3616 samples of COVID positive cases (NCP), and 1345 samples of normal pneumonia cases (CP). Data augmentation is performed as a regularisation and normalisation step, and does not influence the data distribution during training. Data augmentation includes random crop, random horizontal flip, and image value normalisation. The paper also utilises CIFAR-10 as a baseline dataset, to compare results on balanced and unbalanced data.

2.2 Scoring Functions

A scoring function is a function that assesses how informative a new data point would be when added to the training pool. Different underlying motivations exist when designing scoring functions and the following will be evaluated.

Model Uncertainty. Query the instances where the model is least certain of what the output should be. This is one of the simpler and most commonly used framework [15]. For classification tasks, there are several uncertainty scoring functions that utilise the probabilistic output of the model:

Uncertainty Scoring - Confidence. Score the informativeness of the data based on the confidence predicted by model. By sampling these confidences from the unlabelled data, we identify the labels where the model is least confident. Below, y^* refers to the label with the highest confidence.

$$\phi_{LC} = (1 - P_\theta(y^*|x))$$

Uncertainty Scoring - Margin Scoring. Similar to confidence, instead of taking the absolute least confident labels, we measure using the margin of the top two predicted classes of the label. This potentially improves the scoring function as it directly chooses labels where the model is uncertain between two classes, which improves the decision boundaries between the two classes after training. Below, y_0^* is the label with the most confidence and y_1^* is the label with the second most confidence.

$$\phi_M = (1 - (P_\theta(y_0^*|x) - P_\theta(y_1^*|x)))$$

Expected Model Change. Query the instances which would lead to the greatest change to the current model once learned. This would be the instance where, for each possible class designation, we calculate the gradients if that class designation *was indeed* the actual class label, and we take the highest predicted class designation's gradient.

2.3 Sampling Strategies

The sampling strategy refers to the algorithms which determine which instances get picked. The following four methods will be evaluated, as explored in [4].

Top-N Sampling. By utilising the informativeness score, we choose the top-N most uncertain samples. One of the popular, more efficient methods to utilise the uncertainty scores produced by the scoring functions, but lacks any exploration. It is a greedy sampling method that is very efficient as it does not require computing any distance metrics and hence can scale with larger query and unlabelled data sizes.

K-Nearest Neighbours Sampling. We take the K-nearest samples of known data points based on some heuristic, in this case, based on the scoring functions

[9]. The intuition is to exploit the embedding representation produced by the model to help sample from areas where the model is uncertain.

Core-Set Sampling. By modelling the sampling process as an optimisation problem to find the optimal subset of points that best covers the whole dataset, we get the core-set problem. The core-set approach was introduced into the active learning sphere by [12]. This technique requires solving a mixed-integer optimisation problem, but an efficient greedy approximation (known as **K-centre**) is often used in practice.

Mixture Sampling. This strategy relies on mixing top-N and random sampling to allow the model to both greedily sample points that it believes is the most informative points whilst also introducing exploration via. random sampling.

3   Experiments

In this section, we exhaustively measure how the active learning scoring functions and sampling algorithms perform on the balanced and unbalanced datasets.

3.1   Experimental Setup

We utilise the ResNet18 model to benchmark results for the CIFAR-10 dataset, and the MobileNet_v2 model for the COVID-19 dataset. For both datasets, we initialise the active learning cycle with 10% of the full dataset as the starting seed, using a uniform random sampling. We train the model to convergence and sample an additional 10%, and repeat this cycle 3 times until we reach 40% of the original dataset size. We use a 80:10:10 train/val/test split for both datasets. The results are validated with 5 validation runs. For evaluation we report the final test accuracy for each experiment. Our open-source code to run these experiments is available at www.github.com/justincqz/active_learning_cv.

3.2   Results

Scoring Functions. We test each of the three scoring functions against all sampling algorithms, over both datasets separately (Tables 1 and 2). Different scoring functions come out on top for both datasets, and the margins between scoring functions remain similar for both datasets throughout different sampling strategies. The scoring function is not affected by the unbalanced datasets. The tradeoff of performing margin and confidence scoring over expected model change make them the best scoring metrics to use, although there is no clear answer.

Sampling Algorithms. Taking a look at the four sampling algorithms, it is clear that Core-Set comes out on top for both datasets. However, the cost of performing core-set is also the most expensive (requires building the distance matrix over all unlabelled data). However, we see that top-N uncertainty based sampling actually performs worse than random sampling when applied to the

COVID-19 dataset (where the data is imbalanced). Taking a closer look at the sampling distributions in Figs. 1 and 2, we can see that the top-N sampling oversamples from the NC class, whereas core-set samples more images from the NCP class than any other class, explaining its improved performance. This is because K-Nearest neighbours also out-performs random sampling by a smaller margin, but has a similar computational cost as core-set. Core-Set is clearly the best performing metric.

Table 1. Comparison of the scoring functions and sampling strategies with 40% of the CIFAR10 dataset

CIFAR10		
Sampling	Scoring	Test Acc. (%)
Random	–	73.53
Top-N	Margin	75.72
	Confidence	75.87
	EMC	75.16
Mixture Sampling	Margin	78.47
	Confidence	77.77
	EMC	77.44
K-Nearest	Margin	79.79
	Confidence	81.04
	EMC	81.00
Core-Set	**Margin**	**82.02**
	Confidence	81.29
	EMC	81.95

Table 2. Comparison of the scoring functions and sampling strategies with 40% of the COVID-19 dataset

COVID-19		
Sampling	Scoring	Test Acc. (%)
Random	–	86.99
Top-N	Margin	84.50
	Confidence	83.53
	EMC	82.03
Mixture Sampling	Margin	86.23
	Confidence	85.73
	EMC	85.47
K-Nearest	Margin	88.77
	Confidence	87.53
	EMC	85.60
Core-Set	Margin	89.86
	Confidence	**90.87**
	EMC	89.9

4   Discussion

The key take-away of the results of our experiments is that the active learning strategies based on uncertainty and entropy methods perform *worse* than random sampling. This differs from results of prior art where these methods

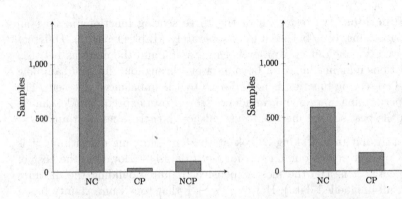

Fig. 1. Samples selected by Top-N   **Fig. 2.** Samples selected by Core-Set

were shown to work well when applied to datasets with a balanced number of classes. Only the strongly diverse sampling algorithms of K-Nearest Neighbours and Core-Set manage to continue to outperform random sampling.

To dive deeper, Fig. 1 shows that the top-N sampling method prioritises NC instances over other classes. One explanation could be that the model is *confidently incorrect*. Since in active learning, when we evaluate the informativeness of an unlabelled instance unknown to the model, the model could completely mislabel those instances. This means that exploitation sampling strategies prioritises the uncertain instances that are within the state space that it is familiar with, leading to the repeated sampling of normal class instances even though a large majority of the training set consists of instances from this class.

On the other hand, the core-set sampling Fig. 2, the sampling strategy prioritises the NCP over the NC instances. This could be hypothesised as when the model improves, so does its understanding of the distinguishing features between the classes. Therefore when modelling the state-space, the model can better distance NCP class feature vectors from NC feature vectors, which in turn allows the core-set sampling strategy to sample from the newly recognised feature space.

Core-Set sampling and other diversity sampling algorithms are more focused on state-space exploration rather than model exploitation. In uneven scenarios, given a small biased training set causes the model to biasly over-predict the majority class, which leads to an oversampling from the majority class in exploitation algorithms. This leads to diversity algorithms showing a **substantial** improvement in the quality of querying in the cases of class imbalance.

5   Conclusion

This paper shows that the assumption of balanced class distributions in the unlabelled data can have a profound impact to the overall results of the experiments, particularly for active learning algorithms which rely on exploitation of the informativeness scores. Our initial results suggest that methods based on diversity sampling should be favoured in this case.

Future benchmarks of active learning algorithms and techniques should include a section which evaluates the performance on unbalanced data, by artificially introducing the unbalanced class distributions in currently available datasets. This is an important factor to consider if active learning was to be applied to a real-world scenario.

References

1. Budd, S., Robinson, E.C., Kainz, B.: A survey on active learning and human-in-the-loop deep learning for medical image analysis. Medical Image Analysis, p. 102062 (2021)
2. Esteva, A., et al.: Dermatologist-level classification of skin cancer with deep neural networks. Nature **542**(7639), 115–118 (2017)

3. Gulshan, V., et al.: Development and validation of a deep learning algorithm for detection of diabetic retinopathy in retinal fundus photographs. JAMA **316**(22), 2402–2410 (2016)
4. Haussmann, E., et al.: Scalable active learning for object detection. In: 2020 IEEE Intelligent Vehicles Symposium (IV), pp. 1430–1435. IEEE (2020)
5. He, K., Zhang, X., Ren, S., Sun, J.: Deep residual learning for image recognition. In: CVPR, pp. 770–778 (2016)
6. Huang, J., Child, R., Rao, V., Liu, H., Satheesh, S., Coates, A.: Active learning for speech recognition: the power of gradients. arXiv:1612.03226 (2016)
7. Larrazabal, A.J., Nieto, N., Peterson, V., Milone, D.H., Ferrante, E.: Gender imbalance in medical imaging datasets produces biased classifiers for computer-aided diagnosis. Proc. Natl. Acad. Sci. **117**(23), 12592–12594 (2020)
8. Lee, C.S., Baughman, D.M., Lee, A.Y.: Deep learning is effective for classifying normal versus age-related macular degeneration oct images. Ophthalmol. Retina **1**(4), 322–327 (2017)
9. Lindenbaum, M., Markovitch, S., Rusakov, D.: Selective sampling for nearest neighbor classifiers. Mach. Learn. **54**(2), 125–152 (2004)
10. Massion, P.P., et al.: Assessing the accuracy of a deep learning method to risk stratify indeterminate pulmonary nodules. Am. J. Respir. Crit. Care Med. **202**(2), 241–249 (2020)
11. Rahman, T., et al.: Exploring the effect of image enhancement techniques on COVID-19 detection using chest x-ray images. Comput. Biol. Med. **132**, 104319 (2021)
12. Sener, O., Savarese, S.: Active learning for convolutional neural networks: A core-set approach. arXiv preprint arXiv:1708.00489 (2017)
13. Settles, B.: Active learning literature survey. Computer Sciences Technical report 1648, University of Wisconsin-Madison (2009)
14. Tran, T., Do, T.T., Reid, I., Carneiro, G.: Bayesian generative active deep learning. In: International Conference on Machine Learning, pp. 6295–6304. PMLR (2019)
15. Wang, D., Shang, Y.: A new active labeling method for deep learning. In: 2014 International Joint Conference on Neural Networks (IJCNN), pp. 112–119. IEEE (2014)
16. Wu, X., Chen, C., Zhong, M., Wang, J., Shi, J.: COVID-al: the diagnosis of COVID-19 with deep active learning. Med. Image Anal. **68**, 101913 (2021)
17. Yoo, D., Kweon, I.: Learning loss for active learning. In: Proceedings of the IEEE/CVF on Computer Vision and Pattern Recognition, pp. 93–102 (2019)

Zero-Shot Domain Adaptation in CT Segmentation by Filtered Back Projection Augmentation

Talgat Saparov[1,3(✉)], Anvar Kurmukov[2,4], Boris Shirokikh[1],
and Mikhail Belyaev[1]

[1] Skolkovo Institute of Science and Technology, Moscow, Russia
[2] Artificial Intelligence Research Institute, Moscow, Russia
[3] Moscow Institute of Physics and Technology, Moscow, Russia
[4] Higher School of Economics, Moscow, Russia

Abstract. Domain shift is one of the most salient challenges in medical computer vision. Due to immense variability in scanners' parameters and imaging protocols, even images obtained from the same person and the same scanner could differ significantly. We address variability in computed tomography (CT) images caused by different convolution kernels used in the reconstruction process, the critical domain shift factor in CT. The choice of a convolution kernel affects pixels' granularity, image smoothness, and noise level. We analyze a dataset of paired CT images, where smooth and sharp images were reconstructed from the same sinograms with different kernels, thus providing identical anatomy but different style. Though identical predictions are desired, we show that the consistency, measured as the average Dice between predictions on pairs, is just 0.54. We propose Filtered Back-Projection Augmentation (FPBAug), a simple and surprisingly efficient approach to augment CT images in sinogram space emulating reconstruction with different kernels. We apply the proposed method in a zero-shot domain adaptation setup and show that the consistency boosts from 0.54 to 0.92 outperforming other augmentation approaches. Neither specific preparation of source domain data nor target domain data is required, so our publicly released FBPAug (https://github.com/STNLd2/FPBAug) can be used as a plug-and-play module for zero-shot domain adaptation in any CT-based task.

1 Introduction

Computed tomography (CT) is a widely used method for medical imaging. CT images are reconstructed from the raw acquisition data, represented in the form of a sinogram. Sinograms are two-dimensional profiles of tissue attenuation as a function of the scanner's gantry angle. One of the most common reconstruction algorithms is Filtered Back Projection (FBP) [11]. This algorithm has an important free parameter called *convolution kernel*. The choice of a convolution kernel defines a trade-off between image smoothness and noise level [10]. Reconstruction with a high-resolution kernel yields *sharp* pixels and a high noise level.

© Springer Nature Switzerland AG 2021
S. Engelhardt et al. (Eds.): DGM4MICCAI 2021/DALI 2021, LNCS 13003, pp. 243–250, 2021.
https://doi.org/10.1007/978-3-030-88210-5_24

In contrast, usage of a lower-resolution kernel results in *smooth* pixels and a low noise level. Depending on the clinical purpose, radiologists use different kernels for image reconstruction.

Modern deep neural networks (DNN) are successfully used to automate computing clinically relevant anatomical characteristics and assist with disease diagnosis. However, DNNs are sensitive to changes in data distribution which are known as *domain shift*. Domain shift typically harms models' performance even for simple medical images such as chest X-rays [13]. In CT images, factors contributing to domain shift include [4] slice thickness and inter-slice interval, different radiation dose, and reconstruction algorithms, e.g., FBP parameters. The latter problem is a subject of our interest.

Recently, several studies have reported a drop in the performance of convolutional neural networks (CNN), trained on *sharp* images while being tested on *smooth* images [1,5,6]. Authors of [9] proposed using generative adversarial networks (GAN) to generate realistic CT images imitating arbitrary convolution kernels. A more straightforward approach simultaneously proposed in [6], [1], and [5] suggests using a CNN to convert images reconstructed with one kernel to images reconstructed with another. Later, such image-to-image networks can be used either as an augmentation during training or as a preprocessing step during inference.

We propose FBPAug, a new augmentation method based on the FBP reconstruction algorithm. This augmentation mimics processing steps used in proprietary manufacturer's reconstruction software. We initially apply Radon transformation to all training CT images to obtain their sinograms. Then we reconstruct images using FBP but with different randomly selected convolution kernels. To show the effectiveness of our method, we compare segmentation masks obtained on a set of paired images, reconstructed from the same sinograms but with different convolutional kernels. These paired images are perfectly aligned; the only difference is their style: smooth or sharp. We make our code and results publicly available, so the augmentation could be easily embedded into any CT-based CNN training pipeline to increase its generalizability to smooth-sharp domain shift.

2   Materials and Methods

In this section, we detail our augmentation method, describe quality metrics, and describe datasets which we use in our experiments.

2.1   Filtered Back-Projection Augmentation

Firstly, we give a background on a discrete version of inverse Radon Transform – Filtered Back-Projections algorithm. FBP consists of two sequential operations: generation of filtered projections and image reconstruction by the Back-Projection (BP) operator.

Projections of attenuation map have to be filtered before using them as an input of the Back-Projection operator. The ideal filter in a continuous noiseless case is the ramp filter. Fourier transform of the ramp filter $\kappa(t)$ is $\mathcal{F}[\kappa(t)](w) = |w|$.

The image $I(x, y)$ can be derived as follows:

$$I(x, y) = \text{FBP}(p_\theta(t)) = \text{BP}(p_\theta(t) * \kappa(t)), \tag{1}$$

where $*$ is a convolution operator, $t = t(x, y) = x \cos\theta + y \sin\theta$ and $\kappa(t)$ is the aforementioned ramp filter.

Assume that a set of filtered-projections $p_\theta(t)$ available at angles $\theta_1, \theta_2, ..., \theta_n$, such that $\theta_i = \theta_{i-1} + \Delta\theta$, $i = \overline{2, n}$ and $\Delta\theta = \pi/n$. In that case, BP operator transforms a function $f_\theta(t) = f(x \cos\theta + y \sin\theta)$ as follows:

$$BP(f_\theta(t))(x, y) = \frac{\Delta\theta}{2\pi} \sum_{i=1}^{n} f_{\theta_i}(x \cos\theta_i + y \sin\theta_i) = \frac{1}{2n} \sum_{i=1}^{n} f_{\theta_i}(x \cos\theta_i + y \sin\theta_i)$$

In fact, $\kappa(t)$ that appears in (1) is a generalized function and cannot be expressed as an ordinary function because the integral of $|w|$ in inverse Fourier transform does not converge. However, we utilize the convolution theorem that states that $\mathcal{F}(f * g) = \mathcal{F}(f) \cdot \mathcal{F}(g)$. And after that we can use the fact that the BP operator is a finite weighted sum and Fourier transform is a linear operator as follows:

$$\mathcal{F}^{-1}\mathcal{F}[I(x, y)] = \mathcal{F}^{-1}\mathcal{F}[\text{BP}(p_\theta * \kappa)] = \text{BP}(\mathcal{F}^{-1}\mathcal{F}[p_\theta * \kappa]) = \text{BP}(\mathcal{F}^{-1}\{\mathcal{F}[p_\theta] \cdot |w|\}),$$

$$I(x, y) = \text{BP}\left(\mathcal{F}^{-1}\{\mathcal{F}[p_\theta] \cdot |w|\}\right).$$

However, in the real world, CT manufacturers use different filters that enhance or weaken the high or low frequencies of the signal. We propose a family of convolution filters $k_{a,b}$ that allows us to obtain a smooth-filtered image given a sharp-filtered image and vice versa. Fourier transform of the proposed filter is expressed as follows:

$$\mathcal{F}[k_{a,b}](w) = \mathcal{F}[\kappa](w)(1 + a\mathcal{F}[\kappa](w)^b) = |w|(1 + a|w|^b).$$

Thus, given a CT image I obtained from a set of projections using one kernel, we can simulate the usage of another kernel as follows:

$$\hat{I}(x, y) = \text{BP}\left(\mathcal{F}^{-1}\{\mathcal{F}[\mathcal{R}(I)] \cdot \mathcal{F}[k_{a,b}]\}\right).$$

Here, a and b are the parameters that influence the sharpness or smoothness of an output image and $\mathcal{R}(I)$ is a Radon transform of image I. The output of the Radon transform is a set of projections. Figure 1 shows an example of applying sharping augmentation on a soft kernel image (Fig. 1(a) to (c)) and vice versa: applying softening augmentation on a sharp kernel image (Fig. 1(b) to (d)).

2.2   Comparison Augmentation Approaches

We compare the proposed method with three standard augmentations: gamma transformation (Gamma), additive Gaussian noise (Noise), and random windowing (Windowing), the technique proposed by [4]. As a baseline method, we train a network without any intensity augmentations (Baseline).

Gamma [12], augments images using gamma transformation:

$$\hat{I}(x,y) = \left(\frac{I(x,y) - m}{M - m} \right)^{\gamma} \cdot (M - m) + m,$$

where $M = \max(I(x,y))$, $m = \min(I(x,y))$ with a parameter γ, such that we randomly sample logarithm of γ from $\mathcal{N}(0, 0.2)$ distribution.

Noise is the additive gaussian noise from $\mathcal{N}(0, 0.1)$ distribution.

Windowing [4] make use of the fact that different tissue has different attenuation coefficient. We uniformly sample the center of the window c from $[-700, -500]$ Hounsfield units (HU) and the width of the window w from $[1300, 1700]$ HU. Then we clip the image to the $[c - w/2, c + w/2]$ range.

FBPAug parameters were sampled as follows. We uniformly sample a from $[10.0, 40.0]$ and b from $[1.0, 4.0]$ in sharpening case and a from $[-1.0, 0]$, b from $[0.1, 1.0]$ in smoothing case.

In all experiments, we zoom images to 1×1 mm pixel size and use additional rotations and flips augmentation. With probability 0.5 we rotate an image by multiply of 90 degrees and flip an image horizontally or vertically.

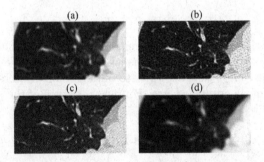

Fig. 1. An example of paired CT slices (top row) and the effect of the augmentation by the proposed method (bottom row). The top row contains original images: a slice reconstructed either with a smooth kernel (a) or sharp kernel (b). The bottom row shows augmented images: the top-left image processed by FBPAug with parameters $a = 30$, $b = 3$ shifting it from smooth to sharp (c); the top-right image processed by FBPAug with parameters $a = -1$, $b = 0.7$ from sharp to smooth (d).

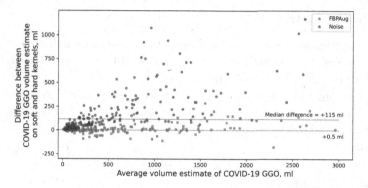

Fig. 2. Bland-Altman plot showing prediction agreement using FBPAug (proposed augmentation, red) and next best competitor (Gaussian noise, blue). Agreement is measured between predictions on paired images reconstructed with soft and hard convolution kernels from *Covid-private* dataset. Difference in image pairs were always computed as $\text{Volume}_{\text{soft}} - \text{Volume}_{\text{sharp}}$. (Color figure online)

2.3 Datasets

We report our results on two datasets: *Mosmed-1110* and a private collection of CT images with COVID-19 cases (*Covid-private*). Both datasets include chest CT series (3D CT images) of healthy subjects and subjects with the COVID-19 infection.

Mosmed-1110. The dataset consists of 1110 CT scans from Moscow clinics collected from 1st of March, 2020 to 25th of April, 2020 [7]. The original images have 0.8 mm inter-slice distance, however the released studies contain every 10th slice so the effective inter-slice distance is 8 mm. *Mosmed-1110* contains only 50 CT scans that are annotated with the binary masks of ground-glass opacity (GGO) and consolidation. We additionally ask three experienced radiologists to annotate another 46 scans preserving the methodology of the original annotation process. Further, we use the total of 96 annotated cases from *Mosmed-1110* dataset.

Covid-Private. All images from Covid-private dataset are stored in the DICOM format, thus providing information about corresponding convolution kernels. The dataset consists of paired CT studies (189 pairs in total) of patients with COVID-19. In contrast with many other datasets, all of studies contain two series (3D CT images); the overall number of series is 378. Most importantly, every pair of series were obtained from one physical scanning with different reconstruction algorithms. It means the slices within these images are perfectly aligned, and the only difference is *style* of the image caused by different convolutional kernels applied. *Covid-private* does not contain ground truth mask of GGO or consolidation, thus we only use it to test predictions agreement.

2.4   Quality Metrics

For the comparison, we use the standard segmentation metric, Dice Score. Dice Score (DSC) of two volumetric binary masks X and Y is computed as $DSC = \frac{2|X \cup Y|}{|X|+|Y|}$, where $|X|$ is the cardinality of a set X.

Furthermore, we perform statistical analysis ensure significance of the results. We use one-sided Wilcoxon signed-rank test as we consider DSC scores for two methods are paired samples. To adjust for multiple comparisons we use Bonferroni correction.

3   Experiments

3.1   Experimental Pipeline

To evaluate our method, we conduct two sets of experiments for COVID-19 segmentation.

First, we train five separate segmentation models: baseline with no augmentations, FBPAug, Gamma, Noise, and Windowing on a *Mosmed* dataset to check if any augmentation results in significantly better performance. *Mosmed* is stored in Nifti format and does not contain information about the kernels. Thus, we use it to estimate the in-domain accuracy for COVID-19 segmentation problem.

Second, we use trained models from the previous experiment to make predictions on a paired *Covid-private* dataset. We compare masks within each pair of sharp and soft images using Dice score to measure prediction agreement for the isolated domain shift reasons, as the only difference between images within each pair is their smooth or sharp style, see Fig. 3.

3.2   Network Architecture and Training Setup

For all our experiments, we use a slightly modified 2D U-Net [8]. We prefer the 2D model to 3D since in the Mosmed-1110 dataset images have an 8 mm inter-slice distance and the inter-slice distance of Covid-private images is in the range from 0.8 mm to 1.25 mm. Furthermore, the 2D model shows performance almost equal to the performance of the 3D model for COVID-19 segmentation [2]. In all cases, we train the model for 100 epochs with a learning rate of 10^{-3}. Each epoch consists of 100 iterations of the Adam algorithm [3].

At each iteration, we sample a batch of 2D images with batch size equals to 32. The training was conducted on a computer with 40GB NVIDIA Tesla A100 GPU. It takes approximately 5 hours for the experiments to complete.

4   Results

Table 1 summarizes our results. First, experiments on *Mosmed-1110* show that segmentation quality almost does not differ for compared methods. The three best augmentation approaches are not significantly different (p-value for

Wilcoxon test are 0.17 for FBPAug vs Gamma and 0.71 for FBPAug vs Windowing). Thus our method does not harm segmentation performance. Our segmentation results are on-par with best reported for this dataset [2]. Next, we observe a significant disagreement in predictions on paired (*smooth* and *sharp*) images for all methods, except FBPAug (p-values for Wilcoxon test for FBPAug vs every other method are all less than 10^{-16}). For FBPAug and its best competitor, we plot a Bland-Altman plot, comparing GGO volume estimates Fig. 2. We can see that the predictions of FBPAug model agree independent of the volume of GGO.

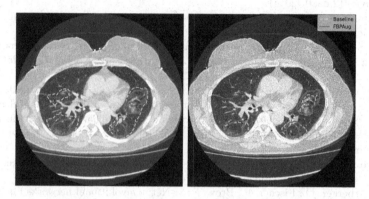

Fig. 3. Example prediction. Left - example image reconstructed with a *smooth* kernel. Right - image reconstructed with a *sharp* kernel from the same sinogram. Red dashed contour - the baseline model (predictions agreement is 0.64), blue contour - FBPAug (predictions agreement is 0.91). (Color figure online)

Table 1. Comparison results. Numbers are mean (std) obtained on 3-fold cross-validation. Results for *Mosmed-1110* are segmentation Dice score compared with ground truth; for *Covid-private* are predictions agreement (between paired images) measured using Dice score.

	Baseline	FBPAug	Gamma	Noise	Windowing
Mosmed-1110	0.56 (0.23)	0.59 (0.22)	**0.61 (0.19)**	0.56 (0.21)	0.59 (0.18)
Covid-private	0.54 (0.27)	**0.92 (0.05)**	0.68 (0.21)	0.79 (0.13)	0.63 (0.23)

5   Conclusion

We propose a new physics-driven augmentation methods to eliminate domain shifts related to the usage of different convolution kernels. It outperforms existing augmentation approaches in our experiments. We release the code, so our flexible and ready-to-use approach can be easily incorporated into any existing deep learning pipeline to ensure zero-shot domain adaptation.

The results have been obtained under the support of the Russian Foundation for Basic Research grant 18-29-26030.

References

1. Choe, J., et al.: Deep learning-based image conversion of CT reconstruction kernels improves radiomics reproducibility for pulmonary nodules or masses. Radiology **292**(2), 365–373 (2019)
2. Goncharov, M., et al.: CT-based Covid-19 triage: deep multitask learning improves joint identification and severity quantification. arXiv preprint arXiv:2006.01441 (2020)
3. Kingma, D.P., Ba, J.: Adam: a method for stochastic optimization. arXiv preprint arXiv:1412.6980 (2014)
4. Kloenne, M., et al.: Domain-specific cues improve robustness of deep learning-based segmentation of CT volumes. Sci. Rep. **10**(1), 1–9 (2020)
5. Lee, S.M., et al.: CT image conversion among different reconstruction kernels without a sinogram by using a convolutional neural network. Korean J. Radiol. **20**(2), 295 (2019)
6. Missert, A.D., Leng, S., McCollough, C.H., Fletcher, J.G., Yu, L.: Simulation of CT images reconstructed with different kernels using a convolutional neural network and its implications for efficient CT workflow. In: Medical Imaging 2019: Physics of Medical Imaging, vol. 10948, p. 109482Y. International Society for Optics and Photonics (2019)
7. Morozov, S., et al.: MosMedData: data set of 1110 chest CT scans performed during the Covid-19 epidemic. Digit. Diagn. **1**(1), 49–59 (2020)
8. Ronneberger, O., Fischer, P., Brox, T.: U-Net: convolutional networks for biomedical image segmentation. In: Navab, N., Hornegger, J., Wells, W.M., Frangi, A.F. (eds.) MICCAI 2015. LNCS, vol. 9351, pp. 234–241. Springer, Cham (2015). https://doi.org/10.1007/978-3-319-24574-4_28
9. Sandfort, V., Yan, K., Pickhardt, P.J., Summers, R.M.: Data augmentation using generative adversarial networks (CycleGAN) to improve generalizability in CT segmentation tasks. Sci. Rep. **9**(1), 1–9 (2019)
10. Schaller, S., Wildberger, J.E., Raupach, R., Niethammer, M., Klingenbeck-Regn, K., Flohr, T.: Spatial domain filtering for fast modification of the tradeoff between image sharpness and pixel noise in computed tomography. IEEE Trans. Med. Imaging **22**(7), 846–853 (2003)
11. Schofield, R., et al.: Image reconstruction: part 1-understanding filtered back projection, noise and image acquisition. J. Cardiovasc. Comput. Tomogr. **14**(3), 219–225 (2020)
12. Tureckova, A., Turecek, T., Kominkova Oplatkova, Z., Rodriguez-Sanchez, A.J.: Improving CT image tumor segmentation through deep supervision and attentional gates. Front. Robot. AI **7**, 106 (2020)
13. Zech, J.R., Badgeley, M.A., Liu, M., Costa, A.B., Titano, J.J., Oermann, E.K.: Variable generalization performance of a deep learning model to detect pneumonia in chest radiographs: a cross-sectional study. PLoS Med. **15**(11), e1002683 (2018)

Label Noise in Segmentation Networks: Mitigation Must Deal with Bias

Eugene Vorontsov[(✉)] and Samuel Kadoury

École Polytechnique de Montréal, Montréal, QC H3T1J4, Canada

Abstract. Imperfect labels limit the quality of predictions learned by deep neural networks. This is particularly relevant in medical image segmentation, where reference annotations are difficult to collect and vary significantly even across expert annotators. Prior work on mitigating label noise focused on simple models of mostly uniform noise. In this work, we explore biased and unbiased errors artificially introduced to brain tumour annotations on MRI data. We found that supervised and semi-supervised segmentation methods are robust or fairly robust to unbiased errors but sensitive to biased errors. It is therefore important to identify the sorts of errors expected in medical image labels and especially mitigate the biased errors.

Keywords: Label noise · Segmentation · Neural networks

1 Introduction

The reference annotations used to train neural networks for the segmentation of medical images are few and imperfect. The number of images that can be annotated is limited by the need for expert annotators and the result is subject to high inter- and intra-annotator variability [1]. Furthermore, the objects targetted by medical image segmentation also tend to be highly variable in appearance. Thus, to make the best of use of limited labeled data, it is important to understand which sorts of errors in reference annotations most affect the segmentation performance of deep neural networks.

Noisy labels can be dealt with by modeling the noise [2–4], re-weighting the contribution of labels depending on some estimate of their reliability [5,6], training on pseudo-labels [5,7], designing noise-tolerant objective functions [8,9], or estimating true labels [10–14].

A generative model of the noise was presented in [4]. A true segmentation map is estimated using this model, and the segmentation model is updated accordingly. For this approach, a good estimate of the noise model must be known. In [3] and [2], it is learned with the limitation that a fraction of the dataset has to

Electronic supplementary material The online version of this chapter (https://doi.org/10.1007/978-3-030-88210-5_25) contains supplementary material, which is available to authorized users.

S. Engelhardt et al. (Eds.): DGM4MICCAI 2021/DALI 2021, LNCS 13003, pp. 251–258, 2021.
https://doi.org/10.1007/978-3-030-88210-5_25

be known to have clean labels. Instead of estimating the noise model directly, the reliability of labels could be estimated instead so that examples with unreliable labels are reweighted to contribute little to the loss function. This was done in [6] by filtering for examples for which gradient directions during training differ greatly from those measured on known clean examples. A similar estimate can be made without requiring clean examples by giving a low weight to examples that tend to produce higher error during training [5]. Alternatively, model predictions [5] (especially those deemed confident by an adversarially trained discriminator [7]) can be used as pseudo-labels in further training iterations.

The choice of objective function also affects robustness to label noise. Mean absolute error (MAE) in particular exhibits some theoretically grounded robustness to noise [9] and is the inspiration for a modified Dice loss that makes Dice less like a weighted mean squared error objective and more like MAE.

True labels can be estimated from multiple imperfect reference labels with expectation minimization (EM) [10–13]. Creating multiple segmentation annotations is typically too expensive but lower quality results can be obtained with crowdsourcing [15]. Indeed, it has been demonstrated that the most efficient labeling strategy is to collect one high quality label per example for many examples and then estimate the true labels with model-bootstrapped EM [14]. The authors state that this is effective when "the learner is robust to noise" and then assume that label errors are random and uniform. This raises the question: which sorts of errors in data labels are deep neural networks robust to?

We show that recent supervised and semi-supervised deep neural network based segmentation models are robust to random "unbiased" annotation errors and are much more affected by "biased" errors. We refer to errors as biased when the perturbation applied to reference annotations during training is consistent. We test recent supervised and semi-supervised segmentation models, including "GenSeg" [16], trained on artificially noisy data with different degrees of bias. Overall, we demonstrate that:

1. All models have robustness to unbiased errors.
2. All models are sensitive to biased errors.
3. GenSeg is less sensitive to biased errors.

2   Segmentation Models

Four different deep convolutional neural networks are tested for robustness to label noise in this work. Network architectures and training are detailed in [16]; the models are briefly described below.

Supervised. The basic segmentation network is a fully convolutional network (FCN) with long skip connections from an image encoder to a segmentation decoder, similar to the U-Net [17]. It is constructed as in [16], with compressed long skip connections, and trained fully supervised with the soft Dice loss.

Autoencoder. The supervised FCN is extended to semi-supervised training by adding a second decoder that reconstructs the input image, as in [16].

Mean Teacher. The supervised FCN is extended to mean teacher training as in [18]. A teacher network maintains an exponential moving average of the weights in a student network. When an input has no reference annotation to train on, the student network learns to match the teacher. A potential limitation of this method is that the reliability of the teacher may depend on the size and richness of the annotated training dataset. The supervised FCN architecture is re-used and hyperparameters were selected as in [16].

GenSeg. GenSeg extends the FCN for tumour segmentation by using image-to-image translation between "healthy" and "diseased" image domains as an unsupervised surrogate objective for segmentation [16]. In order to make a diseased image healthy or *vice versa*, the model must learn to disentangle the tumour from the rest of the image. This disentangling is crucial for a segmentation objective. Importantly, this generative method can learn the locations, shapes, and appearance of tumours conditioned on healthy tissue without relying on tumour annotations.

3   Model Performance on Corrupted Labels

We aim to evaluate the robustness of segmentation models to errors in the reference annotations. To that end, we test different types of perturbations applied to the annotations of the training data. Each perturbation (except for permutation) is applied on the fly—that is, the annotation is perturbed from its reference state each time it is loaded during training (once per epoch). Experiments on these various types of perturbation are presented below, along with some loose intuition on their level of bias.

All experiments were performed on the 2D brain tumour dataset proposed in [16], using the same training, validation, and testing data split. Created from axial slices extracted from the MRI volumes of the 3D brain tumor segmentation challenge (BraTS) 2017 data [19], this dataset includes a set of 8475 healthy slices (without tumour) and a set of 7729 diseased slices (with tumour).

Segmentation performance was evaluated as a Dice score computed on the test set over all inputs combined together. That is, all reference and predicted annotations are stacked together before computing this overlap metric.

3.1   Random Warp

We introduced random errors into annotations during training by randomly warping every tumour mask by an elastic deformation. The deformation map was computed with b-spline deformation with a 3×3 grid of control points. Each time warping was applied, each control point was randomly shifted from its initial grid configuration by some number of pixels sampled from a zero-mean Normal distribution with standard deviation σ. Examples of random warping performed for different values of σ are shown in Fig. 1.

When warping the tumour mask, the per-pixel label error depends on other pixels, so it cannot be simply averaged out. Nevertheless, there is no bias in

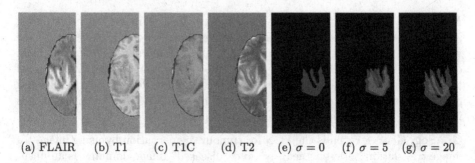

(a) FLAIR   (b) T1   (c) T1C   (d) T2   (e) $\sigma = 0$ (f) $\sigma = 5$ (g) $\sigma = 20$

Fig. 1. Example of warped annotations with different σ (red) vs original (blue). FLAIR, T1, T1C, and T2 are the MRI acquisition sequences that compose the four channels of each input image. (Color figure online)

the displacement of the control points (all sampled from a zero-mean Normal distribution) and there is no bias in the distribution of shapes produced; in a sense, the original tumour shape remains the average case. Thus, we consider warping to be a largely unbiased error.

The relative performance of segmentation models trained on warped tumour masks is shown in Fig. 2a. For different σ, segmentation performance is evaluated relative to the Dice score achieved with no perturbation of the annotations. All models show a linear relationship of percent reduction in Dice score to σ, with no measurable reduction in performance for small deformations at $\sigma = 2$ and a reduction of only between 4% and 6% for unrealistically large deformations at $\sigma = 20$. Interestingly, when annotations were only provided for about 1% of the patient cases, σ had no effect on model performance (Fig. 2b). As expected, the semi-supervised autoencoding (AE), mean teacher (MT), and especially GenSeg segmentation methods outperformed fully supervised segmentation in this case. These results show that state of the art segmentation models are surprisingly robust to unbiased deformations of the tumour masks.

3.2   Constant Shift

We introduced consistent errors into annotations during training by shifting the entire tumour mask by n pixels, creating a consistent misalignment between the target segmentation mask and the input image. Because this error is consistent and the correct (original) annotations cannot be inferred from the distribution of corrupted annotations, we refer to shifting as a biased error. As shown in Fig. 2c, this kind of error strongly affected all models, resulting in about 10% lower Dice scores when $n = 5$. Interestingly, GenSeg showed remarkable robustness to extreme shift errors, compared to other models. GenSeg was the only model to not show a linear relationship between segmentation performance and the amount of shift, n; at an urealistic shift of $n = 30$, GenSeg showed a 27% drop in performance compared to the mean teacher (MT), autoencoding (AE), and purely supervised segmentation (Seg) models which each performed about 53%

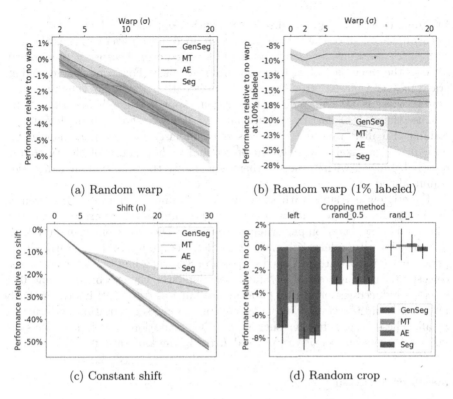

(a) Random warp

(b) Random warp (1% labeled)

(c) Constant shift

(d) Random crop

Fig. 2. Relative segmentation performance for different kinds of errors applied to tumour annotations during training. Compares semi-supervised autoencoding (AE), mean teacher (MT), and GenSeg models, as well as a fully supervised (Seg) network. Performance is relative to each model's peak performance (as Dice score) on clean data. Each experiment was repeated three times. Solid lines: mean; shaded regions: standard deviation. Variance calculations considered both the variance of the presented experiments and of results on clean data.

worse than on error-free annotations. The same trends were observed when 99% of the annotations were omitted. These experiments suggest that segmentation models are sensitive to biased errors in the annotations.

3.3   Random Crop

To further test the effect of bias in annotation errors, we devised three variants of cropping errors. In all cases, we made sure that half of each tumour area is cropped out on averaged. First, we performed a simple a consistent crop of the left side of each tumour ("left"). Second, we cropped out a random rectangle with relative edge lengths distributed in [0.5, 1] as a fraction of the tumour's bounding box ("rand_0.5"); the rectangle was randomly placed completely within the bounding box. Third, we did the same but with relative edge lengths in [0, 1] ("rand_0"). For both "rand_0.5" and "rand_0", edge lengths were sampled according to a linear random distribution with a slope selected empirically so as to ensure that half of the tumour area is cropped out on average, as is the case with "left". We consider "left" as the most biased error because it consistently removes the same part of each tumour from the mask. Following this reasoning, "rand_0.5" is a biased error because it consistently reduces the tumour area—that is, there is always a hole in the tumour—but it is less biased than "left" because there is no part of the tumour that is never shown to models during training. Finally, "rand_0" is the closest to unbiased because it is inconsistent both in which pixels are made incorrect and in how much of the tumour is removed. Indeed, with "rand_0", there is a chance that the tumour annotation is unmodified.

Experimental results with these random cropping strategies are presented in Fig. 2d. Similarly to what we observed with unbiased warp errors and biased shift errors, segmentation performance relative to Dice score on clean data drops the most for all models with the biased "left" cropping strategy. Surprisingly, performance only decreased between 5% and 8% when half of each tumour was consistently missing from the training set. Similarly, the less biased "rand_0.5" strategy reduced segmentation performance but less severely. Finally, the fairly unbiased "rand_0" strategy did not result in reduced segmentation performance at all. These results further suggest that segmentation models are robust to unbiased errors but sensitive to biased errors in the annotations.

3.4   Permutation

Occasionally, annotations and images end up being matched incorrectly during the creation of a dataset. We test the effect of this kind of biased error on the performance of a supervised segmentation network. Before training, we randomly permuted the annotations for a percentage of the data. Permutation was done only one time and maintained throughout training. We compared segmentation performance, as the Dice score, to training on clean data but with the same percentage of data discarded from the training dataset. As shown in Fig. 3, permutation errors reduce segmentation performance far more than if the corrupted data were simply discarded.

4   Limitations and Future Work

We presented the effects of various biased and unbiased errors, applied to the annotation maps in the training subset of data, on segmentation performance. Although it appears that models are more sensitive to biased errors than unbiased ones, it would be prudent to test many more strategies for introducing error. One simple test could be randomly switching the class of each pixel, independently. This sort of error is commonly considered in the literature for classification; it would be interesting to measure whether segmentation models are more robust to it due to the contextual information from neighbouring pixels.

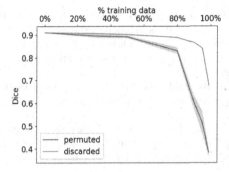

Fig. 3. The segmentation Dice score goes down as the percentage of the dataset for which annotations are permuted, or as the percentage of the dataset that is removed from the training set, goes up. Corrupting data by permuting the labels is more detrimental to model performance than discarding the data.

All tested models used the soft Dice objective since they were trained as in [16]. However, different objective functions have different robustness to label noise [8,9] so it would be prudent to explore these options further. Furthermore, it would be interesting to evaluate how much of the errors that we introduced could be removed with model-bootstrapped EM [14] or accounted for with an explicit model of the expected noise [2,3]. Finally, it would be interesting to estimate how much of the error in intra- or inter-annotator variability in medical imaging is systematic biased error, and thus potentially difficult to reduce.

5   Conclusion

State of the art deep neural networks for medical image segmentation have some inherent robustness to label noise. We find empirically that while they are robust or partially robust to unbiased errors, they are however sensitive to biased errors. We loosely define biased errors as those which most consistently modify parts of an annotation. We conclude then that when considering on annotation quality (e.g. crowdsourcing vs expert annotation) or when working on robustness to label noise, it is particularly important to identify and mitigate against biased errors.

References

1. Vorontsov, E., et al.: Deep learning for automated segmentation of liver lesions at CT in patients with colorectal cancer liver metastases. Radiol.: Artif. Intell. **1**(2), 180014 (2019)
2. Patrini, G., Rozza, A., Krishna Menon, A., Nock, R., Qu, L.: Making deep neural networks robust to label noise: a loss correction approach. In: Proceedings of the IEEE Conference on Computer Vision and Pattern Recognition, pp. 1944–1952 (2017)

3. Sukhbaatar, S., Bruna, J., Paluri, M., Bourdev, L., Fergus, R.: Training convolutional networks with noisy labels arXiv preprint arXiv:1406.2080 (2014)

4. Mnih, V., Hinton, G.E.: Learning to label aerial images from noisy data. In: Proceedings of the 29th International Conference on Machine Learning (ICML-12), pp. 567–574 (2012)

5. Karimi, D., Dou, H., Warfield, S.K., Gholipour, A.: Deep learning with noisy labels: exploring techniques and remedies in medical image analysis. Med. Image Anal. **65**, 101759 (2020)

6. Mirikharaji, Z., Yan, Y., Hamarneh, G.: Learning to segment skin lesions from noisy annotations. In: Wang, Q., et al. (eds.) DART/MIL3ID -2019. LNCS, vol. 11795, pp. 207–215. Springer, Cham (2019). https://doi.org/10.1007/978-3-030-33391-1_24

7. Nie, D., Gao, Y., Wang, L., Shen, Đ.: ASDNet: attention based semi-supervised deep networks for medical image segmentation. In: Frangi, A.F., Schnabel, J.A., Davatzikos, C., Alberola-López, C., Fichtinger, G. (eds.) MICCAI 2018. LNCS, vol. 11073, pp. 370–378. Springer, Cham (2018). https://doi.org/10.1007/978-3-030-00937-3_43

8. Wang, G.: A noise-robust framework for automatic segmentation of COVID-19 pneumonia lesions from CT images. IEEE Trans. Med. Imaging **39**(8), 2653–2663 (2020)

9. Ghosh, A., Kumar, H., Sastry, P.S.: Robust loss functions under label noise for deep neural networks. In: Proceedings of the AAAI Conference on Artificial Intelligence, vol. 31, no. 1 (2017)

10. Dawid, A.P., Skene, A.M.: Maximum likelihood estimation of observer error-rates using the EM algorithm. J. R. Stat. Soc.: Ser. C (Appl. Stat.) **28**(1), 20–28 (1979)

11. Whitehill, J., Wu, T.-F., Bergsma, J., Movellan, J., Ruvolo, P.: Whose vote should count more: optimal integration of labels from labelers of unknown expertise. In: Advances in Neural Information Processing Systems, vol. 22, pp. 2035–2043 (2009)

12. Welinder, P., Branson, S., Perona, P., Belongie, S.: The multidimensional wisdom of crowds. In: Advances in Neural Information Processing Systems, vol. 23, pp. 2424–2432 (2010)

13. Zhou, D., Liu, Q., Platt, J.C., Meek, C., Shah, N.B.: Regularized minimax conditional entropy for crowdsourcing arXiv preprint arXiv:1503.07240 (2015)

14. Khetan, A., Lipton, Z.C., Anandkumar, A.: Learning from noisy singly-labeled data arXiv preprint arXiv:1712.04577 (2017)

15. Ørting, S., et al.: A survey of crowdsourcing in medical image analysis arXiv preprint arXiv:1902.09159 (2019)

16. Vorontsov, E., Molchanov, P., Beckham, C., Kautz, J., Kadoury, S.: Towards annotation-efficient segmentation via image-to-image translation arXiv preprint arXiv:1904.01636 (2019)

17. Ronneberger, O., Fischer, P., Brox, T.: U-Net: convolutional networks for biomedical image segmentation. In: Navab, N., Hornegger, J., Wells, W.M., Frangi, A.F. (eds.) MICCAI 2015. LNCS, vol. 9351, pp. 234–241. Springer, Cham (2015). https://doi.org/10.1007/978-3-319-24574-4_28

18. Perone, C.S., Cohen-Adad, J.: Deep semi-supervised segmentation with weight-averaged consistency targets. In: Stoyanov, D., et al. (eds.) DLMIA/ML-CDS -2018. LNCS, vol. 11045, pp. 12–19. Springer, Cham (2018). https://doi.org/10.1007/978-3-030-00889-5_2

19. Bakas, S., et al.: Identifying the best machine learning algorithms for brain tumor segmentation, progression assessment, and overall survival prediction in the brats challenge arXiv preprint arXiv:1811.02629 (2018)

DeepMCAT: Large-Scale Deep Clustering for Medical Image Categorization

Turkay Kart[1]($\boxtimes$), Wenjia Bai[1,2], Ben Glocker[1], and Daniel Rueckert[1,3]

[1] BioMedIA Group, Department of Computing, Imperial College London,
London, UK
t.kart@imperial.ac.uk
[2] Department of Brain Sciences, Imperial College London, London, UK
[3] Institute for AI and Informatics in Medicine, Klinikum rechts der Isar,
Technical University of Munich, Munich, Germany

Abstract. In recent years, the research landscape of machine learning in medical imaging has changed drastically from supervised to semi-, weakly- or unsupervised methods. This is mainly due to the fact that ground-truth labels are time-consuming and expensive to obtain manually. Generating labels from patient metadata might be feasible but it suffers from user-originated errors which introduce biases. In this work, we propose an unsupervised approach for automatically clustering and categorizing large-scale medical image datasets, with a focus on cardiac MR images, and without using any labels. We investigated the end-to-end training using both class-balanced and imbalanced large-scale datasets. Our method was able to create clusters with high purity and achieved over 0.99 cluster purity on these datasets. The results demonstrate the potential of the proposed method for categorizing unstructured large medical databases, such as organizing clinical PACS systems in hospitals.

Keywords: Deep clustering · Unsupervised learning · Categorization · DICOM sequence classification · Cardiac MRI

1 Introduction

Highly curated labelled datasets have recently been emerging to train deep learning models for specific tasks in medical imaging. Thanks to these fully-annotated images, supervised training of convolutional neural networks (CNNs), either from scratch or by fine-tuning, has become a *dominant* approach for automated biomedical image analysis. However, the data curation process is often manual and labor-intensive as well as requiring expert domain knowledge. This time-consuming procedure is simply not practical for each single task in medical imaging, and therefore, automation is a necessity.

Electronic supplementary material The online version of this chapter (https://doi.org/10.1007/978-3-030-88210-5_26) contains supplementary material, which is available to authorized users.

© Springer Nature Switzerland AG 2021
S. Engelhardt et al. (Eds.): DGM4MICCAI 2021/DALI 2021, LNCS 13003, pp. 259–267, 2021.
https://doi.org/10.1007/978-3-030-88210-5_26

The first step of data curation in medical imaging typically starts from data cleaning where desired images are extracted from a hospital image database such as a PACS system. Due to the nature of such image databases in hospitals, these systems often record important attributes such as image sequences in an unstructured fashion as meta-data in the DICOM header of the images. Meta-data in the DICOM standard, the most widely adapted format for data storage in medical imaging, may seem as a reliable option for automated annotation but it is often incorrect, incomplete and inconsistent. This represents a major challenge for data curation. Gueld et al. [8] analyzed the quality of the DICOM tag *Body Part Examined* in 4 imaging modalities at Aachen University Hospital and found that, in 15% of the cases, the wrong information had been entered for the tag because of the user-originated errors. Misra et al. [11] reported that labelling with the user-defined meta-data containing inconsistent vocabulary may introduce human-reporting bias in datasets, which degrades the performance of deep learning models. Categorization can be even more difficult for images stored in other formats, e.g. NIfTI in neuroimaging, where meta-data is limited and/or simply not available for image categorization.

To categorize medical images in a realistic scenario, designing fully supervised methods would require a prior knowledge about the data distribution of the entire database, accounting for long-tailed rare classes and finally devoting significant effort to accurately and consistently obtaining manual ground-truth. In this work, we propose a different paradigm by efficiently using abundant unlabelled data and perform unsupervised learning. Specifically, we demonstrate that large-scale datasets of cardiac magnetic resonance (CMR) images can be categorized with a generalizable clustering approach that uses basic deep neural network architectures. Our intuition is that categorization of unknown medical images can be achieved if clusters with high purity are generated from learned image features without any supervision. Our approach builds on a recent state-of-the-art method, DeepCluster [4].

Our main contributions are the following: (i) we show that pure clusters for CMR images can be obtained with a deep clustering approach; (ii) we investigate end-to-end training of the approach for both class-balanced dataset and highly imbalanced data distributions, the latter being particularly relevant for medical imaging applications where diseases and abnormal cases can be rare; (iii) we discuss the design considerations and evaluation procedures to adapt deep clustering for medical image categorization. To the best of our knowledge, this is the first study to perform simultaneous representation learning and clustering for cardiac MR sequence/view categorization and evaluating its performance on a large-scale imbalanced dataset (n = 192,272 images).

2 Related Work

A number of self-supervised and unsupervised methodologies have been explored to train machine learning models with abundant unlabelled data. In self-supervised learning (SSL), a pretext task is defined to train a model without ground-truth. While several studies have been explored in the context of self-supervision [3,7], domain expertise is typically needed to formulate a pretext task unlike our work. Similar to self-supervised learning, different strategies of unsupervised learning have been implemented with generative networks [6] and deep clustering [20] to learn visual features. In this study, we focus on unsupervised deep clustering approaches at large scale. Although this has been investigated in a number of studies for natural images [4,5], various attempts in medical imaging have explored them with only limited amount of curated data in contrast to our methodology.

Moriya et al. [12] extended the JULE framework [20] for simultaneously learning image features and cluster assignments on 3D patches of micro-computed tomography (micro-CT) images with a recurrent process. Perkonigg et al. [15] utilized a deep convolutional autoencoder with clustering whose loss function is a sum of reconstruction loss and clustering loss to predict marker patterns of image patches. Ahn et al. [1] implemented an ensemble method of deep clustering methods based on K-means clustering. Pathan et al. [14] showed clustering can be improved iteratively with joint training for segmentation of dermoscopic images. Maicas et al. [10] combined deep clustering with meta training for breast screening.

One related approach to our study is the "Looped Deep Pseudo-task Optimization" (LDPO) framework proposed by Wang et al. [19]. LDPO extracts image features with joint alternating optimization and refine clusters. It requires a pre-trained model (trained on medical or natural images) at the beginning to extract features from radiological images and then fine-tunes the model parameters by joint learning. Therefore, the LDPO framework starts with a priori information and strong initial signal about input images. On the contrary, our model is completely unsupervised and trained from scratch with no additional processing. In addition, we do not utilize any stopping criteria, which is another difference from LPDO [19].

3   Method

Our method builds upon the framework of DeepCluster [4]. The idea behind their approach is that a CNN with random parameters θ provides a weak signal about image features to train a fully-connected classifier reaching an accuracy (12%) higher than the chance (0.1%) [13]. DeepCluster [4] combines CNN architectures and clustering approaches, and it proposes a joint learning procedure. The joint training alternates between extracting image features by the CNN and generating pseudo-labels by clustering the learned features. It optimizes the following objective function for a training set $X = \{x_1, x_2, ..., x_N\}$:

$$\min_{\theta,W} \frac{1}{N} \sum_{n=1}^{N} \ell(g_w(f_\theta(x_n)), y_n) \tag{1}$$

Here g_w denotes a classifier parametrized by w, $f_\theta(x_n)$ denotes the features extracted from image x_n, y_n denotes the pseudo-label for this image and l denotes the multinominal logistic loss [4]. Pseudo-labels are updated with new cluster assignments at every epoch. To avoid trivial solutions where output of the CNN is always same, the images are uniformly sampled to balance the distribution of the pseudo-labels [4].

In this study, we keep parts of DeepCluster [4] such as VGG-16 with batch normalization [18] as the deep neural architecture and K-means [9] as the clustering method, and then we adapt the rest for cardiac MR image categorization, illustrated in Fig. 1. To begin with, we add an adaptive average pooling layer between the VGG's last feature layer and the classifier. In DeepCluster [4], PCA is performed for dimensionality reduction which results in 256 dimensions whereas we preserve the original features. These features are ℓ_2-normalized before clustering. DeepCluster [4] feeds Sobel-filtered images to the CNN instead of raw images. In contrast, our method uses raw cardiac MR images in our experiments. We utilize heavy data augmentations including random rotation, resizing and cropping with random scale/aspect ratio for both training and clustering. Lastly, we normalize our images with z-scoring independently instead of using global mean and standard deviation.

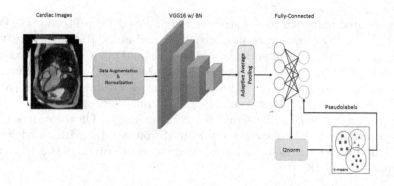

Fig. 1. Entire processing pipeline of our method based on DeepCluster [4]

We utilize the UK Biobank cardiac MR dataset which is open to researchers and contains tens of thousands of subjects. The whole dataset contains 13 image sequences/views, including short-axis (SA) cine, long-axis (LA) cine (2/3/4 chamber views), flow, SHMOLLI, etc. [16]. These images are in 2D, 2D + time or 3D + time. UK Biobank employs a consistent naming convention for different cardiac sequences and view-planes. We generated ground-truth labels using this naming convention and classified images into 13 categories [2]. To investigate the effect of class distribution on our methodology as well as the training stability, we designed three experiment settings using subsets of the entire dataset: (i) a subset of 3 well-balanced classes (LA 2/3/4 chamber views), and (ii) the large dataset of and (iii) the smaller dataset of high class imbalance of 13 classes. In these datasets, 2D images at t = 0 were saved in PNG format for faster loading and training. If the images are in 3D + time, every single slice in z direction at

$t = 0$ were saved. Total numbers were 47,637 images in the dataset (i), 192,272 images in the dataset (ii), 23,943 images in dataset (iii). Example images are illustrated in Fig. 1, and the class distributions are reported at the Table 1 in the supplementary material.

4 Results and Discussion

In our experiments, we followed a systematic analysis of the proposed methodology. We want to answer these four questions below:

1. Is it feasible to categorize uncurated large-scale cardiac MR images based on their cluster assignments?
2. How does the class balance affect deep clustering for medical images?
3. How stable is training given there are no clear stopping criterion?
4. How should we interpret the evaluation metrics?

Experiment Settings: For training, we set the total number of epochs as 200. Our optimizer was stochastic gradient descent (SGD) with momentum 0.9 and weight decay of $1e-5$. Our batch size was 256 and initial learning rate was 0.05. In the literature, there is a large body of empirical evidence which indicates that over-segmentation improves the performance of a deep clustering method [4]. Based on this evidence, we set the number of clusters to be 8 times of number of classes in the datasets, which corresponded to 24 for the dataset of 3 well-balanced classes, and 104 for the datasets with 13 classes.

Evaluation Metrics: We used normalized mutual information (NMI) [17] and cluster purity (CP) [17] to evaluate the clustering quality of our models.

$$NMI(X,Y) = \frac{2I(X;Y)}{H(X) + H(Y)} \tag{2}$$

Here I is the mutual information between X and Y and H is the entropy. For our experiments, we calculate two NMI values: NMI against the previous cluster assignments $(t-1)$ and NMI against ground-truth labels.

$$CP(X,L) = \frac{1}{N} \sum_k \max_j |x_k \cap l_j| \tag{3}$$

Here N is the number of images, X are the cluster assignments at epoch t and L is the ground-truth labels.

Accurate interpretation of our metrics, CP and NMI, is important. CP has a range from 0 to 1, which shows poor and perfect clusters, respectively. As the number of clusters increases, CP generally tends to increase until every image forms a single cluster, which achieves perfect clusters. In addition, we utilize NMI which signifies the mutual shared information between cluster assignments and labels. If clustering is irrespective of classes, i.e. random assignments, NMI has a value of 0. On the other hand, if we can form classes directly from cluster assignments, then NMI has a value of 1. The number of clusters also affects the NMI value but normalization enables the clustering comparison [17]. In our experiments, we did not employ any stopping criteria; thus, we always used the last model. In addition, during the training, we did not use NMI between cluster assignments and ground-truth, or cluster purity for validation.

Table 1. Performance of our method for different data configurations after 200 epochs

Dataset	Balanced Classes	# of Images	# of Classes	# of Clusters	NMI t vs t-1	NMI t vs labels	Cluster Purity
(i)	✓	47,637	3	24	0.675	0.519	0.997
(ii)	✗	192,272	13	104	0.782	0.605	0.991
(iii)	✗	23,943	13	104	0.745	0.609	0.994

Discussion: Metrics and loss progression throughout the training are given at Fig. 2. Results of our deep clustering method, which are calculated from features at the 200th epoch, are given at Table 1. Our method is able to reach a clustering purity above 0.99 for both class balanced and imbalanced datasets, which shows the feasibility of the deep clustering pipeline to categorize large-scale medical images without any supervision or labels. The class imbalance does not affect overall performance but balanced classes provide a more stable purity throughout the training. We also show that a relatively smaller dataset can be enough for efficient clustering with high cluster purity.

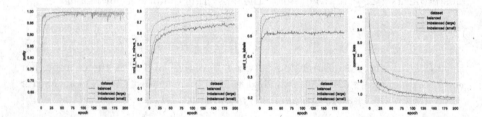

Fig. 2. Training metrics of our method for different data configurations

Additionally, we want to extend the discussion about deep clustering at [4] to medical imaging in a realistic scenario. One major challenge in deep clustering is the lack of a stopping criterion. Supervised training with labelled data as a stopping criterion could be utilized but this usually requires the prior knowledge of classes, which may not be possible to have beforehand at an unstructured hospital database. Pre-defined threshold-based methods on evaluation metrics, e.g. NMI and purity from adjacent epochs [19], could be another option but their robustness has yet to be proven. This is why it is important to investigate whether the training diverges. For this aim, we trained the dataset (iii) with 1000 epochs to observe the training stability. As we can see from Fig. 3, although we observed some fluctuations in metrics from time to time, they were stable throughout the training, which is similar to the observation at [4].

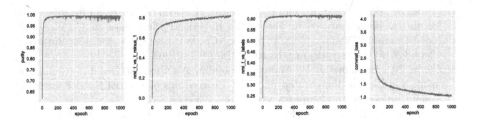

Fig. 3. Training stability and metrics for 1000 epochs

Lastly, we observed that changes in NMI and CNN loss could indicate changes in clustering quality. Normally, we expect to see a steady increase in NMI and a steady decrease in CNN loss during the training. A sudden decrease in NMI and/or a sudden increase in CNN loss may be a sign of worse clusters generated. However, steady decrease in CNN loss does not necessarily mean better cluster purity. Therefore, we think that it is beneficial to closely monitor not one but all metrics for unusual changes as well as to consider other metrics of clustering.

5   Conclusion

In this work, we propose an unsupervised deep clustering approach with end-to-end training to automatically categorize large-scale medical images without using any labels. We have demonstrated that our method is able to generate highly pure clusters (above 0.99) under both balanced and imbalanced class distributions. In future work, expanding the evaluation, adapting deep clustering approaches to other clinical tasks and improving their robustness and generalizability are some of interesting avenues that could be explored.

Acknowledgement. This work is supported by the UK Research and Innovation London Medical Imaging and Artificial Intelligence Centre for Value Based Healthcare. This research has been conducted using the UK Biobank Resource under Application Number 12579.

References

1. Ahn, E., Kumar, A., Feng, D., Fulham, M., Kim, J.: Unsupervised feature learning with k-means and an ensemble of deep convolutional neural networks for medical image classification. arXiv preprint arXiv:1906.03359 (2019)

2. Bai, W., et al.: Automated cardiovascular magnetic resonance image analysis with fully convolutional networks. J. Cardiovasc. Magn. Reson. **20**(1), 65 (2018). https://doi.org/10.1186/s12968-018-0471-x

3. Bai, W., et al.: Self-supervised learning for cardiac MR image segmentation by anatomical position prediction. In: Shen, D., et al. (eds.) MICCAI 2019. LNCS, vol. 11765, pp. 541–549. Springer, Cham (2019). https://doi.org/10.1007/978-3-030-32245-8_60

4. Caron, M., Bojanowski, P., Joulin, A., Douze, M.: Deep clustering for unsupervised learning of visual features. In: Ferrari, V., Hebert, M., Sminchisescu, C., Weiss, Y. (eds.) Computer Vision – ECCV 2018. LNCS, vol. 11218, pp. 139–156. Springer, Cham (2018). https://doi.org/10.1007/978-3-030-01264-9_9

5. Caron, M., Bojanowski, P., Mairal, J., Joulin, A.: Unsupervised pre-training of image features on non-curated data. In: Proceedings of the IEEE/CVF International Conference on Computer Vision, pp. 2959–2968 (2019)

6. Donahue, J., Krähenbühl, P., Darrell, T.: Adversarial feature learning. arXiv preprint arXiv:1605.09782 (2016)

7. Gidaris, S., Singh, P., Komodakis, N.: Unsupervised representation learning by predicting image rotations. arXiv preprint arXiv:1803.07728 (2018)

8. Gueld, M.O., et al.: Quality of DICOM header information for image categorization. In: Siegel, E.L., Huang, H.K. (eds.) Medical Imaging 2002: PACS and Integrated Medical Information Systems: Design and Evaluation, vol. 4685, pp. 280–287. International Society for Optics and Photonics, SPIE (2002)

9. Johnson, J., Douze, M., Jégou, H.: Billion-scale similarity search with GPUs. IEEE Trans. Big Data **7**, 535–547 (2019)

10. Maicas, G., Nguyen, C., Motlagh, F., Nascimento, J.C., Carneiro, G.: Unsupervised task design to meta-train medical image classifiers. In: 2020 IEEE 17th International Symposium on Biomedical Imaging (ISBI), pp. 1339–1342. IEEE (2020)

11. Misra, I., Zitnick, C.L., Mitchell, M., Girshick, R.: Seeing through the human reporting bias: visual classifiers from noisy human-centric labels. In: 2016 IEEE Conference on Computer Vision and Pattern Recognition (CVPR), pp. 2930–2939 (2016)

12. Moriya, T., et al.: Unsupervised segmentation of 3d medical images based on clustering and deep representation learning. In: Medical Imaging 2018: Biomedical Applications in Molecular, Structural, and Functional Imaging, vol. 10578, p. 1057820. International Society for Optics and Photonics (2018)

13. Noroozi, M., Favaro, P.: Unsupervised learning of visual representations by solving jigsaw puzzles. In: Leibe, B., Matas, J., Sebe, N., Welling, M. (eds.) ECCV 2016. LNCS, vol. 9910, pp. 69–84. Springer, Cham (2016). https://doi.org/10.1007/978-3-319-46466-4_5

14. Pathan, S., Tripathi, A.: Y-Net: biomedical image segmentation and clustering. arXiv preprint arXiv:2004.05698 (2020)

15. Perkonigg, M., Sobotka, D., Ba-Ssalamah, A., Langs, G.: Unsupervised deep clustering for predictive texture pattern discovery in medical images. arXiv preprint arXiv:2002.03721 (2020)

16. Petersen, S.E., et al.: UK biobank's cardiovascular magnetic resonance protocol. J. Cardiovasc. Magn. Reson. **18**(1), 1–7 (2015)

17. Schütze, H., Manning, C.D., Raghavan, P.: Introduction to Information Retrieval, vol. 39. Cambridge University Press, Cambridge (2008)

18. Simonyan, K., Zisserman, A.: Very deep convolutional networks for large-scale image recognition. arXiv preprint arXiv:1409.1556 (2014)

19. Wang, X., et al.: Unsupervised joint mining of deep features and image labels for large-scale radiology image categorization and scene recognition. In: 2017 IEEE Winter Conference on Applications of Computer Vision (WACV), pp. 998–1007. IEEE (2017)
20. Yang, J., Parikh, D., Batra, D.: Joint unsupervised learning of deep representations and image clusters. In: Proceedings of the IEEE Conference on Computer Vision and Pattern Recognition, pp. 5147–5156 (2016)

MetaHistoSeg: A Python Framework for Meta Learning in Histopathology Image Segmentation

Zheng Yuan[✉], Andre Esteva, and Ran Xu

Salesforce Research, Palo Alto, CA 94301, USA
zyuan@salesforce.com

Abstract. Few-shot learning is a standard practice in most deep learning based histopathology image segmentation, given the relatively low number of digitized slides that are generally available. While many models have been developed for domain specific histopathology image segmentation, cross-domain generalization remains a key challenge for properly validating models. Here, tooling and datasets to benchmark model performance *across* histopathological domains are lacking. To address this limitation, we introduce MetaHistoSeg – a Python framework that implements unique scenarios in both meta learning and instance based transfer learning. Designed for easy extension to customized datasets and task sampling schemes, the framework empowers researchers with the ability of rapid model design and experimentation. We also curate a histopathology meta dataset - a benchmark dataset for training and validating models on out-of-distribution performance across a range of cancer types. In experiments we showcase the usage of MetaHistoSeg with the meta dataset and find that both meta-learning and instance based transfer learning deliver comparable results on average, but in some cases tasks can greatly benefit from one over the other.

Keywords: Histopathology image segmentation · Transfer learning · Meta learning · Pan-cancer study · Meta-dataset

1 Introduction

For cancer diagnosis and therapeutic decision-making, deep learning has been successfully applied in segmenting a variety of levels of histological structures: from nuclei boundaries [1] to epithelial and stromal tissues [2], to glands [3,4] across various organs. It's generalizability that makes it effective across a wide variety of cancers and other diseases. Admittedly, the success relies largely on the abundance of datasets with pixel level segmentation labels [5–9].

Few-shot learning is of particular importance to medicine. Whereas traditional computer vision benchmarks may contain millions of data points, histopathology typically contains hundreds to thousands. Yet in histopathology images, different cancers often share similar visual components. For instance,

© Springer Nature Switzerland AG 2021
S. Engelhardt et al. (Eds.): DGM4MICCAI 2021/DALI 2021, LNCS 13003, pp. 268–275, 2021.
https://doi.org/10.1007/978-3-030-88210-5_27

adenocarcinomas, which occur in glandular epithelial tissue, contain similar morphological structure across many organs where they can arise [10]. Thus models that distill transferable histopathological features from one cancer can potentially transfer this knowledge to other cancers. To utilize different histopathology datasets collectively, benchmarks and tooling that enable effective learning across domains are strongly desired to support more accurate, and more generalizable models across cancers.

The key question is how to formulate the learning-across-task setup for histopathology segmentation? Naturally meta-learning [11–13] is the best reference as for its precise effectiveness to handle limited data availability. It is widely used in few shot classification with a canonical setup: a task of K-way-N-shot classification is created on the fly by sampling K classes out of a large class pool followed by sampling N instances from each of the K classes. Then a deep neural network is trained by feeding batches of these artificial tasks. Eventually during inference the whole network is shared with new tasks (composed by K classes never seen during training) for refinement.

While this setup is ubiquitous in meta classification, we find that it is difficult to extend to the meta segmentation problem. First, a task of segmenting histopathology images should justify medical validity (e.g. cancer diagnosis) before even created. One cannot generate factitious tasks by randomly combining K layers of pixels based on their mask label, as oppose to the routine in meta classification. For example, based on a well-defined Gleason grading system, a prostate cancer histopathology image usually requires to be classified into 6 segments for each pixel. Meanwhile for another histopathology image in nuclei segmentation, researchers in general need to classify each pixel as either nuclei or others. Notwithstanding each case exhibits a valid medical task in its own right, criss-crossing them just as in the canonical setup to form a new task is not medically sound. Moreover, the underlying assumption of meta classification is that shared knowledge must exist across any K-way classification tasks. It is generalizable among tasks by a composite of any K classes, as long as the number of classes involved is K. Generally, we will not observe this "symmetrical" composite in segmentation task space. In the same example, the first task is to segment 6 classes pixel wise whereas the other is to segment 2 classes. Therefore, the knowledge sharing mechanism in the deep neural also needs to be adjusted to reflect this asymmetry.

In this paper, we introduce a Python framework MetaHistoSeg to facilitate the proper formulation and experimentation of meta learning methodology in histopathology image segmentation. We also curate a histopathology segmentation meta-dataset as the exemplar segmentation task pool to showcase the usability of MetaHistoSeg. To ensure the medical validity of the meta dataset, we build it from existing open-source datasets that are (1) rigorously screened by world-wide medical challenges and (2) well-annotated and ready for ML use.

MetaHistoSeg offers three utility modules that cover the unique scenarios in meta learning based histopathology segmentation from end to end:

1) Data processing functions that normalize each unique dataset pertaining to each medical task into a unified data format.
2) Task level sampling functions (the cornerstone of the meta learning formulation in segmentation) for batch generation and instance level sampling functions provided as a baseline.
3) Pre-implemented task-specific heads that are designed to tail customized backbone to handle the asymmetry of tasks in a batch, with multi-GPU support.

We open-source both MetaHistoSeg and the meta-dataset for broader use by the community. The clear structure in MetaHistoSeg and the accompanying usage examples allow researcher to easily extend its utility to customized datasets for new tasks and customized sampling methods for creating task-level batches. Just as importantly, multi-GPU support is a must in histopathology segmentation since a task level batch consists of fair number of image instances, each of which is usually in high resolution. We also benchmark the performance of meta learning based segmentation as compared with the instance based transfer learning as a baseline. Experiments show that both meta-learning and instance based transfer learning and deliver comparable results on average, but in some cases tasks can greatly benefit from one over the other.

2   MetaHistoSeg Framework

MetaHistoSeg offers three utilities: task dataset preprocessing, task or instance level batch sampling, task-specific deep neural network head implementation.

2.1   Histopathology Task Dataset Preprocessing

MetaHistoSeg provides preprocessing utility functions to unify the heterogeneity of independent data sources with a standard format. Here we curate a meta histopathology dataset to showcase how knowledge transfer is possible via meta learning among different segmentations tasks. Following the tasks in the dataset as examples, users can easily create and experiment with new tasks from customized datasets.

The meta-dataset integrates a large number of histopathology images that come from a wide variety of cancer types and anatomical sites. The contextual information of each data source, the preprocessing method and the meta information of their data constituents are detailed as follows.

– *Gleason2019*: a dataset with pixel-level Gleason scores for each stained prostate cancer histopathology image sample. Each sample has up to six manual annotations from six pathologists. During preprocessing, we use the image analysis toolkit SimpleITK [14] to consolidate multiple label sources into a single ground truth. The dataset contains 244 image samples with resolution of 5120×5120 and each pixel belongs to one of 6 Gleason grade grades. The data source was a challenge [5] hosted in MICCAI 2019 Conference.

- *BreastPathQ*: a dataset of patches containing lymphocytes, malignant epithelial and normal epithelial cell nuclei label. This is an auxiliary dataset in the Cancer Cellularity Challenge 2019[6] as part of the 2019 SPIE Medical Imaging Conference where the original task is to evaluate patch as a single score. In our context, we use the dataset for segmentation. Since the annotations only contain the centroid of each cell nuclei, we generate the segmentation mask by assuming each cell is a circle with a fixed radius. The dataset contains 154 samples and each pixel belongs to one of 4 classes.
- *MoNuSeg*: a dataset of pixel-level nuclei boundary annotations on histopathology images from multiple organs, including breast, kidney, liver, prostate, bladder, colon and stomach. This dataset comes from the nuclei segmentation challenge [7] as an official satellite event in MICAII 2018. It contains 30 samples and each label has 2 classes.
- *Glandsegmenatation*: a dataset of pixel-level gland boundary annotations on colorectal histopathology images. This data source comes from the gland segmentation challenge [8] in MICAII 2015. The dataset contains 161 samples and each label has 2 classes.
- *DigestPath*: a dataset of colon histology images with pixel-level colonoscopy lesion annotations. The data source [9] is part of MICCAI 2019 Grand Pathology Challenge. It contains 250 samples and each pixel belongs to one of the 2 classes. Although the original challenge contains both Signet ring cell detection and Colonoscopy tissue segmentation task, we only consider the latter in our context for the obvious reason.

2.2   Task and Instance Level Batch Sampling

MetaHistoSeg implements this core data pipeline of meta learning. It abstracts task level batch creation as a dataloader class *episode_loader*. Since *episode_loader* essentially unrolls the entire task space, researchers can customize their sampling algorithm just by specifying a probability distribution function. This enables users to quickly switch between training frameworks, empowering them to focus on model design and experimentation rather than building data pipelines. It also encapsulates instance level batch creation in dataloader class *batch_loader* as a baseline. The sampling schemes are as follows,

- Task level sampling: we sample a task indexed by its data source and then sample instances given the task to form an episode. Then it is split into support and query set. Here a batch is composed of several such episodes.
- Instance level sampling: we first mix up instances from different data sources as a pool and sample instances directly. Noting that data source imbalance can be a problem here, we dynamically truncate each data source to the same size before mixing up. We refresh the random truncation in each epoch.

Figure 1 shows how the preprocessing and batch sampling functions in MetaHistoSeg can be used to construct the data pipelines. Each data source is color coded. In meta learning setting, a batch is organized as episodes, each of which comes from the same data source. In instance based learning, a batch is organized as instances, which comes from mixed up data sources.

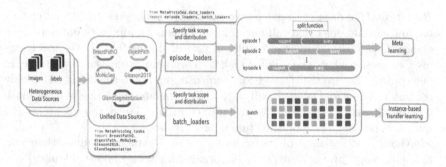

Fig. 1. MetaHistoSeg diagram: utility functions enable fast construction of data pipelines for meta-learning and instance based learning on the meta-dataset.

2.3 Task-Specific Heads and Multi-GPU Support

Since the tasks sampled in a batch usually predict different number of segments, we pre-implement the last layer of a neural network as task-specific heads and route the samples of a task only to its own head during forward propagation (FP). This feature frees researchers from handling task asymmetry in meta segmentation. Meanwhile, note that the default multi-GPU support in pytorch (nn.DataParallel) requires a single copy of network weights. This conflicts with the meta learning scenario, where two copies of weight parameters are involved in its bi-level optimization. Thus we re-implement multi-GPU FP process.

3 Experiments

We use MetaHistoSeg to benchmark MAML [11] on the histopathology image segmentation meta dataset and compare it with instance based transfer learning as a baseline. For each data source in the meta-dataset, we fix it as a test task and train a model using some subset of the remaining data sources, using both MAML and instance based transfer learning.

3.1 Implementation Details

For data augmentation, we resize an input image with a random scale factor from 0.8 to 1.2, followed by random color jittering (with 0.2 variation on brightness, contrast and 0.1 variation on hue and saturation), horizontal and vertical flipping (0.5 chance) and rotation (a random degree from −15 to 15). The augmented image is ultimately cropped to 768×768 before feeding into the neural networks.

During training, we use 4 Nvidia Titan GPU (16G memory each) simultaneously. This GPU memory capacity dictates the maximum batch size as 4 episodes with each consisting of 16 image samples. During meta learning, each episode is further split into a support set of size 8 and a query set of size 8.

When forming a batch, we use MetaHistoSeg.episode_loader to sample data in a bi-level fashion: first among data sources then among instances.

In both methods we choose U-Net [15] as the backbone model given its effectiveness at medical image segmentation tasks. Training is performed with an Adam optimizer and a learning rate of 0.0001 for both methods. For MAML, we adapt once with a step size of 0.01 in the inner loop optimization. The maximum training iteration is set to 300000 for both settings.

We use the mean Intersection Over Union (mIoU) between predicted segmentation and ground truth as our performance metric:

$$\text{mIoU} = \frac{1}{N} \sum_i \frac{P_i \cap T_i}{P_i \cup T_i} \tag{1}$$

where P_i and T_i are the predicted and ground truth pixels for class i, respectively, across all images in evaluation, and N is the number of classes.

3.2   Results

Table 1 summarizes the mIoU scores for both methods where each data source is treated as a new task, and models are trained on some subset of the remaining data sources. We enumerate over the other data sources as well as their combination to form five different training sets - the five columns in the table.

Table 1. mIoU performance on each new task (row) refined from pretrained models with different predecessor tasks (column)

New task	Training tasks									
	All others		BreastPathQ		MoNuSeg		Gland segmentation		DigestPath	
	MAML	TransferL	MAML	TransferL	MAML	TransferL	MAML	TransferL	MAML	TransferL
BreastPathQ	**0.301**	0.282	NA	NA	**0.302**	0.287	**0.326**	0.300	0.285	**0.299**
MoNuSeg	0.669	**0.676**	**0.682**	0.636	NA	NA	0.691	**0.694**	0.639	**0.653**
Gland segmentation	**0.557**	0.556	**0.540**	0.539	0.563	**0.573**	NA	NA	0.535	**0.553**
DigestPath	**0.632**	0.628	0.609	**0.613**	**0.607**	0.599	**0.624**	0.617	NA	NA

As shown in the table, MAML and instance based transfer learning deliver comparable performance across tasks, with MAML outperforming the other in 9 of the 16 settings. Of note, for a number of tasks, one of the two performs noticeably better than the other. However, we don't observe a consistent advantage of one methodology over the other on all testing data sources. We hypothesize that the suitability of a knowledge sharing methodology highly depends on the interoperability between the predecessor tasks and the testing task. For example, when evaluating data source MoNuSeg as a new task, meta learning outperforms transfer learning with BreastPathQ as predecessor task while the reverse is true with GlandSegmentation or DigestPath as predecessor tasks. This suggests that BreastPathQ might share more task level knowledge with MoNuSeg than GlandSegmentation and DigestPath.

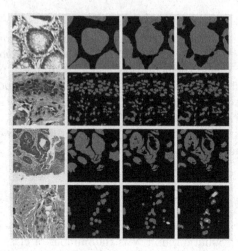

Fig. 2. Meta-dataset task examples. Top to bottom: GlandSegmentation, MoNuSeg, digestpath and BreastPathQ. Left to right: original image, ground truth, segmentation by MAML and segmentation by instance based transfer learning.

Figure 2 depicts the visual comparison of two knowledge sharing methodologies. Each row is a sample of each data source while each column is original histopathology images, the ground truth masks, segmentation results from MAML and instance based transfer learning respectively. Note that for BreastPathQ (the fourth row), the raw label is standalone centroid of each nuclei and we augment them into circles with a radius of 12 pixels to generate the segmentation masks. Yet we don't impose this simplified constraint on predictions. Therefore the results are not necessarily isolated circles. We also observe that for breastPathQ, both methodologies sometimes make predictions that falsely detect (green region) a long tail class. This is due to the innate class imbalance in the data source and can be alleviated by weighted sampling. Another interesting observation is that sometimes pathologists can make ambiguous annotations. As the third row shows, there is an enclave background in the tissue while human labeler regards it as the same class as the surrounding tissue, perhaps out of medical consistency. Whereas it also makes sense in the prediction results the two methodologies still predict it as background. Overall, as shown in these figures, MAML and transfer learning produce similar qualitative results.

4   Conclusions

In this work, we introduced a Python based meta learning framework MetaHistoSeg for histopathology image segmentation. Along with a curated histopathology meta-dataset, researchers can use the framework to study knowledge transferring across different histopathological segmentation tasks. To enable easy adoption of the framework, we provide sampling functions that realize the standard sampling procedures in classical knowledge transferring settings. We also benchmark against the meta dataset using MAML and instance based transfer learning. Based on experiment results, MAML and transfer learning deliver comparable results, and it is worthwhile to attempt each when fitting models. However, it remains unclear how interoperability of the testing task and predecessor task(s) in the training set precisely determine meta learning and transfer learning effectiveness. Also, we observe there isn't always performance gain when we add more predecessor task sources. It concludes that a naive combination of task-level training data may not be beneficial. This addressed observation points to a future research goal of explainable interoperability between tasks.

References

1. Xing, F., Xie, Y., Yang, L.: An automatic learning-based framework for robust nucleus segmentation. IEEE Trans. Med. Imaging **1**(1), 99 (2015)
2. Al-Milaji et al.: Segmentation of tumor into epithelial vs. stromal regions. In: CONFERENCE 2016, LNCS, vol. 9999, pp. 1–13. Springer, Heidelberg (2017)
3. Manivannan, A., et al.: Segmented the glandular structures by combining the hand-crafted multi-scale image features and features computed by a deep convolutional network. Med. Image Comput. Comput. Assist Interv. **16**(2), 411–8 (2013)
4. Chan, L., et al.: HistoSegNet: histological tissue type exocrine gland endocrine gland. Transp. Vessel. Med. Image Comput. Comput. Assist Interv. **16**(2), 411–8 (2013)
5. Nir, G., Hor, S., Karimi, D., Fazli, L., et al.: Automatic grading of prostate cancer in digitized histopathology images: learning from multiple experts. Med. Image Anal. **1**(50), 167–80 (2018)
6. Peikari, M., Salama, S., et al.: Automatic cellularity assessment from post-treated breast surgical specimens. Cytom. Part A **91**(11), 1078–1087 (2017)
7. Kumar, N., Verma, R., Sharma, S., et al.: A dataset and a technique for generalized nuclear segmentation for computational pathology. IEEE Trans. Med. Imaging **36**(7), 1550–1560 (2017)
8. Sirinukunwattana, K., Snead, D.R.J., Rajpoot, N.M.: A stochastic polygons model for glandular structures in colon histology images. IEEE Trans. Med. Imaging **34**(11), 2366–2378 (2015). https://doi.org/10.1109/TMI.2015.2433900
9. Li, J., et al.: Signet ring cell detection with a semi-supervised learning framework. In: Chung, A.C.S., Gee, J.C., Yushkevich, P.A., Bao, S. (eds.) IPMI 2019. LNCS, vol. 11492, pp. 842–854. Springer, Cham (2019). https://doi.org/10.1007/978-3-030-20351-1_66
10. American Cancer Society: Cancer Facts and Figures 2020. American Cancer Society, Atlanta, Ga (2020)
11. Finn, C., Abbeel, P., Levine, S.: Model-agnostic meta-learning for fast adaptation of deep networks. In: Proceedings of the International Conference of Machine Learning (2017)
12. Mishra, N., Rohaninejad, M., et al.: A simple neural attentive metalearner. In: Proceedings of the International Conference on Learning Representations (2018)
13. Munkhdalai, T., Yu, H.: Meta networks. In: Proceedings of the International Conference on Machine Learning, pp. 2554–2563 (2017)
14. Beare, R., Lowekamp, B., Yaniv, Z.: Image segmentation, registration and characterization in R with SimpleITK. J. Stat. Softw. **86**(8) (2018)
15. Ronneberger, O., Fischer, P., Brox, T.: U-Net: convolutional networks for biomedical image segmentation. In: Navab, N., Hornegger, J., Wells, W.M., Frangi, A.F. (eds.) MICCAI 2015. LNCS, vol. 9351, pp. 234–241. Springer, Cham (2015). https://doi.org/10.1007/978-3-319-24574-4_28

Author Index